*New York Recentered*

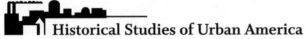 **Historical Studies of Urban America**

*Edited by* Lilia Fernández, Timothy J. Gilfoyle, Becky M. Nicolaides,
    and Amanda I. Seligman
James R. Grossman, Editor Emeritus

RECENT TITLES IN THE SERIES

*The Gateway to the Pacific: Japanese Americans and the Remaking of San Francisco*
    Meredith Oda

*Bulls Markets: Chicago's Basketball Business and the New Inequality*
    Sean Dinces

*Newsprint Metropolis: City Papers and the Making of Modern Americans*
    Julia Guarneri

*Evangelical Gotham: Religion and the Making of New York City, 1783–1860*
    Kyle B. Roberts

*Crossing Parish Boundaries: Race, Sports, and Catholic Youth in Chicago, 1914–1954*
    Timothy B. Neary

*The Fixers: Devolution, Development, and Civil Society in Newark, 1960–1990*
    Julia Rabig

*Chicago's Block Clubs: How Neighbors Shape the City*
    Amanda I. Seligman

*The Lofts of SoHo: Gentrification, Art, and Industry in New York, 1950–1980*
    Aaron Shkuda

*The Newark Frontier: Community Action in the Great Society*
    Mark Krasovic

*Making the Unequal Metropolis: School Desegregation and Its Limits*
    Ansley T. Erickson

*Confederate Cities: The Urban South during the Civil War Era*
    Andrew L. Slap and Frank Towers, eds.

A complete list of series titles is available on the University of Chicago Press website.

# New York Recentered

*Building the Metropolis from the Shore*

KARA MURPHY SCHLICHTING

The University of Chicago Press
Chicago and London

PUBLICATION OF THIS BOOK HAS BEEN AIDED
BY A GRANT FROM THE BEVINGTON FUND.

The University of Chicago Press, Chicago 60637
The University of Chicago Press, Ltd., London
© 2019 by The University of Chicago

Published 2019
Printed in the United States of America

28  27  26  25  24  23  22  21  20  19      1  2  3  4  5

ISBN-13: 978-0-226-61302-4 (cloth)
ISBN-13: 978-0-226-61316-1 (e-book)
DOI: https://doi.org/10.7208/chicago/9780226613161.001.0001

Library of Congress Cataloging-in-Publication Data
Names: Schlichting, Kara Murphy, author.
Title: New York recentered : building the metropolis from the shore / Kara Murphy
    Schlichting.
Other titles: Historical studies of urban America.
Description: Chicago ; London : The University of Chicago Press, 2019. | Series: Historical
    studies of urban America | Includes bibliographical references and index.
Identifiers: LCCN 2018051008 | ISBN 9780226613024 (cloth : alk. paper) |
    ISBN 9780226613161 (e-book)
Subjects: LCSH: New York (N.Y.)—History. | City planning—New York (State)—New
    York—History—19th century. | City planning—New York (State)—New York—
    History—20th century. | Cities and towns—New York (State)—New York—Growth. |
    Urban ecology (Sociology)—New York (State)—New York—History.
Classification: LLC F128.47.S35 2019 | DDC 974.7/1—dc23
LC rcord available at https://lccn.loc.gov/2018051008

*This book is lovingly dedicated to my mother,*
*Mary Murphy Schlichting*

# Contents

# Introduction

[In t]his unified metropolis of the East—the greatest New York, if one may so call it . . . the overflowing of big towns toward each other must strike forcibly whoever travels observantly. . . . the omni-present suburban villas, improved residential parks, beach properties, trolley car stations and clanging trolleys, telephone pay stations, newsboys hawking late editions of the metropolitan "yellows"—these assure the traveler that he has not left the city universe behind. . . . It is the city of the future.
—Frederick Coburn, "The Five-Hundred Mile City"[1]


Before F. Scott Fitzgerald's famous 1925 novel *The Great Gatsby* was about a man, it was about a place. Fitzgerald's original title was *Among the Ash Heaps and Millionaires*, which captured the diversity of the edge of greater New York City, a place of palatial estates but also a landscape of marshes, like Flushing Meadows, the site of the novel's ash dump.[2] Fitzgerald's fictionalized account of New York's wealthy in the Roaring Twenties unfolds in the real spaces of Manhattan and the borough of Queens, although East and West Egg were imaginary composites of Long Island's exclusive North Shore. The ash heap of Flushing Meadows is also an introduction to the ecological fabric of the city's environs. Shallow bays lined with extensive mudflats and salt marshes characterized greater New York's coast, an environment that influenced not just Fitzgerald's story but regional patterns of expansion and metropolitan development. In *New York Recentered: Building the Metropolis from the Shore*, I argue that the New York metropolis of today was built over a hundred years of development in the diverse spaces of its edge.

If the skyscraper symbolizes Manhattan, the greater metropolis is captured by a more subtle yet nevertheless equally salient feature—its coast. Focusing on this waterway as a place of connection rather than a boundary brings to light how environmental issues of the shore, topography, and riparian land-use patterns unified the region. Urban growth occurred in a discernible geographic region best identified as the coastal metropolitan corridor. It includes the Bronx and Queens on the upper East River and the adjacent counties along the Sound. Westchester County in New York and Fairfield County in Connecticut form the Sound's north shore. Queens and Nassau counties on Long Island form the opposite coast. I sometimes take a broader, county-level view that tracks inland but always with an eye to the specific environments and places of the coast.

While New York has been studied extensively, the city merits a new look through the spaces of its environs. Regionally situated actors across greater New York directed much of the metropolis's formation. Manhattan's role as an economic engine and its concentrated wealth and population dominate histories of the city. This interest is logical, but the nineteenth-century city also has a rich history beyond Manhattan. The histories of the Bronx and Long Island, for example, tend to be told as twentieth-century stories, but their nineteenth-century evolutions are more than prequels to twentieth-century suburbanization, slum clearance, and urban decline. This book shifts the historical gaze from Manhattan to the city's dynamic environs to reassess the avenues of power and engines of growth that shaped greater New York. This shift frees greater New York from two interconnected myths concerning the twentieth-century city. The first is the myth that urbanization is inherently unidirectional and linear, radiating outwards from an urban core. A foundation of this myth is that areas beyond the center city were undifferentiated and unimportant until they were "transformed" by suburbanization.[3] The second is that professional planners and the powerful administrative center that Manhattan represented were solely responsible for shaping large-scale development. Even places that power brokers considered peripheral, in terms of both geography and consequence, had their own histories and planning.

*From City to Region*

In 1874 New York City, confined to Manhattan for two centuries, stretched beyond its island shores and annexed its first mainland territory. Annexation of southern Westchester County—the future South Bronx—initiated a transformative era of expansion that accelerated changes in urban form. In 1886 city reformer and early planner George Waring announced that jurisdictional expansion combined with urbanization heralded a new type of city. While the city's coastal environs had their own distinct history of development, they were increasingly knit into greater New York's networks. Waring argued that "to limit [the city's] population, its industries, and its achievements to what we now find on Manhattan Island" was a misleading view of the city. Rather, "each great metropolis should be credited with the natural outgrowth of the original nucleus."[4] Waring identified metropolitan connectivity in the wider canvas of what became the five-borough city in 1898. Metropolises emerged nationwide in this period: Chicago became the premier city in the urban hierarchy of the Midwest, and Boston matured as the hub of New England. New York led them all, becoming the nation's chief city.

At midcentury New York Harbor grew to be one of the world's greatest industrial regions, and Manhattan Island was its epicenter. Commerce and industry crowded lower Manhattan's business districts and its waterfront. Walt Whitman famously captured the vitality of the port at midcentury, exalting the busyness of "mast-hemm'd Manhattan" and the harbor churning with the wake of steamers, tugs, schooners, and sloops. In 1860 New York ranked first in the nation in the value of its manufacturers; its neighboring ports of Brooklyn and Newark, New Jersey, ranked fifth and sixth, respectively.[5] New York led the nation's sugar refining and garment industries, produced a third of its printing and publishing, and manufactured an array of machines, engines, and specialty goods. A vast network of advertising and merchandising firms, financial markets and credit facilities supported this prodigious manufacturing economy. New York's economic supremacy attracted immigrants and American migrants; by 1850 over half a million people lived on the twenty-three-square-mile island. The greater city's population rose at an amazing pace, doubling between 1820 and 1840, again between 1840 and 1860, and for a third time between 1860 and 1890, by which time Manhattan alone housed over 1.4 million people.[6] Entrepreneurs built an extraordinary system of rapid transit—the largest in the world by the end of the century—to increase the city's accessibility. From midcentury on, population and economic growth combined with Manhattan's geography (the island is thirteen miles long and but two miles across at its widest) to intensify expansionist pressures.

No precedent could prepare New Yorkers for the speed and scale of metropolitan growth in the second half of the nineteenth century. Speculation, the relentless search for new areas to develop, and municipal leaders' devotion to the city's growth drove expansion.[7] In 1895, the city annexed additional mainland territory from Westchester County. Voters in Manhattan, Brooklyn, Queens, and Staten Island also approved consolidation with the city. Greater New York officially came into existence on January 1, 1898. Overnight, the city grew from sixty square miles to a staggering three hundred square miles.

Consolidation is perhaps the best-known date in the history of New York's growth. Yet Waring identified the treatment of a common future between the city and its hinterlands more than a decade earlier, while regional residents grappled with the effect of expansion for decades after.[8] In reality, the transition zones between rural, urban, and suburban remained thin, diverse, and difficult to characterize in greater New York. Political boundaries also often proved incompatible with the translocal world of the region. In 1870, the city's Department of Parks commissioners expressed frustration that New York's did not include the East Bronx but stopped at the Bronx River. They deemed the river a "purely arbitrary" border that would "lead to some embarrassment on the part of the department," since planning drainage for only half the mainland's drainage basin would not effectively address sewage pollution.[9] Consolidation turned areas considered rural or suburban into "urban" neighborhoods overnight, but even expanded jurisdictions could not contain the extent of the mutually constructed and shared world of the region.

Multiple definitions of *region* were at work from 1840 to 1940. Conceptions of the region changed as its public works infrastructure and social and economic networks expanded. Landscape architect Frederick Law Olmsted and public works czar Robert Moses both famously conceptualized the regional city through park plans, although the region they surveyed changed dramatically in the fifty years between their work. Greater New York in the 1870s, when Olmsted planned the newly annexed South Bronx, was not the same point of reference in 1924 when Moses was appointed head of the Long Island State Park Commission. Changes in population distribution and the amount of open space in the New York metropolitan area dramatically affected ideas of the region. By 1929 the groundbreaking *Regional Plan of New York and Its Environs* defined greater New York as a five-thousand-square-mile region. Not only did the meaning of the region change, multiple definitions now existed. Conceptions of the region varied depending on the frame used to define it, be it hydrology, communities, real estate patterns, or parkland. In the same year that the regional plan was unveiled, for example, a local planning body situated in the coastal corridor offered a different definition of the metropolitan area. The City of New Rochelle's planning commission identified a smaller region of "interrelated and inter-dependent" cities along the Sound, claiming "in each . . . questions are constantly arising that cannot be solved by one community alone. They have a common stake."[10] While regional definitions evolved, classifying the extent and character of greater New York remained a central goal of city officials and professional planners.

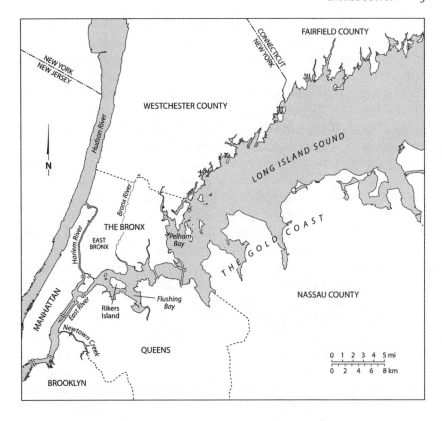

I.1. Map of greater New York along the upper East River and Long Island Sound. *Source*: Map by Bill Nelson.

In *New York Recentered* I reinterpret the process of metropolitan expansion from the periphery, emphasizing the spaces beyond Manhattan long considered "peripheral." In fact, they were central to metropolitan growth. Even the concept of the urban periphery or edge is largely based on the traditional one-dimensional binary of core/periphery, city/suburb, or urbanity/nature. New Yorkers use the particular terminology of the "outer boroughs" to describe the four boroughs beyond Manhattan. Yet like *periphery*, terms with the prefix *out-*—outer boroughs, outskirts, and outlying—create a bifurcation that positions such spaces as subordinate to a city center. *Edge* can similarly conjure subordination to a center, but it is a term that also usefully suggests contrast and demarcates difference.

In *New York Recentered* I consider four types of edge spaces: the borderlands of urbanized landscapes, political boundaries, the wider range of the city's environs, and the environmental edge of the coast. In addition, I use the terms *threshold* and *liminal* and the geographic concepts of *borderland*, *environs*, and *hinterland* to capture the dynamic, integrated region. A *borderland* is a district near a border or an area of overlap between two places, a useful term to consider land-use patterns that span political boundaries; *threshold* denotes a place of beginning and *liminal* an intermediate state or transitional quality. These terms evoke the complexity with which core and edge were tied together and overlapped. While city-centric biases are inherent in the terms *environs* and *hinterland*, terms often used to define metropolitan regions, both convey the geographic breadth of a region adjoining or beyond, yet connected to, a major metropolitan center.

The rise of the regional city begs several important questions. How did the city's coastal environment shape urban expansion? In turn, how did growth rearrange the natural world? Who counts as a city builder, and what constitutes planning in the coastal corridor? What constituted the ligaments that tied edge and core together? How did various levels of governance give rise to (or fragment) regionalism, the perspective that city and hinterland needed to be jointly planned in the service of a shared public interest? The answers to these questions constitute this book's four major contributions to the scholarship on New York and metropolitan history. First, *New York Recentered* identifies a nonprofessional planning and city-building pattern that emerged alongside professional channels of expansion. Second, the power of property fostered home rule and shaped large-scale patterns of development. Government and planning shaped urban growth, but not alone. When and where planning and public-sector leadership were absent, property owners developed beaches, parks, roads, and amusement parks that despite their variety helped expand the metropolis and change urbanism in key ways. Third, local government was a significant battleground for the implementation of state and federal policies, part of a search for an effective level of government oversight. Finally, while regional development shaped environmental change, nonhuman nature played a formative role in shaping urban expansion.

## City Builders

In *New York Recentered* I approach the regional city as an evolving series of temporary environments with different users who had their own visions and practices. Three groups shaped regional growth: government officials, planning experts, and nonprofessional private citizens and their local communities. Both appointed and elected officials played significant roles in this history. Andrew Haswell Green, who promoted a vanguard regionalist perspective in the nineteenth century, and Robert Moses, who forwarded it in the twentieth, held powerful appointed positions in city and state government. Elected officials such as governor Alfred E. Smith importantly backed Moses's vision of top-down, centralized planning. Planning experts worked in both the private and public sectors, for the Westchester County Park Commission and the private-sector Regional Plan Association, organized to promote the landmark 1929 *Regional Plan of New York and Its Environs*. While formal plans and planning institutions mattered, diverse nonprofessionals were also involved.[11] Private citizens substantiated new land-use patterns at the city's limits and shaped the regional city that straddled it. The interaction between citizens, governments, and professional planners allowed for a reevaluation of two common tenets of urban planning: the assumption that planning was a progressive force, and the acceptance of planners' views as normative. At times communities demanded government intervention, but they also made their own localized, nonprofessional interventions in urban form.

The modern planning profession did not appear until the early twentieth century, but early antecedents abound. Planning historians have identified a range of conscious attempts to shape long-term development comprehensively by a diverse group of civic elites, self-interested businessmen, landscape architects, and bourgeois reformers.[12] As planning professionalized between 1905 and 1915, the field's new leaders framed growth as a problem and asserted their authority to solve it. Even as planners interacted, cooperated, and fought each other in interpreting, enacting, and opposing various plans and planning standards, they remained uniformly committed to expanding their new authority over local control. Governments embraced the profession, hiring professional planners and establishing public-sector regional planning bodies.[13]

Robert Moses, New York City's infamous power broker, rose to prominence as city and state government created powerful planning authorities and committed resources to structure urbanization. While Moses was not a designer, he was phenomenally successful at implementing plans. Moses has fascinated the public and historians alike since he rose to prominence in New York state government in the 1920s. His unmatched building record, genius at amassing power, and boundless vision of growth and dedication to it, as well as his arrogance and his bullying, are all appealing fodder for critique. While his work was generally applauded by government officials and the press during his first three decades of active public life, the

defining assessment of Moses's career was long Robert Caro's scathing biography *The Power Broker: Robert Moses and the Fall of New York* (1974).[14]
Historians and journalists still debate Moses's legacy. With this quandary
in mind, I reassess the influences on and contingencies of Moses's career.
Focusing on his prewar work, I consider Moses as a conservationist particularly dedicated to preserving the shore, although his building indubitably had environmental consequences for it. I also examine his celebrated
parks, beaches, and bridges in addition to his vision of democracy, which
ultimately benefited a broad public interest in the face of privatism, localism, and special interests that he deemed antidemocratic.[15]

The history of park planning is vital to the history of metropolitan development, decentralization, and suburbanization. Literature on park development has considered public parks to be a phenomenon of centralized
authority at the city and state levels as well as occurring in relative proximity to the urban core itself. To the modern viewer, New York's Central Park
and Boston, Massachusetts's, metropolitan park system fit this framing, but
historians Catherine McNeur and Michael Rawson argue that parks reveal
a desire to control borderlands during rapid, widespread urban growth.[16]
In *New York Recentered* I continue to examine parks and parkways as central structural components of urban decentralization. Regional parks were
long-term commitments to publicly owned space, a commitment that government officials wielded as a powerful tool to shape growth.

I define *city planning* as including the activities of private citizens and
local communities alongside the plans of bureaucrats and planners. When
planning bodies came to consider exurban spaces, residents indubitably
responded; such responses remind us to look beyond end-product plans
to see the vested interests and unresolved dialectical processes that gave
rise to the modern metropolis. *New York Recentered* reframes growth as a
complex political process in part shaped by the litany of modest choices by
local actors and filled with compromises, incremental accomplishments,
and unanticipated consequences. Local custom augmented official plans;
public interest intersected with private enterprise. While the private sector did not usually possess formal plans, private development exhibited a
considerable degree of deliberateness, order, and regularity.[17] In aggregate,
the choices of locals, people who were neither planners nor politicians,
overwrote public plans or influenced the shape and timing of public infrastructure.

The diverse nonprofessional city builders active on the urban fringe
and its hinterland resist categorization. City-building activists included
the wealthy and the working class, people originally from Manhattan and

those raised in the outskirts of the Bronx. This list includes figures such as William Steinway and Louis Comfort Tiffany as well as William Kells, whose family ran a dance hall at Steinway's trolley park, and black real estate investor Solomon Riley. While Steinway and Tiffany are well-known figures, Kells and Riley are not. Diversity, however, did not inhibit cumulative influence. No planner or private citizen could accept responsibility for any corner of the city without participating to some degree in the joint shaping of the region. Locals negotiated geographies of both proximity and difference, at times encouraging closer ties through public works but at other moments defining their interests in contradistinction to regionalism through antigrowth positions. Recognizing the unpredictable dynamism of localism turns the periphery into a richly populated and politically contested territory rather than a space defined solely by official plans.

*Property*

Public and private, property played a multifaceted role in regional growth. Three important aspects of property influenced public-sector development. The first two, the property values associated with parks and the government takings and condemnation law used to build them, constituted an unprecedented government intervention into the traditionally private realm of the real estate market. The third, the Public Trust Doctrine (PTD), is the unique legal framework that polices the divide between public and private property on America's coasts.

Parks were associated with one of the most powerful forces of development in the city: real estate profits. Real estate profits accrued from the development of open space drove the breakneck pace of urbanization on nineteenth-century Manhattan. Yet real estate interests also viewed open spaces as useful tools to bolster property values and neighborhood reputations.[18] Parks and tax revenues went hand in hand, and the benefits of rising real estate motivated government acquisition of land. Profits motivated Westchester officials to accumulate parkland in the 1920s. And twentieth-century commissions were powerfully equipped to do so. The park commissions that planned the Bronx River Parkway, Westchester's county park system, and Long Island and Connecticut's state park systems all enjoyed expanded powers for takings. The title of condemned land vested immediately upon notice to owners.

This expedited vesting contrasted from condemnation in most places. Judges usually appointed commissions to assess the value of lands and title did not vest until after a court of claims fixed the award and the state accepted the amount. Condemnation empowered park commissions to avoid rapid jumps in value and to acquire land where land records were inaccurate or incomplete or when nominal owners were unable to grant valid title even if willing.[19] Government takings were essential to the quick acquisition of vast regional parks in the 1920s and 1930s. New York State's management of the PTD, however, proved a formidable challenge to public beaches. Because of the nation's PTD, which derived from Roman civil and English common law, the government was required to preserve public use of tidelands and protect against private monopoly.[20] Wealthy estate owners, however, worked to block recreationalists from the shore, elevating their rights as property owners over the public interest of the PTD. These divergent views inspired battles over property and public rights in the coastal corridor.

As fights on Long Island over public beach access reveal, private property shaped regional development patterns. Property was an investment tool for benefactors-turned-planners such as William Steinway, who acquired a controlling interest in Long Island City's Fifth Ward to build a company town. In aggregate, distinct patterns of property use formed local property regimes that shaped the first wave of urbanization in the regional city. Residential property patterns in bungalow colonies gave rise to blue-collar suburbs; estate building created elite enclaves. The legal rights of property ownership, particularly the right to exclude, were also wielded as a tool to control development. The segregation of the urban core, the spatial separation of the upper class from lower classes and racial segregation, was recreated on the city's outer reaches. Planners billed their work as a public good, yet at times they endorsed exclusionary land use rather than challenging the effects of the real estate market. This practice underscores the power of property regimes to shape the regional city.

*Regional Government and Its Challenges*

In addition to environmental and topographical factors, political considerations make the northern corridor of metropolitan New York distinct from other sections, such as the Hudson River Valley and northern New Jersey. I examine the northern corridor of metropolitan New York, an important part of the regional city, but only a part of it. Between 1840 and 1940, annexation, consolidation, village government organization, and new public agencies made greater New York America's most jurisdictionally complex metropolis.[21] A comprehensive history of the entire metropolitan area, inland hinterlands, or New Jersey is beyond the scope of this book. As the authors of the *Regional Plan of New York and Its Environs* explained in 1929, greater New York is really a "twin region or two intimately related sub-regions" divided by the Hudson River. New Jersey dominates the 2,883-square-mile region to the west of the Hudson, and its urban core is not Manhattan but Newark and Jersey City. The case studies in *New York Recentered* are drawn from the region to the east, an area of 2,232 square miles in New York State and 413 square miles in southwestern Connecticut.[22]

Regionalism demands the treatment of city and hinterland growth jointly and in consideration of a shared public interest; localism rejects this perspective.[23] Localism was a powerful, alternative shaper of the landscape design, transportation infrastructure, and real estate regimes that produced the coastal corridor. Interest groups wary of regionalism focused on the bifurcation of edge communities from the city and demanded "buffer" parks to separate city and suburban land use in the 1920s. Localism, when combined with home-rule government, empowered communities to reject the government-endorsed regionalism of state plans. Derived from colonial traditions, home-rule government was based on the assumption that state or regional government could not know what was best for a locality. The classic judgment of home rule was that it prevented well-orchestrated regional growth. In fact, home rule empowered local communities. Home-rule government was a significant battleground for the meaning and implementation of state and county policies concerning urban growth. The success of home-rule governments in blocking public works in Connecticut and on Long Island's North Shore reveals a largely untold process of state building, and challenges thereto, from the bottom up.[24]

*The Nature of Greater New York*

For residents of greater New York, thinking about the city's islands, coasts, and waterways was a precondition of thinking about growth. New York is a city of islands at the mouth of the Hudson River. It is a city built across an archipelago of three dozen landmasses of which Manhattan, Long Island, and Staten Island are the largest. From New York's colonial beginnings at the southern tip of Manhattan, the harbor and its waterways were a circulatory system connecting the archipelago and the mainland. As Michael Rawson has argued, "the wholesale rearrangement of the natural world lay at the heart of the nation's first great period of city building" in the early-to-mid-nineteenth century. From Boston to Chicago, New Orleans to New York, people found new ways of relating to their natural environments through the configuration of urban form.[25] From the mid-nineteenth century on, New York's power brokers worked to unify their city of islands. Government and residents alike accommodated themselves to the physical parameters, advantages, and disadvantages of this environment.[26] New York City's waterways and tides, islands, and estuary system were parameters that shaped the form and function of the regional city. The gaze of city builders spanned this coastal geography and considered its ecological fabric—so, too, should the historian's gaze.

New York's sprawling coastal zone includes the broad, deep waters of the upper and lower bays of New York Harbor, the lower Hudson River and Long Island Sound, and the Atlantic coasts of New Jersey and Long Island. New York sits at the center of an estuary system of more than sixteen hundred square miles; New York Harbor alone encompasses 585 miles of waterfront. In New York Recentered I tell the history of the northeast section of New York Harbor and the Sound. Environmental factors make the corridor distinct and discoverable. Before urbanization, the fine fluvial outwash of tidal channels sculpted the corridor's wetlands and mudflats while salt marsh plants captured sedimentation and maintained this environment. Thousands of acres of hydrologically connected wetlands covered the shores of the upper East River and Sound. The two waterways are in fact a single system: the East River, which runs between Manhattan and Long Island, is an eighteen-mile tidal strait of the Sound.

The diverse environmental issues addressed in New York Recentered speak to the complexity of human interactions with New York Harbor, its tributary waterways, and its estuaries. Although the tools used to assess ecological health and the ideas that structured their assessments changed over time, New Yorkers were keenly aware of environmental issues. They dealt with issues of drainage, topography and soil characteristics, climate and tides; assigned cultural and economic values to tidal marshes; promoted cultural ideas about green space and leisure; and worried about water quality, sewage, and dumps. Nature was imbued with meanings concerning health, wealth, leisure and social goals, and the need for human intervention and "improvement." Each interpretation revealed the community from which it came, be it professional planners, elites, working-class recreationalists, or Progressive politicians. All reflected different visions of what a restructured environment should look like.[27]

The political and environmental histories of regional New York are entwined. The cultural value and use of parks; their location, design, and roles as public institutions; and their "publicness," both "in the sense of democratic access and democratic control," all have political dimensions.[28] Planners and government officials wanted to restructure both the natural world and the urban form they built on it to realize their social and economic ambitions for the modernizing city.[29] Concerns over sanitation and public recreation were the central issues attending the preservation of natural landscapes in the nineteenth-century city. Apart from recreation, city officials valued the city's waterways as a sink for waste and its marshes as potential landfill sites and new real estate through land making. Landfill fundamentally reshaped New York Harbor's coastal ecology. In 1905 the city's Department of Street Cleaning declared that the city's "ecosystems of wetlands . . . afford an almost unlimited supply of dumping ground. . . . The possibilities of land reclamation are almost boundless."[30] Moses ferociously championed this approach in the 1920s and 1930s. Government agencies altered the environment through extensive marsh drainage and landfill, channelizing rivers, and taking private land for parks and parkways. By the 1930s, however, environmental concerns also inspired consideration of both sides of the waterline and the long-standing practice of using the city's natural spaces as a sink for waste. Environmental issues shaped politics in terms of motivating politicians' agendas and, in a larger sense, in terms of the features of urban government. Both types of politics motivated development goals, and achieving these goals required political might. City, county, and state governments amassed huge sums of public money to fund this work. And such work expanded the public sector to create regulatory oversight and enabled power brokers such as Moses entry into the project of city building.

The coast zone is not the only environment that underwent change because of turn of the century urbanization. Historians have examined the environmental transformations of New York's rise through the history of Central Park; by means of the construction of its vast water supply system; by looking to the ties between city and state and the ecological change wrought by the construction of the Erie Canal; and through the city's maritime economy and marine life.[31] The coast is a particularly useful frame with which to study environmental change in a city of islands because the city's real estate market, recreational networks, shipping and manufacturing interests, and public works authorities all had vested interests in coastal access. Shoreline development was integral to metropolitan change in any port. Coastline was a finite type of real estate along the nation's largest port. Competition between different users over the shore reveals the complexity of coastal urbanization and its environmental transformations.

*New York Recentered* integrates pieces of the region too often dismissed as disparate as interdependent pieces of metropolitan growth. Its series of regionally bound case studies includes a diverse set of places, from the company town of Steinway Village to the park systems of the Bronx to the 110-square-mile estate district of the North Shore. The spaces, while diverse, share a common history. Each experienced more than one of the chief development trends that spanned the regional city and helped make it a more concrete entity. For example, the Bronx's vanguard park system helped drive annexation and the idea of a greater New York City and inspired Progressive Era regional park plans. Coastal reclamation, which was systematized in the regional park programs, became essential to site construction for the New York World's Fair in Flushing Meadows. Shared development patterns and the environment, both its features and challenges, fostered intraregional connections that gave rise to the coastal metropolitan corridor.

In 1907 Henry James captured the complexity, tensions, and utter originality of the emerging regional city. In *The American Scene*, he observed that New York City appeared "as an heir whose expectations are so vast and so certain," at a moment in which its "whole case must change and her general opportunity, swallowing up the mainland, become a new question altogether."[32] Private investors, small-time builders, and bureaucrats came to share a regional perspective that situated the city and its environs as a contiguous space with a shared future. The modern metropolitan area was the result not of a master plan but the culmination of a century's worth of regional projects. James's observation reflected a mounting recognition that New York was more than a single urban center: it was the civic heart of a great system of industrial and commercial centers and suburbs. This perspective, emergent in the late 1800s, made possible the realization of a modern regional city by the mid-twentieth century.

# 1

## Benefactor Planning

### *Barnum's Bridgeport and Steinway's Queens*

In the early 1860s residents of Bridgeport, Connecticut, rarely enjoyed the panoramic views and cooling breezes afforded by their city's shoreline. At high tide the waters of Long Island Sound submerged Bridgeport's beach. Low tide exposed slick, seaweed-covered boulders. This environment discouraged pedestrians and proved inaccessible to carriages. Seizing the moment one low tide, the famous showman P. T. Barnum explored this terrain on horseback. Barnum scouted the shore with the hope of identifying a potential public drive route. As Bridgeport's greatest benefactor and booster, the showman hoped to capitalize on the summer excursionist traffic of the New York, New Haven and Hartford Railroad, which ran along the shore. A first-class waterfront park, a rarity in the developing corridor between New York and New Haven, might induce "strangers who came to spend the summer with us . . . to make the city their permanent residence."[1] The showman probably hoped that a park would raise real estate values, for he owned substantial investment properties across the city. On the shores of the Sound, Barnum might simultaneously build a park, increase the value of his property, and expand the size and economic scope of a city.

Less than ten years later and sixty miles to the east, a similarly shrewd businessman also capitalized on the undeveloped coastline of greater Long Island Sound. On a rainy November Saturday in 1870, William Steinway clad himself in a pair of "great India rubber boots" and tramped through the extensive salt meadows of Long Island City, Queens.[2] The marsh fronted on the upper half of the East River, a tidal strait of the Sound. Steinway's piano manufacturing firm had outgrown the production capacity of its Fifty-Third Street factory in Manhattan.[3] Queens's marshland offered Steinway an expansive area on which to build a new manufacturing settlement.

Steinway's and Barnum's work on the periphery—both on the edge of the city and on the edge of its hinterland—highlights the importance of local processes in regional patterns of growth. At a time when professional city planning did not exist as a distinct area of city governance, both men effectively created nonprofessional but nonetheless comprehensive city plans. Steinway and Barnum typified the class of civic-minded capitalists and propertied New Yorkers whom historian David Scobey identifies as having "amassed the capital, expertise, state power, and cultural authority necessary" to control the city's real estate and public works.[4] The success that Steinway and Barnum achieved underscores the critical role land-ownership and private enterprise played in shaping the New York metropolitan area.

Both Steinway and Barnum enjoyed public influence as leading citizens of nineteenth-century America. Steinway was born in Germany in 1835. His father, Henrich Steinweg, moved the family to New York in the 1840s, Americanized his name, and established the celebrated piano manufacturing firm. By midcentury New York City was already a manufacturing behemoth; it outranked every other industrial center in the nation in terms of the value of its manufactures. In 1860 the city housed more than 4,300 specialty goods manufacturers. In a city crowded with specialty firms—more than thirty produced pianos alone—Steinway & Sons surpassed all competition.[5] William directed the marketing that made the Steinway piano a symbol of refinement. Charismatic, driven, and meticulous, he gained international celebrity from his prestigious role in manufacturing and the city's music world.

Barnum's reputation eclipsed even Steinway's renown. Born in Bethel, Connecticut, in 1810, Barnum gained national attention in the 1830s as a showman and opened his famed American Museum in lower Manhattan in 1841. Barnum's museum and circus, like Steinway's instruments and music venues, captured the nation's imagination and bolstered his fame. The showman often bragged of his celebrity, repeating president Ulysses S. Grant's comment that he was "the best known man in the world."[6] Their celebrity, wealth, and connections to powerful investors uniquely empowered Barnum and Steinway to shape the metropolitan landscape.

Barnum and Steinway experimented in urban form to cultivate local growth as well as a sense of regional connectivity. Localized urban development paradoxically served as a major force that shaped the entire metropolis. The binaries of city-suburb and urban-rural that by tradition bifurcate the city and its environs fail to capture this dynamic. While both entrepreneurs invested substantial time and effort in local projects, they did so with an eye to the different regional networks of which their locales were a part. In the search for a manufacturing site, Steinway looked to be "away from the city, and yet within easy access of it."[7] Barnum saw similar value in Bridgeport's position in New York's hinterlands, just on a larger geographic scale. He moved to the City on the Sound, fifty-nine miles from the city, for its proximity and ease of access to New York, what he believed to be greater Bridgeport's two essential "elements of prosperity."[8] The periphery was a unique place to experiment in urban form because it could be at once accessible to regional patterns of urbanization yet distinctive.

Steinway's manufacturing town and Bridgeport framed the Sound's coastal metropolitan corridor in the late nineteenth century. The relationship between core and edge existed at multiple scales simultaneously: Steinway and Barnum were at work on two different edges in the same urban context. The city's jurisdictional limits formed one edge. But this edge was impermanent and shifted multiple times in the late nineteenth century. Until New York City annexed mainland territory, the future South Bronx, in 1873, its limits were coterminous with Manhattan Island. Long Island, the site of the future boroughs of Queens and Brooklyn, was not incorporated into the city's limits until 1898. The farthest reach of the city's nineteenth-century hinterland formed a second edge, although it was more a transitional zone than a specific boundary line. In the second half of the century, the city's influence encompassed New York Harbor, including both sides of the East River, and north shore of the Sound as far east as Bridgeport. In the 1840s and 1850s, Barnum declared Bridgeport to be "about the proper distance from the great metropolis" for real estate and manufacturing investment.[9] The showman grasped that because of the advantages of its harbor and railroads, Bridgeport was positioned to become the northeastern anchor of greater New York's orbit. In the 1870s, the Steinway settlement developed on the eastern shore of New York Harbor, just beyond municipal boundaries but still within the city's economic sphere on the upper East River. In the words of a guidebook, the coastal corridor of the Sound "well-nigh" represented "a continuous extension of New York City."[10] Bridgeport and the Steinway settlement mark the multiple scales of urbanism, that of the city outgrowing municipal limits and the rise of an urban hinterland.

Steinway and Barnum's work in underdeveloped edge spaces offered new and different opportunities to create model urban settlements. Bridgeport and the Steinway settlement represent the tradition of model or company towns, the successors to planned antebellum mill villages, plantations, and communitarian experiments.[11] A different version of the history of the growth of the hinterland of New York City could focus on how agricultural and nonagricultural land uses came into contact as the city expanded. The Sound's marine economy was robust. There was ship-building, whaling, and fishing (lobster, scallop, blue crab, flounder, striped bass, and bluefish) all along the coast along with some of the nation's most productive oyster farms. The edges of the city and its hinterland were not a blank slate awaiting urbanization.

Barnum's Bridgeport and Steinway's settlement were progenitors of a new urban landscape—a recentered city. Model company towns became satellite manufacturing nodes around major cities in the second half of the nineteenth century. Manufacturing needs motivated Steinway's move, but once he was in Queens, he built more than factory space. During the late 1800s Queens became home to several planned communities and in-dustrial villages; the Steinway settlement was the largest and best known and left the most enduring mark on the borough's urban fabric.[12] While not a manufacturer, Barnum nevertheless took it upon himself to shape Bridgeport as a model industrial town replete with waterfront villa district, park, and planned industrial core. His desire to create a new space came out of his personal vision for the best path of urban growth. Reflecting on his investments in 1871 he declared, "I wanted to build a city. I 'had . . . Bridgeport on the brain.'"[13]

Steinway's and Barnum's city building aligned with the judgment from civic leaders and citizens in the city center and its surrounding territory that cities could no longer grow organically. Roiling change characterized New York in the second half of the nineteenth century. The multifaceted modern city required self-conscious "building."[14] Barnum was an early voice for intervention in laissez-faire urban growth, arguing for large-scale public works as a way to stabilize and encourage economic growth. "Con-servatism may be a good thing in the state, or in the church," he argued, "but it is fatal to the growth of cities."[15] While Steinway did not make such explicit declarations, in effect his investments in Queens aligned with this progrowth philosophy. Both men explained their public works investments as largess, but in serving the public good they also enjoyed personal re-turns. Steinway and Barnum linked their personal reputations to their city building and courted regional growth. For example, the piano manufac-turer hoped to isolate his workers in Queens and forestall strikes, but he also assumed the city would eventually outgrow Manhattan and absorb his settlement.

These benefactors-cum-developers embodied the ways in which leading citizens originated city-hinterland relationships for greater New York in the era before professional city planning. Steinway and Barnum invested in private-sector projects such as workers' housing, factories, and commercial amusements. They made little distinction between private-sector development and privately funded public works projects such as municipal infrastructure, since both types of investments were necessary for urban growth. Both men also enjoyed influential public-sector positions with which they shaped growth. Steinway was appointed to lead New York City's Rapid Transit Commission, which put forward the city's first subway and elevated railroad plan, and Barnum served in state and city government. While private and public interests often collided in urbanization, the work of Steinway and Barnum makes clear that private enterprise and municipal goals intersected and overlapped as well. Moving between the roles of public works officials, benefactors, and private speculators—and at times consolidating these positions—Steinway and Barnum were city builders.

*Favorable Features for Settlement*

The coastal environs of the upper East River and Long Island Sound offered favorable conditions for experimental urban settlements. Coastwise networks built from trade and travel along the coast had linked the upper East River and Sound since colonial settlement in the seventeenth century. Yet settlement was localized in nature. Long Island blocked the Sound from direct Atlantic trade, and it never developed a port large enough to compete with New York or Boston.[16] Maritime networks encouraged connectivity, but the coast's environment presented challenges to urbanization. The bays of the upper East River and Sound were more likely to be shallow and shoaling rather than deep, which discouraged commercial port development. The boggy ground of the vast tidal marshes of the river and Sound could isolate settlements and encouraged localized development. The final feature that made New York's hinterland conducive to localized experiments in urban form was the lack of large or strong municipal governments.

That both Barnum and Steinway invested in coastal sites reflected the long-standing preeminence of the maritime economy and coastal trade in the region. Steam and rail transportation transformed the Port of New York into a global port while also bolstering regional shipping and manufacturing. Steinway developed on the urban periphery of New York Harbor, while Bridgeport grew as the eastern anchor of the Sound's coastal environs. The two settlements bookended a segment of the city's coastal hinterland in the second half of the nineteenth century. By the middle of the century, transportation networks connected the Sound's southern ports to each other and New York City in a distinctive coastal region. Farmers, industrial producers, and fishing interests first shaped this network; the development of coastal railroads ushered in an era of urbanization linked to the emergence of the New York metropolis. The Long Island shore of the Sound lacked a main transit corridor, but the New York–New Haven Railroad ran along the Sound's northern shore. The rail line paralleled the Post Road, which since the colonial era had linked Boston and New York and the intermediary coastal towns. At the end of the nineteenth century, the conglomerate New York, New Haven, and Hartford Railroad (NYNH&H), with its steamboat lines to Long Island, Brooklyn, and Manhattan, constituted the principal tendon of greater New York's connectivity.

Like so many nineteenth-century American cities, Bridgeport's fortunes changed when the railroad arrived at midcentury. Bridgeport became the terminus of three railroads, the Housatonic (opened 1840), the New York–New Haven (opened 1848), and the Naugatuck (opened 1849). Before the railroads arrived, Bridgeport had epitomized the agricultural settlements that characterized the Sound's shoreline. A harbor port at the mouth of the Pequonnock River, Bridgeport occupied a coastal plain backed by a series of terraces that rose fifty feet to present a commanding panorama of the Sound. In the first decades of the 1800s, the city of Bridgeport did not exist; it was but a rural port of two hundred and fifty farmers and fishermen. By 1827 the settlement stretched seven blocks along the wharves of the Pequonnock and only four blocks inland before giving way to open lots.[17] In 1836, residents declared Connecticut's borough government system too limited for their needs and incorporated as the city of Bridgeport with a population of around 3,300.[18] *City* had referred to the type of government created, not to the size of the settlement. The settlement became city-like in form and function following the opening of the railroads, when Bridgeport became the market for central Connecticut farm products as well as the manufactured goods produced in the Housatonic and Naugatuck river valleys, the hubs of Connecticut enterprise. In the 1840s, the decade railroads arrived, Bridgeport's population grew by 130 percent; in the following decade it grew by another 76 percent to number more than thirteen thousand people by 1860.[19]

Across the Sound and to the southwest, the shores of Queens on Long Island were similarly rural. Established in 1683, Queens County contained no cities until 1870. Truck farming for city markets, nurseries, and wholesale flower farms characterized the majority of eastern Queens into the twentieth century. Development was minimal. The limited suburbanization that occurred followed the Flushing Railroad from Hunter's Point to Flushing beginning in the 1850s. The limited urbanization that occurred concentrated in the west along the East River. In 1870 Long Island City incorporated on this waterfront, the result of a home-rule drive from industrialists in the Hunter's Point district on Newtown Creek, the city's southern limit. Five years after incorporation, the city's population was 15,500.[20] Long Island City's five wards covered only seven square miles, but even after incorporation, the city's existing settlements, while close to each other, remained independent places. An observer reported that Long Island City was made up of "three communities—Astoria, Ravenswood and Hunter's Point—so distinct and separate that in common parlance their connection with each other was generally ignored."[21] The Steinway settlement covered but four hundred acres, half a square mile, in Astoria. In Long Island City, like the rest of Queens, urban development was piecemeal and small scale. Steinway used this neighborhood autonomy to his advantage.[22]

Long Island City's fiscal problems and hands-off approach to government further encouraged local autonomy. A chronic shortage of revenue, due to graft and the unwillingness of residents to pay taxes, stymied public works. The city accumulated a crushing debt of more than a million dollars by 1876; in contrast, the entire state of New York had an indebtedness of only ten million at the time. In early August 1882, Long Island City defaulted on the interest due on its bonds.[23] Overwhelmed by debt, the city was incapable of and uninterested in municipal oversight for its northernmost ward. Although he never incorporated his settlement as an independent municipality, the fragmentation of Queens development and the corruption of Long Island City government left Steinway free to assemble a company town there.

The material nature of Queens's wetlands also encouraged such localism. Marshes such as Lubbert's Swamp in East Astoria, the mile-wide, three-mile-long Flushing Meadows, and the Ravenswood Swamp isolated settlements from one another. Until the mid-nineteenth century, the territory that became Long Island City had few roads or settlers because of its tendency to flood. The industrialization of Hunter's Point led to the reclamation of Lubbert's Swamp after the Civil War, but farm towns isolated by tidal marsh remained the norm across Queens.[24] Steinway made the most of northern Queens's isolation. Steinway looked for a locale that was isolated from labor politics and government supervision. Steinway &

Sons battled with workers over unions, the eight-hour day movement, and wage increases for five decades. Steinway wanted to move his factory to "a place outside the city . . . to escape the machinations of the anarchists and socialists."[25] Queens's boggy waterfront had dissuaded intensive land use beyond agriculture, and these same conditions made the territory ideally isolated for a company town.

Greater New York's expansive estuaries shaped agricultural and then urban development in the waterfront settlements of the city's environs. The eastern shore of Bridgeport's Pequonnock River, like much of the coastal marshes on the Sound, had long been agricultural. The thick salt marsh cordgrass that grew in the Sound's tidelands was harvested and sold as hay and feed by local farmers.[26] Salt marshes were also plentiful along the upper East River in Long Island City. Tidal wetlands stretched between the mainland and Berrian Island. Directly to the east, marshes spread around Luyster Creek to ring the elevated bluff of Luyster Island. The Steinway settlement's location in the Fifth Ward was so rural it was not included in the Beers real estate atlas of the developed territories of Long Island in 1873. When Steinway first purchased Luyster's Island, the future site of his factory, woodlot, salt meadows, and fields lined the shore. A handful of buildings scattered along the district's only road, Old Bowery Bay Road, were the extent of the region's development.[27] Steinway and Barnum made the most of the low real estate value of marshes in agricultural and urban markets. Marshland was generally considered unproductive beyond the salt hay industry. Since marshes made poor farmland, marshland was inexpensive and usually constituted the least expensive land in port cities. Developers interested in deep navigational channels and dock space had little use for shallow tidelands. Steinway and Barnum hoped to turn such inexpensive land into profits.

Barnum and Steinway disparaged coastal New York and Connecticut's marshy environment. Steinway described his new territory as "waste lands, and vacant lots"; the New York Times employed the same description for the marshes of Queens.[28] Soon after arriving in Bridgeport, Barnum began a campaign calling for the improvement of the city's waterfront. He eventually funded landfill operations on the eastern shore of the Pequonnock and claimed he turned a "filthy, repulsive, mosquito-inhabited and malaria-breeding marsh into a charming sheet of water." The declaration of property as "wasteland" in its natural state was strategic; if they claimed to start with worthless property, Barnum and Steinway could then boast of personally "improving" the land and elevating the value of local real estate with their subsequent projects. The investors aimed to signal the

**1.1.** Steinway, Queens. Steinway built his manufacturing settlement on this undeveloped expanse of low-lying tidal marsh and fields. *Source*: Courtesy of the Greater Astoria Historical Society.

start of a new era of urban growth in their adopted settlements. As environmental historian Ted Steinberg observes in his ecological history of greater
New York, the word *improvement* signified property development to bring
profits.[29] Urban development could make valuable real estate out of such
idle, untapped landscapes. Coastal landfill was a defining characteristic of
waterfront property development in greater New York. Extensive landfill
in the Upper Bay estuaries and land underwater began in the early 1900s,
shrinking the bay to three-quarters of its 1845 size by the 2000s.[30] Barnum
and Steinway had vested interests in describing the coastal environment
as a malarial marsh, since painting the unimproved shoreline in desolate,
unhealthy terms made reclamation that much more impressive.

*Manufacturing Cities*

Upon arrival in their adopted towns, Steinway and Barnum dedicated
their formidable personal influence and financial resources to open
streets, finance rapid transit, public parks, and residential developments,
and attract industry. Both urban experiments incorporated the attributes of both company and model towns. The existence of a principal
industry, the town builder's status as a dominant property owner, and a
community populated primarily by the employees of a single company
or group of companies with wide-ranging control over daily life characterized company towns. Company towns emphasized the demands of
industry, economic growth, and labor control over conceptual order. In
model towns, symbolic form was a means to social and physical planning.[31] Neither Steinway nor Barnum expressed a formal urban ideology or commissioned comprehensive designs. Yet both experimented in
urban-industrial form and social planning, part of a national trend that
inspired an avalanche of civic, political, and intellectual interest in the
second half of the century.[32]

Barnum's midcentury arrival was deemed "the first great boom for the
celebrity of Bridgeport"; because of him, the *New York Tribune* claimed,
"people of wealth and culture wanted to live in Bridgeport, and they
came."[33] The showman boasted that he commissioned "buildings of a
novel order" in hopes they would "serve as an advertisement of my various enterprises" in Bridgeport.[34] Barnum had his first Bridgeport home,
Iranistan, patterned on George IV's oriental Brighton Pavilion, replete
with domes and minarets. He deliberately oriented Iranistan in full view
of the New York–New Haven Railroad tracks and scheduled one of his

**1.2.** Barnum captured the public's attention in Bridgeport. The NYNH&H Railroad steams by his property. *Source*: Samuel Orcutt, *A History of the Old Town of Stratford and the City of Bridgeport, Connecticut*, vol. 2 (New Haven, CT: Tuttle, Morehouse, and Taylor, 1886), 843.

circus performers, dressed in "oriental costume," to outfit an elephant with a plow and till his lawn according to commuter timetables.[35] This stunt, he crowed, was a "capital investment" in his brand of entertainment and personal fame. Widely incongruous with the area's colonial, Queen Anne, italianate, and Gothic homes, Iranistan captured the attention of the national press. *Frank Leslie's Illustrated Newspaper* lauded the home as a "thing of beauty . . . a marvel of wonder and an honor to all America."[36] Newspapers from Milwaukee, Wisconsin, to Fayetteville, North Carolina, regularly reported on Barnum's doings in Bridgeport.

The King of Humbug fashioned himself as Bridgeport's ultimate booster. Barnum frequently used his circus's fame to further not just his personal celebrity but to publicize what he deemed his "pet" city.[37] In 1881, while his five-hundred-man and seventy-car circus train wintered in town, he donated elephants to push a derailed engine back on track. In December 1888, he assisted the Bridgeport Public Works Department in "testing" a new bridge with twelve elephants weighing thirty-six tons and a crowd of two hundred spectators.[38] The image of Barnum and his elephants on the Stratford Avenue Bridge exemplifies Barnum's merging

of city improvements and his celebrity. The popular press repeated and endorsed Barnum's assessment of his stature as Bridgeport's leading citizen.
As far away as Missouri, newspapers praised Barnum's "nature as an organizer of men and systems" as the reason for his significant influence over
urban growth.[39] "He could not live in a town without being the source and
center of the forces that uprise to improve it," *Century Illustrated Magazine*
declared. "If ever a city can point to one man as its preeminent benefactor,
that city is Bridgeport, and that benefactor was P.T. Barnum."[40] Reputation
aside, the showman evinced a sincere interest in facilitating urban growth
in greater Bridgeport.

Steinway's and Barnum's real estate investments empowered them to
influence the nature of development. Steinway endeavored to control as
much contiguous property as possible. In twelve separate purchases between 1870 and 1872, Steinway acquired more than four hundred acres
north of the Long Island City neighborhood of Astoria, including the
former Pike estate and Luyster farm. Barnum took a different approach,
diversifying his holdings rather than consolidating them, yet he amassed
comparable influence over real estate in his adopted city. Bridgeport's
land records reveal that Barnum invested in every ward of the city. This
activity included real estate leases and sales as well as the purchasing of
mortgages. For example, in a single volume of the city's land records for
1855 and a portion of 1856, Barnum was recorded as a grantor (seller) in
forty-four separate transactions and a grantee (buyer) in eighteen separate transactions.[41] He acquired additional commercial buildings downtown and a winter camp for his circus on the western edge of town, and he
opened Recreation Hall on Main Street, home to business space as well as
community amenities including a theater, gymnasium, and skating rink.[42]
His purchases included high-end properties such as multiple estates in the
elite Golden Hill district. He also invested in working-class housing. In
the 1850s he speculated in worker housing and factory complexes in East
Bridgeport. And from the 1860s to the 1890s he invested in an elite villa
and park district fronting the Sound.[43]

Steinway and Barnum turned themselves into city builders by privately
initiating public works, beginning with streets. The lack of a strong municipal public works program enabled Barnum to personally instigate an
aggressive agenda within the city. He began his Bridgeport city-building
career as an unofficial street planner around Iranistan in Golden Hill. To
advance his real estate investments, the showman declared "there was
other work to do," in other words, opening streets. He had Iranistan Avenue and connecting streets built and straightened, and he widened others,

including Hanover Street, at his own expense. Barnum explained his approach to street opening in exclusive Golden Hill and beyond: "When land was dull and people did not seem disposed to buy" properties he wished to sell, he would pay the expense of laying out a new street; "Then everybody rushed in to get corner lots."[44]

As a private street builder, Barnum criticized local interests, such as one farmer who opposed him, claiming he blocked "my way, his own way, and the highway." In his autobiography, Barnum characterized his neighbors as "old fogys" who complained "we don't believe in these improvements of Barnum's. What's the use of them. . . . The new street will cut the pasture or mowing-lot in two. . . . It was bad enough to have the railroad go through . . . but this new street business is all bosh!" Barnum dismissed his critics as overwrought fiscal conservatives shortsighted for their inability to see that public improvements and personal gains could go hand in hand. Claiming that local property owners "looked upon me as a restless, reckless innovator, because I was trying to remove the moss from everything around them, and even from their own eyes," Barnum fashioned himself a reputation as a progressive benefactor dedicated to the greater good. He presented his street opening in distinction to "grasping farmers," who encroached on and fenced off public land.[45] While street building was expensive and might cut into private property, Barnum believed the ends would justify potential individual losses by encouraging economic growth in the city at large.

Steinway built streets to encourage both commercial and residential development in his settlement. In November 1871, the manufacturer walked his Queens farmland with architect and engineer Henry Reck, who surveyed for streets. While the settlement was never comprehensively designed, Steinway and Reck arranged a street plan in advance of development. Although most of the sixty platted blocks were not built right away, the plan revealed Steinway's awareness that not just jobs but public works and utilities would attract residents.[46] He invested in both his factory and town simultaneously. On Friday, June 20, 1873, Steinway recorded a typical visit to his growing settlement: "to farm, where it is splendid, Mill & foundry in fine running order. . . . Steinway Ave. making good progress. Cowing is driving wells. . . . Engine & boiler are tested in the factory yard."[47] Steinway charted foot by foot the progress as his factory building walls went up—a private project—and excavation work set drainage on Steinway Avenue—a privately funded public works project. He also oversaw the public works of marsh landfill and street grading and curbing.[48] Across the East River in New York City, a variety of municipal commissions

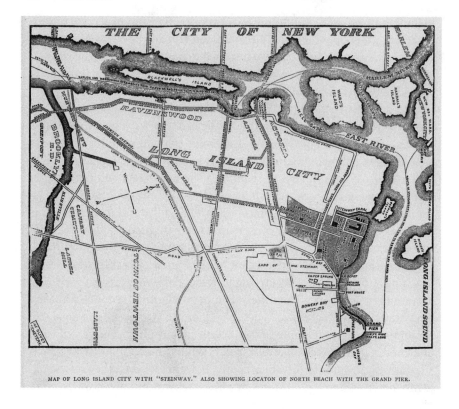

MAP OF LONG ISLAND CITY WITH "STEINWAY." ALSO SHOWING LOCATON OF NORTH BEACH WITH THE GRAND PIER.

**1.3.** Map of Steinway property ca. 1874. This map includes the horsecar routes of the Steinway & Hunter's Point Railroad, the predecessor of the Steinway Railway that ran from the Steinway & Sons factory to the piers of the Bowery Bay resort district. *Source*: J. S. Kelsey, *History of Long Island City, A Record of Its Early Settlement and Corporate Progress* [ . . . ] (New York: Long Island Star, 1896), 44. Courtesy of the Brooklyn Historical Society.

oversaw the infrastructure work that Steinway pursued as a private citizen. In the late 1860s, the state legislature authorized the Board of Commissioners of Central Park to create a street plan for northern Manhattan and even southern Westchester County, the future Bronx. While pubic interest and private enterprise often conflicted over public works, collision and competition were not guaranteed. In the Steinway settlement, private enterprise established a public works program.

Steinway enjoyed greater autonomy than Barnum in shaping the built environment because he did not have to contend with other property owners. At first Steinway used personal funds to build streets, but over time the work of leveling land and laying out roads became Steinway & Sons' business. Steinway eventually incorporated the Astoria Homestead Company to oversee his extensive real estate holdings in January 1889. He personally provided funds to establish the company and eventually amassed all but fourteen of its 2,500 shares.[49] Capitalized at a million dollars, the company enabled Steinway to realize improvement projects without using personal or Steinway & Sons funds. The Astoria Homestead Company sold bonds to pay for projects such as street curbs. By the mid-1890s Steinway held controlling interests in the company as well as his brother George's public improvement corporation, the Ravenswood Improvement Company. He eventually streamlined operations and consolidated the firms. Urbanization would take place on his terms, on his property and streets, and it would bring him profits.[50]

Barnum began his city building in Bridgeport proper but eventually decided to establish a new city where he claimed there were no "conservatives" to battle.[51] Like Steinway, he sought direct control over the physical aspects of the urban environment and set his sights beyond Bridgeport's municipal boundaries. In 1851 Barnum partnered with a prominent Bridgeport landowner, the lawyer William H. Noble, to develop 124 acres of Noble's family's estate on the east side of the Pequonnock River. Playing off the reputation of the larger city directly across the river, they renamed the area East Bridgeport. Barnum and Noble filled tidal plats and laid out twenty-six blocks in a wide grid. Like Steinway, they chose not to hire a landscape architect but directed the laying out of the neighborhood themselves. The investors established an industrial district as well as a planned working-class residential neighborhood.[52] Barnum oversaw the laying out of the industrial core, what he believed to be the ideal combination of factories and worker housing. He recognized that he effectively acted as a planner, although he did so a generation before the professionalization of urban planning.

Steinway and Barnum stimulated urban growth by investing in manufacturing. Their city building created a manufacturing periphery for greater New York. Bridgeport grew from a rural outpost to a manufacturing center and railroad hub. Steinway attracted additional industrial development to his village. He diversified the industry of his settlement and also invested in regional connectivity via rapid transit and bridge and tunnel projects. The Steinway settlement and East Bridgeport demonstrated their benefactors' personal designs to secure returns on their investments and attract capital to their model cities.

Barnum did not plan a one-shop town or seek the level of control associated with company towns. He did, however, play a central role in Bridgeport's industry. Barnum believed that Bridgeport's advantages as a port and railway hub needed only to be supplemented by development projects. But unlike the industrial towns of interior Connecticut, Bridgeport was not tied to the state's major sources of capital, Hartford and New Haven. The city was also removed from the precision machinery industries of the Connecticut Valley. To secure the success of his settlement, Barnum convinced manufacturing concerns to relocate to East Bridgeport. In 1852, a year into their city building, Barnum and Noble built and leased the first factory in the district. Four years later, Wheeler & Wilson relocated its sewing machine operation from central Connecticut to East Bridgeport. Barnum asserted that the arrival of Wheeler & Wilson "was a fresh impulse towards the building up of the new city and the consequent increase of the value of the land belonging to my estate."[53] In 1867, the Union Metallic Cartridge Company and the Winchester Repeating Arms Company opened factories as well.[54]

In the late nineteenth century, New York City was the manufacturing capital of America. Industrial production flourished in the city's environs from Newark, New Jersey, on New York Harbor's western shore, to Yonkers on the Hudson River, to Sound ports. Long Island City had more than one industrial center; in the 1870s and 1880s, oil refineries, asphalt factories, and light manufacturing industrialized Hunter's Point, followed by chemical factories and gas plants. Newtown Creek became infamous for the pollution released from Hunter's Point. Port cities on the Sound produced hats, carriages, locks and keys, buttons, screws, nuts, bolts, and iron that they shipped, along with agricultural products from the New England interior, to New York. Norwalk, Connecticut, foundries and hatting concerns prospered while Port Chester, New York, blossomed as a manufacturing and financial center, but neither matched East Bridgeport's pace and scale of urbanization and industrialization, nor the national renown it earned by the 1870s.[55] East Bridgeport and Steinway came of age as industrial nodes of the metropolitan region.

By 1871 the sprawling manufacturing complex of Steinway & Sons was under construction along the northwestern shore of Queens. While Steinway had deemed the landscape a waste of marshland, he maximized the malleable environment of the coastal edge. The company built a wharf and bulkheads and expanded the existing canal infrastructure on Luyster Creek, deepening the natural creek bed.[56] Making the most of his coastal site, Steinway built a basin to hold lumber floated to his sawmill to prevent cracking by keeping logs moist. A new dock, tide gate, and

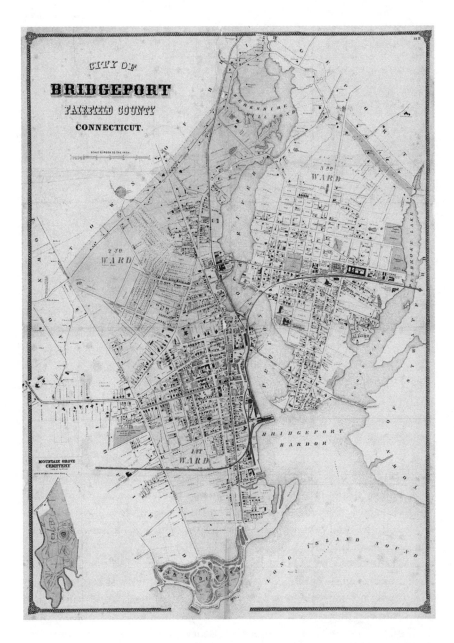

**1.4.** The city of Bridgeport in 1867. Crescent Avenue paralleled the NYNH&H Railroad through the center of East Bridgeport, clearly visible curving across the Pequonnock River. Crescent Avenue reflects Barnum's interest in controlling industrial development and population density. *Source*: City of Bridgeport, Fairfield County, Connecticut, Published by Beers, Ellis & Soule, 1867, lithography, black printer's ink and watercolor on wove paper, 2012.312.125, the Connecticut Historical Society.

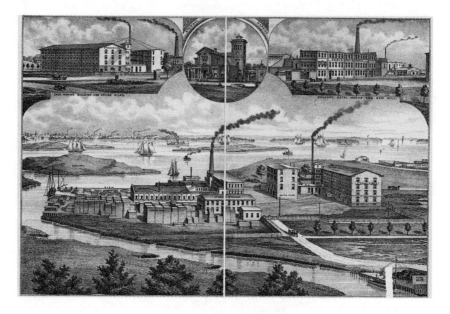

**1.5.** Steinway & Sons in the 1880s. Upper Manhattan, pictured on the horizon, was a two-mile sail or steam across the East River. While on the East River, this portion of the Queens coast was sometimes referred to as part of Long Island Sound. *Source:* Steinway & Sons Pianoforte Company, Metal Works and Lumber Yard, ca. 1875. Courtesy of the Queens Borough Public Library, Archives, Ephemera Prints Collection.

three-hundred-foot bulkhead enclosed the large basin. By 1873 the company had vertically integrated the operations of piano making. It included the basin and sprawling lumberyard, a steam-powered sawmill, an iron and brass foundry, and boiler and engine houses.[57] In 1879, Steinway & Sons employed four hundred workers at its Queens factory and six hundred at the Manhattan factory. Manhattan workers continued to piece together instruments, while Steinway settlement workers, with space to spread out, sawed lumber, fired the kiln to dry wood veneers, and heated iron presses to cast piano cases.[58]

Steinway not only built a modern factory complex in Queens but also invested Steinway & Sons funds to draw additional manufacturers to his company town. Until 1893, Steinway used the "Accommodation" account within Steinway & Sons to lend money to companies like the Astoria Silk Works, the Astoria Veneer Mills, and the Daimler Motor Company. While this approach tangled the books of the piano manufacturing firm with the investor's streetcar and housing companies, the practice allowed Steinway to transfer capital across his projects. Such overlaps were commonplace.

When he electrified his streetcar lines, the power plant also powered his piano factory. The Astoria Homestead Company used the Steinway & Son's Fourteenth Street office in Manhattan and the Steinway & Hunter's Point Railroad Company building in Queens as headquarters. Daimler Motor's boat works shared the Steinway & Sons boat basin.[59] Daimler Motor exemplified Steinway's tendency to merge personal investments and settlement growth. Founded in 1888 in Germany, Daimler Motor held patents on the first practical internal combustion engine powered by gasoline. Steinway signed a contract for the rights to manufacture the engine in America and opened the American Daimler Motor Company. The company headquartered at 931 Steinway Avenue, along with the Steinway Land Office and, for a time, the Ravenswood Improvement Company. The diversification of manufacturing interests might have appeared to a casual observer to be a departure from Steinway's company town goals. Ultimately, however, Steinway's overlapping of companies was more revealing of his aims for the settlement than his apparent diversification. The settlement remained a one-investor town where Steinway combined manufacturing and his personal ideas for urbanization.

Barnum's and Steinway's activities in public transit fit the mold of real estate promoters. Both diversified their interests from speculative real estate to invest in transportation infrastructure because they understood accessibility would probably increase the value of their holdings. Barnum and Noble secured a charter from the Connecticut state legislature to build a bridge across the Pequonnock River and make East Bridgeport accessible to Bridgeport's business district.[60] Down the Sound, Steinway leased, bought, and merged lines to outmaneuver competitors and gain control of Long Island City's horsecar and streetcar railroads in the 1880s. The first line opened in the Steinway settlement in the mid-1870s. But franchises lapsed, bankruptcies were frequent, and investors who built without proper permits were restrained by the county court. Steinway targeted these chronically ailing traction lines. In April 1883, he bought his first line, the Long Island Shore Company Railroad, which served Steinway village. He had begun investing in transit a decade earlier when he had bought $10,000 worth of the company's bonds to convince the railroad to open the line to his settlement.[61] Steinway finally consolidated all the streetcar railroads of Long Island City in 1892, incorporating the Steinway Railway Company to manage a network of four lines and ferry connections.[62] In the same decades, rapid transit was transforming Manhattan, and Steinway knew it would facilitate urbanization in Queens. The *Times* credited the Steinway Railway as "one of the potent factors" of growth in greater New York.[63]

Steinway invested in the local and regional accessibility of his settlement, improving communication not just across Long Island City but across the East River. For over two decades he invested in bridge and tunnel projects, beginning with Blackwell's Island Bridge and joining the board of directors in July 1875.[64] Steinway stood by the project despite nearly two decades of delays because he expected that a bridge from Manhattan across Blackwell's Island to Queens would "vastly increase the value of [his] property and that of Steinway + Sons in Long Island City."[65] Steinway understood that improved regional transit might draw the city's borderlands into closer contact with the commercial networks and capital of the urban core.

A single day in May 1887 encapsulated the intensity of Steinway's interest in transportation infrastructure and his willingness to diversify his investments. His day began with a visit to Manhattan's Public Works Department to see proposed Blackwell's Island Bridge plans. He then lunched with an investor interested in building an East River tunnel. In the afternoon he discussed the rates of ferriage for his Steinway and Hunter's Point Railroad with Astoria Ferry executives.[66] Steinway recognized that an East River tunnel to Long Island City would link his streetcars to Manhattan's lines. The New York City Board of Aldermen's Committee on Docks and Committee on Railroads alleged such a tunnel would detrimentally siphon business from Manhattan to Long Island.[67] Steinway might have agreed with this conclusion but not with the provincial perspective that framed such a shift as a loss. The industrialist understood that areas once considered distant parts of the urban fringe were but segments of a connected, regional city that required improved communication. In late 1895, after a meeting with his fellow investors, Steinway expressed confidence that the tunnel would bring "vast benefit for St+S. and especially my lands in Queens County."[68] While economic depression and an explosion slowed construction, the first Queens–Manhattan subway would eventually run through the Steinway Tunnel, as it became known, in 1915.

Private enterprise and public interest frequently aligned for Steinway and Barnum. They served the public interest indirectly, through private projects such as street and park building that had public benefits, or directly, in appointed or elected positions. During the years that Barnum privately opened streets and invested in East Bridgeport, he also enjoyed political influence. The showman was elected four times to the Connecticut General Assembly in the 1860s and 1870s and subsequently served as Bridgeport's mayor in 1875. As mayor, Barnum called for public baths and an improved water supply and drainage system.[69] As a member of state and municipal government, Barnum blurred the line between private citizen and public servant, presenting his private growth goals for the city as part of the public interest. Following his term as mayor, Barnum joined the city's park commission, promoting large parks with beaches and bandstands. In the 1890s, Mayor Hugh J. Grant named Steinway head of the rapid-transit commission tasked with laying out routes for elevated and underground railways for Manhattan and the future borough of the Bronx.[70] The commission was characteristic of city elites' nineteenth-century efforts to shape urbanization through public appointments. While private enterprise and real estate interest sometimes caused conflict between individual investors and municipal plans, they could also intersect to benefit both public and private sectors.

City-building investments paid dividends. By July 1859, real estate in Barnum and Noble's model neighborhood—not including territory donated as streets, parkland, school, and church property—was valued at $1,200,000, almost four times its 1851 worth.[71] Steinway also claimed impressive returns on real estate investments. In 1870 Steinway had paid $311.25 for 43,560 square feet of land; by 1874 he was selling 2,500 square feet for $900. Even though the depression of the 1870s slowed development, by 1880 vacant lots had nevertheless risen in price to $300–$800. According to a finance report by R. G. Dun and Company in 1878, Steinway & Sons owned three thousand lots in Astoria worth about $250,000, although the company claimed the acreage was worth as much $500,000.[72]

*Attracting Residents with Homes and Parks*

While vested interests and a degree of paternalism drove both Steinway's and Barnum's approaches, both the Steinway settlement and Bridgeport became more than sources of profit for the benefactor-entrepreneurs. Both settlements offered worker housing and more than token cultural amenities for inhabitants. Both men tied their pursuit of wealth to the community by providing community amenities that could also bolster the prestige of their adopted cities. As Bridgeport's self-proclaimed benefactor and the driving vision of Steinway & Sons' company town, Barnum and Steinway forwarded personal planning agendas that established public space and landscape architecture designs as central to urban form.

To ensure growth, Steinway and Barnum needed workers to man factories as well as working-class renters and homebuyers. In East Bridgeport, Barnum and Noble offered inducements to attract working-class residents. In the Steinway settlement, the Astoria Homestead Company oversaw not only public works but also contributed to the private market of city building—housing. As early as 1872, Steinway had expressed interest in developing worker housing, adding to his notes on factory construction, "I think it is necessary to build some houses for the men to live."[73]

Company towns varied considerably in their policies toward worker housing. George Pullman, railroad car manufacturer, opened the nation's most famous company town in 1881 outside Chicago. In an attempt to make workers more docile, Pullman controlled all aspects of life in his town, from barring alcohol to overseeing housing.[74] The construction of a new town and modern worker housing was thus an essential part of the manufacturer's industrial investments. Steinway similarly hoped to isolate and control his workforce, but he distinguished his town from Pullman's by offering workers the chance to buy the homes he built. In 1883 Steinway explained the link he saw between strikes, worker morality, and New York City's housing to the Senate Committee on the Relations between Labor and Capital. He pointed to the tenement house as contributing to the troubles between labor and capital: "I think it is a great cause of dissatisfaction among the workingmen—the bad places that they have to live in, and the high rents they have to pay."[75] He boasted that his settlement allowed him to "withdraw [his] workmen from contact with [the] temptations of city life in the tenement districts" to environments where "they would be more content and their lot would be a happier one."[76] Steinway self-servingly conflated worker happiness with improved housing and creating a docile labor market, and he found Queens to be advantageous to these goals.

Unlike the classic company town of Pullman, residency in Steinway was not restricted to employees, and ads ran frequently to reach a wide audience of potential buyers.[77] Settlement residents could own their own homes, a departure from the traditional landlord-employer model of company towns. Steinway did not manage workers' housing as a landlord, but selling homes to workers nevertheless gave him power over the community's real estate: it meant that he, not an independent subdivider, controlled the type of housing built and profited from sales. Steinway commissioned housing that ranged from four-bedroom single cottages to three-bedroom row houses to larger villas for manager employees. Simultaneously wooing potential workers and boosting his real estate interests, Steinway emphasized that "no less than sixty of the men employed in [the] Steinway & Sons factory own their own homes." The 1900 federal census confirmed this boast.[78] Advertisements were addressed to manufacturers and mechanics "desirous of living in comfortable and cheery homes, with city comforts—in a thoroughly healthy and desirable neighborhood—in the vicinity of, and in easy access to, all parts of the City of New York." The advertisement ran frequently in the *Long Island City Daily Star* in the 1890s and was also printed as a German-language circular Steinway & Sons Land Office used to attract immigrant workers.[79]

Barnum and Noble also invested in housing. They sold low-cost houses on weekly, monthly, or quarterly installments. Advertising in the *Bridgeport Standard*, Barnum promised to loan anyone who could furnish in cash, labor, or material one-fifth of the construction costs the remaining funds necessary, at 6 percent interest, to buy a home. This payment scheme made homeownership possible for workers earning modest wages. Monthly payments equaled local rents.[80]

Both Steinway and Barnum invested in aesthetic details for their residential districts. For their single and multifamily worker housing, Barnum and Noble had hired the Bridgeport-based, nationally renowned housing design firm Pallisser, Pallisser, and Company.[81] The company was popular for its Queen Anne–style architectural details: overhanging eaves; towers topped with spindles; intricate designs of wood shingles, terra-cotta tiles, and bricks; and fancifully painted balustrades. These architectural details combined with East Bridgeport's wide streets, evocative of elite suburban neighborhoods, to contrast favorably with downtown Bridgeport's narrow streets and dense commercial blocks. To increase neighborhood desirability and land values, Barnum and Noble donated land for Washington Square Park and funded its Pallisser, Pallisser gazebo.[82] Gracious italianate and Queen Anne villas ringed the park, and these were backed by more modest worker housing. In the Steinway settlement, shutters and crown molding ornamented the homes, details often absent from the utilitarian homes of company towns. While Steinway also claimed to have established a villa landscape for his company town, in reality his village mixed rural and residential landscape features. Rather than an elite enclave, the company built about forty-five homes, most of them attached row houses, just a few blocks south of the firm's industrial campus. Weedy lots and the wispy, just-planted trees that lined the street were visual reminders of the settlement's newness and spoke to the area's ongoing transformation from rural community to manufacturing town.[83] The edge was a continuum, a transitional zone that knit together urban, suburban, environmental, and industrial features. For a speculator such as Steinway, the underlying message of the transitioning landscape was that substantial profits might be realized if growth continued.

Barnum and Steinway frequently claimed that their city building served the public good, but the philanthropic and economic goals of these benefactor-entrepreneurs were inherently entwined. Testifying to the Senate Committee on the Relations between Labor and Capital, Steinway credited his removal to Astoria as the "only thing" that could "give the workingmen a chance to live as human beings ought to live." He boasted that in the Steinway settlement "there is no sickness or anything of that kind there, and they are all feeling comfortable and happy." Taking a page from Barnum's publicity tactics, Steinway flattered himself when he testified that other manufacturers should emulate his approach. Barnum and Noble also declared largesse in their city building. They claimed that they regulated density, selling alternate lots to prevent overcrowding. This tactic also helped the landowners turn a profit. If restricting density were their only goal, they would have simply created larger lots. Barnum publicly insisted that rational development of municipal infrastructure was essential to economic growth. But such framing served a purpose as Barnum crafted his celebrity. He merged the role of philanthropist and speculator into a single persona, fostering the role of the benefactor-cum–city builder. Barnum was known to say, "with a twinkle in his eye," that his real estate speculation in East Bridgeport "may properly be termed a profitable philanthropy."[84]

As benefactors-turned-planners, Steinway and Barnum provided cultural amenities as well as public spaces in the dedication of community services and the preservation or augmentation of the natural environment. Steinway received praise from the press for letting ninety acres to the Association for Improving the Condition of the Poor, which supplied the lots as farms to the poor whom it served. Integral to the company town was the allocation of Steinway & Sons funds to provide a full range of German cultural services. The firm paid for the construction of a school, the salary of a German-language and music teacher, a free kindergarten, and a free circulating library. In 1881 Steinway also opened a public bathhouse with fifty dressing rooms on Bowery Bay, making the most of the open waterfront. Steinway expected such investments would keep workers happy and productive, an approach typical of company town entrepreneurs. This work became the prelude to Steinway's subsequent establishment of a waterfront resort district. Both the Steinway settlement and Bridgeport benefited economically from their coastal settings, but the shoreline remained in its natural state and a place of informal recreation. Building a city park or resort could serve to enhance surrounding real estate values. Barnum and Steinway looked to unlock this potential.

In the years preceding Barnum's park-building campaign, the showman invested in public spaces that preserved greater Bridgeport's natural landscape features. Contemporaries pointed out Barnum's dedication to environmental beautification in the growing city, claiming that between 1850 and 1890, he planted anywhere from three thousand to ten thousand trees across Bridgeport. The *Horticulturalist* praised his "love for rural ornament and disposition to add, both in a public and private way some lasting contributions to the subject of rural art and decoration" in Bridgeport.[85] In 1849, Barnum established eighty-acre Mountain Grove Cemetery. Mountain Grove was part of the midcentury phenomenon of the rural cemetery movement. Urban residents used parklike spaces on the edges of American cities as public pleasure grounds. On just such territory Barnum oversaw the laying out of ornamental trees and shrubs across the rolling hills. Copings, fences, and hedges ornamented burial plots.[86] Although the wealthy made up the majority of Mountain Grove Cemetery Association membership, Barnum donated free plots for members of the Bridgeport Fire Department, Civil War veterans, and the poor, specifying that they should be scattered across the cemetery, "not grouped like a potter's field."[87] Such work spoke to Barnum's dedication to preserving a democratic spirit in Bridgeport's open spaces, but the best articulation of his democratic vision of public planning was his final project along the shore.

In the 1860s, Bridgeport proper had no official parks. The only official public space was created as a public highway half a century earlier because the borough charter lacked any provision to create a park or square. In fact, Bridgeport's lack of municipal planning proved fortuitous in the creation of a waterfront park at the end of the century. In 1853, the Common Council surveyed the city south of State Street to the Sound for industrial development, but the mayor failed to sign the street plan, and the project stalled. If the streets had been laid out, they would have made it impossible to build a park along the shore.[88] Because these streets were never opened, the beach west of Bridgeport's public wharves remained farmland. In 1857, Mayor P. C. Calhoun addressed the Common Council, making the first legislative reference to public grounds or parks for Bridgeport. Calhoun announced he regretted that there was "scarce a foot of public ground within the city limits . . . and if such an improvement is ever to be made, no time should be lost." The mayor complimented East Bridgeport as being "wiser in their generation" and "so wonderfully developed" a gracious park when the city was laid out the decade before.[89] Barnum and Noble had laid out Washington Square, greater Bridgeport's first public park. Barnum's experimental town surpassed its predecessor and became the ideal to be emulated.

Barnum declared Bridgeport's lack of parks unacceptable. It was, he said, an "absurdity," and almost criminal, "that a beautiful city like Bridgeport, lying on the shore of a broad expanse of salt water, should so cage itself in." He proposed the beach southwest of downtown be preserved as parkland.[90] The plan, based on the natural features of the shore, foregrounded the ecologic boundaries of Bridgeport Harbor over the city's municipal borders. Barnum began to plan a park straddling the Bridgeport-Fairfield line. Thirteen acres of the proposed Seaside Park lay in the South End district of Bridgeport; the rest lay in Fairfield. Not everyone agreed with Barnum. Bridgeport landowners unwilling to sell their land refused to cooperate with Barnum's plan; the town of Fairfield rejected the plan. It fell to private initiative to realize the park. Gathering the support of prominent, public-spirited citizens including Nathaniel Wheeler and William Noble, with whom he had developed East Bridgeport, Barnum orchestrated public purchase and private donations to create a park.[91]

With characteristic bravado, Barnum took credit for Seaside Park's design, which he judged to be an "aristocratic arrangement."[92] In actuality, the park's winding drives and pastoral lawns were the work of Frederick Law Olmsted and Calvert Vaux. In an act of public service, Nathaniel Wheeler had hired the firm at personal expense.[93] Olmsted and Vaux began their partnership with the construction of Central Park in 1858. The designers won immediate acclaim for the park. Seaside was one of three public recreational grounds that Olmsted and Vaux subsequently designed for smaller cities in the early to mid-1860s. That New York's most influential landscape architects designed Seaside and that its aesthetic principles echoed their celebrated design for Central Park heightened Bridgeport's cachet. The park helped link the city to the prestige of the metropolitan region of which it was a part.

Like Central Park, Seaside featured a grand plaza entrance and open greens framed by stands of trees and rolling ground offering a variety of views. Olmsted valued the material nature of Bridgeport's shoreline. He declared, "there is no park in the country that meets the purpose of a fine dressy promenade as well or as cheaply as Seaside Park may be made to do."[94] The park was developed in three sections from east to west between 1865 and 1918. Olmsted and Vaux designed a promenade and lawns for the eastern point of the park in 1865. New York City civil engineer General E. R. Viele suggested the seawall design around the front of the park, the principal engineering work needed to realize the park. Laborers moved the rocks and boulders that had stymied Barnum's travels along shore. According to Olmsted, the "distinguishing feature" of the fifty-nine-acre park was its "system of driving and walking courses from which, on one side, can be enjoyed the sea breeze and an outlook upon the Sound."[95]

With Seaside, Bridgeport became a vanguard in the parks movement

**1.6.** Seaside Park. This 1867 map presents a design that aligns with the designers' description of their work. The easternmost corner of the park was first developed from 1865 to 1879. The mid and western sections were finished in the 1880s and 1918, respectively. The completed crescent-shaped park covered three hundred acres of shorefront. *Source*: City of Bridgeport, Fairfield County, Connecticut, Published by Beers, Ellis & Soule, 1867, lithography, black printer's ink and watercolor on wove paper, 2012.312.125, the Connecticut Historical Society.

to reconnect urban ports to their maritime environmental amenities. National newspapers and landscape architecture periodicals lauded Seaside as an example that New York City, San Francisco, and Charleston should emulate in their treatment of waterfront public spaces. Travel guides remarked on its charming views of the city skyline, the impressive villas nearby, and the Sound's broad expanse, fresh breezes, and perpetual panorama of sails and steamers.[96] "We go to the park for recreation and whatever influences us while there may be considered a part of it[. A] park should be measured not by its boundaries, but by its features and power to influence," a writer for *Landscape Architecture* explained. "If one walks on Seaside Park in Bridgeport it is the waters beyond him with the passing vessels and the dim shore of Long Island in the far distance that influences him more than the long narrow strip of land. . . . A landscape picture is a feature which may produce an inspiration or cause an influence"—in Seaside's case, the shore.[97]

The Steinway settlement's East River shoreline remained just as undeveloped as Bridgeport's shore in the 1880s. Rocky beaches, marsh grass, and mudflats stretched along the nearly seven miles of riverfront from the westernmost corner of Queens to the Sound. Along this shoreline, Bowery Bay was a popular destination for bathing parties, but it lacked infrastructure apart from the Steinway & Sons' baths. The shore was also difficult to access, since the closest horse-trolley stop was nearly a mile away. Steinway saw an opportunity in the underdeveloped leisure market of northwestern Queens and went about building a resort extension of his company town. The state had already authorized a proprietary "beneficial enjoyment grant" that allowed Steinway to develop the tidelands of Bowery Bay, including fifty-seven acres of lands under water.[98] In the United States, the Public Trust Doctrine requires that state governments preserve public use of tide-washed beaches and lands underwater, although development grants that preserve this access are allowed. New York State had long made commercial grants to promote dock building and commerce on the waterfront. In 1850, however, the state's land board instituted beneficial enjoyment grants in areas declared nonessential to commerce. These new grants allowed grantees to develop tidelands in any way that might "benefit" their needs, such as adding buildings or piers. Steinway used his to build a coastal resort.

Steinway partnered with George Ehret, a leading member of New York's German community, to form the Bowery Bay Building and Improvement Company and build a trolley park in 1886. In a typical overlap of business interests, Steinway headed the company, the rail interests, and was the principal investor at Bowery Bay. In the summer of 1887, Steinway extended his horsecar line to the amusement district with the expectation that a shore resort would profitably fill his horse- and streetcar railroads.[99] Ehret owned the prosperous Hell Gate Brewery in Yorkville and a network of forty-two saloons that sold his beer exclusively.[100] He looked forward to excursionists consuming his beer in dance halls and beer gardens. The improvement company built a seawall to replace the rocky shoreline as well as 130 bathhouses, drives, and picnic grounds. During the resort's first seasons, the Bowery Bay Building and Improvement Company ran some concessions, such as the Hotel Tivoli, directly, maximizing profits and control. Steinway recorded in his diary that following the resort's opening, investors became "very enthusiastic to increase capital and annex neighboring ground."[101] Over time, independent proprietors leased company plots or opened new amusements, and the resort flourished.

Barnum and Steinway established residences near the landscape amenities they financed to underscore their role in these progressive public works. Steinway's Long Island City property included a former country estate. Located on the area's only elevated terrain, the granite estate fronted the bay and overlooked the new factory complex to the west. Although the majority of the tidal environment surrounding Luyster Creek was filled, the creek remained, separating the estate's gardens, lawns, and surrounding woodlot from workers' homes along Wolcott Avenue, Steinway Avenue, and Albert Street.[102] Steinway installed himself in the mansion during the summer, overlooking the factory and the shore. In 1869, Barnum moved to the center of the South End across from Seaside Park. His show home there, Waldemere—Woods by the Sea—abutted Seaside on the north. In Marina Park, a thirty-acre subdivision adjacent to Seaside and Waldemere, Barnum built the villas Wavewood, Petrel's Nest, and Cottage Grove for his daughters in 1886. A local newspaper boasted that the homes surrounding Seaside gave "the whole scene the appearance of a grand art design, with a rich border of lace work gliding into purity itself in the natural and sublime picture of the coming and receding tides."[103]

Barnum blurred distinctions between private property and public spaces in developing Bridgeport. Barnum welcomed Bridgeporters to treat Waldemere as a public environmental amenity by opening the estate, as he had Iranistan, to the public. Through the landscaping around his home, Barnum created functional and visual continuity between the South End villa district and the park's public landscape. He ornamented his lawns with vases, statuary, and fountains. His stables and vegetable gardens sat opposite the mansion, on another property, so that Waldemere appeared a seamless extension of Seaside. The large flag hoisted from the home's tower when he served as mayor signaled when he was in residence, inviting any citizen to call. This accessibility, a national landscape journal noted, was an act of "real good-hearted beneficence": "very few rich men we know ever offered their suburban grounds free to public use and enjoyment like this."[104] As a journalist for *Frank Leslie's Illustrated Newspaper* observed, "the line between the public and private development and access [was] very faint."[105]

Seaside Park and the Bowery Bay resort epitomized the coastal rebuilding underway in the region. By the end of the nineteenth century, the tidal marshes of the metropolitan coastal corridor were disappearing. *The Real Estate Record and Builders' Guide* had encouraged the process since the 1860s, declaring New York City's growing population would "eventuate" the drainage of the marshes surrounding the city to make land for development, advance real estate prices, and save citizens from "miasmatic vapor arising" from marshes thought to be "impregnated with disease."[106] Barnum and Steinway declared the shore wasteland to emphasize the value of their improvements, but both invested in recreational spaces that made the most of the waterfront. Olmsted valued the Sound's coast in its natural state but believed landscape designs were necessary to realize public enjoyment of the shore. In the 1880s, Olmsted pointed to Seaside in a letter in *Century Illustrated Monthly Magazine* as a worthy example of municipal efforts to preserve waterfront vistas for the public.[107] In the coming century, the interconnected trends of coastal leisure and suburban living made New York's aesthetically pleasing and healthful coastlines increasingly valuable commodities. In their city building, Barnum and Steinway recognized the value of both the urban edge and the coastal edge, where land met sea, as places ripe for experimentation.

*Conclusion*

Steinway's company town and Barnum's projects in Bridgeport demonstrate the power of individuals to shape the form and function of the midsize nineteenth-century city. Private citizens could so significantly affect urbanism on the edge in part because of the relatively manageable scale of development there. Influential investors certainly shaped New York City proper, but its magnitude and density prevented the holistic city building possible on its periphery and in its environs. By 1890 the city, home to 1.5 million people, stretched forty square miles across Manhattan and the future south Bronx. In contrast, Long Island City covered seven square miles and housed 30,500 people; Bridgeport covered fifteen square miles and housed 48,800 residents.[108] The comparatively modest size of these urban centers made it possible for individuals to use personal economic and ideological goals to structure urban form.

Steinway and Barnum effectively spearheaded local efforts to imagine

1.7. Waldemere and Seaside Park. Waldemere was the last of the four homes Barnum occupied in Bridgeport. Barnum consciously integrated his home with Seaside. *Left*, Walkways connected the park and Barnum's estate, which was decorated with fountains and statuary that drew strollers from the park. *Right*, Waldemere. "Residence of P.T. Barnum," views of Bridgeport and Vicinity, albumen print, ca. 1875. *Source*: Robert N. Dennis Collection of Stereoscopic Views. Stereoscopic Views of Bridgeport, Connecticut. Miriam and Ira D. Wallach Division of Art, Prints and Photographs, New York Public Library.

the regional city and its hinterland networks. Both men intervened in city growth with forethought, making carefully considered investment choices that reflected interest in comprehensive, long-term development. Yet neither city builder pursued the type of unified comprehensive plan that became the centerpiece of municipal development and professional planning in the twentieth century. Neither Steinway's nor Barnum's planning projects outlasted their years of personal oversight. In 1893 Steinway & Sons began untangling family and business expenses, selling Daimler Motor and the Steinway Railway, although Steinway himself pursued public works and speculative investments until his death in 1896.[109] Notwithstanding the limitations inherent in the city building of singular individuals, Steinway's and Barnum's work proved the power of urban investors to realize comprehensive public works and coordinate public- and private-sector development. Furthermore, they accomplished their city building following the panics of 1873 and 1893, which constrained municipal public works budgets and limited the feasibility of large infrastructure projects such as street grading. The severity of these depressions made their successful private development even more striking.

In both Bridgeport and the Steinway settlement, political boundaries mattered less than the networks of manufacturing, transit, and finance that knit Manhattan, its peripheral ring of industrial sites, and satellite cities into a regional entity. Steinway established his Queens factory to escape Manhattan's labor politics, but he ultimately courted growth to link his settlement to the city center's economic networks, growth that offset the paternalism of his company town. While New York City could not overcome the bifurcation of space inherent in its geography, regional development increased the connectivity between Manhattan and adjacent developing districts such as Long Island City. Steinway did not live to accrue the financial benefits that twentieth-century bridge and tunnel projects brought to Long Island. He had anticipated, however, urbanization in northwestern Queens and the need to link the area to midtown Manhattan.[110]

Annexation followed urban development in both Queens and Bridgeport. When Steinway moved to Queens, the East River divided the urban core of Manhattan from New York's environs. By the end of the century, the river ceased to delineate a municipal boundary. In 1897, residents of the western half of Queens voted in favor of annexation to New York City. The county's eastern residents rejected annexation, choosing instead to form the new, suburban county of Nassau in 1899. This split reflected the different development of the districts that made up Queens. In approving annexation, residents in Long Island City acknowledged their place in the regional network of New York. In greater Bridgeport, residents who supported annexation likewise recognized that urbanism had erased nineteenth-century city limits. When Barnum began investing in East Bridgeport and Seaside, in the South End, the former and half of the latter lay beyond city limits. East Bridgeport residents had in fact seceded from Bridgeport in 1839, fearing financial hardship because of the $150,000 in bonds the city sold to fund a municipal horsecar railroad. But in 1864, residents voted to reincorporate, and additional adjacent districts followed suit in 1889. In the intervening period, residents in the town of Fairfield who lived near Bridgeport's South End voted for annexation to leave their rural seat and attach their fortunes to the city. By 1890 population growth and territorial annexation increased the population of Barnum's adopted city by 550 percent to 48,800.[111]

As New York developed economically and geographically in the second half of the nineteenth century, spaces once considered distant shed this distinction, and the perceived differences between the city center and its environs grew fainter. What emerged was a vision of the region as a connected, if extremely complicated, whole. From his strategic plowing in sight of the NYNH&H, Barnum underscored Bridgeport's expanding ties to New York City. Steinway promoted a similar regionalism linking Long Island City and Manhattan. Steinway represented the powerful interests of real estate and industry of the city center, but he also looked beyond the confines of Manhattan to embrace a metropolitan-level point of view. The physical changes Steinway and Barnum wrought in the landscape obscured the once clear boundaries of their urban experiments and led to annexation drives that joined urban core and environs. Barnum's and Steinway's city building facilitated the rise of greater New York and a more concrete metropolitan hinterland along the Sound.

# 2

# Laying Out the Trans-Harlem City

By 1865 New York City park commissioner Andrew Haswell Green came to the conclusion that the city had outgrown Manhattan Island. In a report for the Board of Commissioners of Central Park (BCCP), Green argued that the city's future should include its mainland environs of Westchester County north of the Harlem River. He articulated a river-spanning future for New York. Green reasoned that lower Westchester was "so intimately connected with and dependent upon the City of New York, that unity of plan for improvements on both sides" of the Harlem was "essential."[1] Improvement could not "be well done, if it even can be done at all," by more than one municipal government.[2] In the subsequent three decades, Green lobbied first for control of public works across lower Westchester and then for annexation to expand the city's boundaries. Green's work reflected a new vision, an environmentally informed regionalism and the desire for government authorities to address the forces of growth regardless of corporate limits. Green is best known for orchestrating the consolidation of the five-borough city in 1898. Yet he acknowledged the city's regional qualities decades earlier in his work north of the Harlem. Consolidation in fact culminated Green's thirty-year campaign to turn New York into a regionally scaled city.

Urban development and large-scale plans on the urban periphery of the Harlem River reconfigured the relationship between the city and its northern environs. While intensive commercialization and industrialization came to dominate Manhattan's waterfront by the second half of the nineteenth century, the waterfront of what became the outer boroughs of Queens and the Bronx followed a different trajectory. While these shores remained largely undeveloped, enabling William Steinway to build a company town in Queens, it also enabled Green and a diverse group of planners, residents, real estate boosters, and industrial developers to experiment in planning in the future South Bronx. As head of the city's park planning between the mid-1850s and 1870 and as city comptroller from 1871 to 1876, Green's influence on the city's nineteenth-century form was matched only by the effect of Manhattan's 1811 grid plan.

Green first called for comprehensive planning power over northern Manhattan and southern Westchester as a member of the BCCP in the 1860s. As a state-empowered authority, the BCCP enjoyed relative independence from municipal politics.[3] The state legislature supported Green's vision and authorized the expansion of the commission's powers, and it tasked the commission with the planning of streets, parks, and bridges. Expansion of the commission into the territory opened the way for annexation. The city annexed lower Westchester in two popular referendums in 1874 and 1895, forming the city's only mainland wards and what became the borough of the Bronx. Green's work on New York's northern edge is a window into the rise of the public-sector work of city, county, and state agencies that oversaw park, dock, and street building and how landowners and neighborhood boosters courted public works. The expansion of planning powers and municipal boundaries north of the Harlem River altered how both locals and Manhattan-based municipal officials envisioned the city's environs.

The rise of the regional city in the South Bronx occurred in three periods in the second half of the nineteenth century, each with a distinct understanding of the territory as the trans-Harlem, the Annexed District, and the North Side. The first articulation of the region was as the *trans-Harlem*, a term used by boosters to underscore that development spanned the Harlem River.[4] After the city acquired lower Westchester in 1873, the trans-Harlem became known as the Annexed District. Finally, in the 1880s, locals renamed the territory the North Side as part of a campaign to combat what they felt was unequal investment and inferior management compared with the city's Manhattan wards. While chronologically divisible into these three distinctive moments of development, the boundary area in the city's environs where urban land use tapered off was a constantly shifting geographic category. These peripheral landscapes were an evolving series of temporary environments that had different users who had their own visions and uses. As a result, they are hard to map. The impermanent nature of the periphery allowed individuals to chart and rechart the course of growth. They configured the relationship between city center and edge and made the trans-Harlem borderland a distinctive region. The importance of the city's northern edge lay not in its demarcation of a boundary but in its position as part of an urban continuum where city planning transcended the bifurcation of core and edge. The traditional core-edge binary obscures the spectrum of interrelated metropolitan spaces. The idea of the trans-Harlem is a step toward conceptualizing the metropolitan region as a whole.

City officials and boosters of the regional city endeavored to reshape the material nature of the cityscape. They approached environmental reform in two ways: through environmental infrastructure and with ideas of "nature" as a healthful resource. Nature became an environmental amenity; this interest led to concerns over pollution due to the dispersal of industry and population. Debates about the physical nature of the urban fringe demonstrate how the city's coastal edge was both an ecological system and a cultural and political landscape. The harbor environment included both sides of the high-tide line—riparian land and lands underwater. Developers looked to control the material characteristics of the coast through infrastructure. Landscape architects, engineers, and street commissioners approached the urban edge as a laboratory for regional planning. These city builders focused on regional environmental boundaries in contradistinction to laissez-faire urbanization and development that overlooked the conditions of environment and topography. Parks, channelized rivers, and street systems expanded the urban fabric into rural hinterlands. If the territory could escape the grid, perhaps it could overcome Manhattan's moral and environmental failings.

Coordinating officials such as Green led the planning charge to demand comprehensive plans for new streets, a mainland park system, and Harlem River improvements. Urban growth sparked debates over environmental reform and regional planning. Part of the debate centered on who should shape growth: state or federal agencies, officials and park designers in city-wide departments, or local officials headquartered in the trans-Harlem? These government bodies embraced regionalism, but locals did not always welcome projects sponsored by "outsiders." While the evolution of the trans-Harlem illuminates how city institutions such as Green's park board penetrated and organized lower Westchester, it also reveals the ways in which Westchesterites responded to the city's arrival. Residents also acquired a voice in streets and park plans, effectively implementing public works through authorities situated on the urban periphery.

*The "Trans-Harlem" City*

While municipal boundaries defined Manhattan Island as New York City proper in the 1860s, city officials increasingly considered the city in

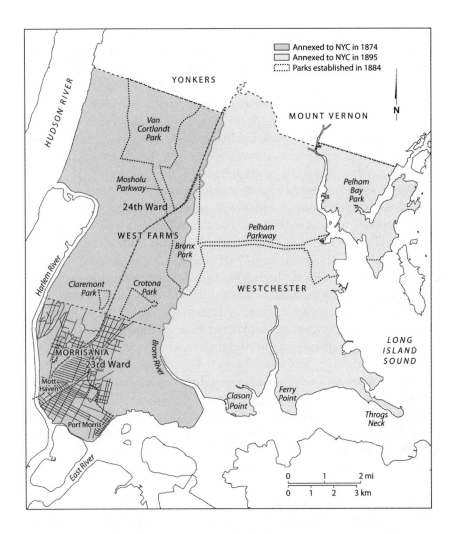

Legend:
- Annexed to NYC in 1874
- Annexed to NYC in 1895
- Parks established in 1884

YONKERS

MOUNT VERNON

N

HUDSON RIVER

Van Cortlandt Park

Mosholu Parkway

24th Ward

WEST FARMS

Bronx Park

Pelham Bay Park

Pelham Parkway

Harlem River

Claremont Park

Crotona Park

WESTCHESTER

LONG ISLAND SOUND

MORRISANIA

23rd Ward

Bronx River

Mott Haven

Clason Point

Ferry Point

Throgs Neck

Port Morris

East River

0   1   2 mi
0   1   2   3 km

2.1. The trans-Harlem. This map shows the expansion of New York City's influence and governance over the southern peninsula of Westchester County. The *trans-Harlem* generally referred to the districts of Mott Haven, Port Morris, and Morrisania. This area became the center of the Twenty-Third Ward and the southern half of the Annexed District in 1874. The Bronx River formed the Twenty-Third and Twenty-Fourth Wards' eastern boundary. Boosters in the mainland wards took up the name the "North Side" to refer to both the Twenty-Third and Twenty-Fourth Wards but also, at times, Westchester County territory east of the Bronx River, around Pelham Bay Park and Parkway. In 1895, New York City annexed the territory about this parkland between the Bronx River and Long Island Sound. *Source*: Map by Bill Nelson.

relation to the territory north of the Harlem River. Between 1865 and 1874 the state legislature expanded the city's public works powers across lower Westchester. Two distinct versions of the trans-Harlem emerged from this era of municipal planning. City leaders such as Green envisioned a modern cityscape built from comprehensive plans created by citywide authorities headquartered on Manhattan. At the same time, local politicians and residents envisioned urbanization guided by home-rule initiatives and voter consensus. A new regional perspective emerged from these complementary, if often competing, visions of the trans-Harlem city.

Westchesterites and city bureaucrats agreed that the systematic coordination of urban improvements would protect private property interests and the public interest in the trans-Harlem. A headline typical of the proplanning *Real Estate Record and Builders' Guide*, the leading trade publication for the industry, proclaimed "No Piecemeal Improvements Wanted."[5] Yet no municipal- or state-level governmental body existed to institute a comprehensive city plan. The closest thing to a plan for growth was the city's street grid, the Commissioners' Plan of 1811. At the time, New York remained largely within its unplanned colonial footprint; City Hall in lower Manhattan marked its northern edge. After four years of study between 1807 and 1811, the state-appointed commissioners theorized that a grid's uniformity and continuity would guarantee that all parts of the city would be equally and rationally treated. An additional benefit was that once set, the municipal government would not need to tend to the plan. The commissioners expected the grid would provide Manhattan with a master plan to shape future growth. The plan set a geometric grid of twelve avenues, 155 cross streets, and two thousand blocks across eleven thousand acres. Yet the commissioners never imagined the future density of Manhattan nor the rate of growth their plan would spur. They misjudged by nearly a century how long it would take for buildings to fill Manhattan. By the 1860s it was apparent that that milestone would be met before the century's end.[6]

Critiques of the street plan proliferated in the 1860s as a vocal contingent of citizens and city officials declared war on the grid. While the grid had fostered unparalleled growth, its critics alleged it failed as an urban design. Frederick Law Olmsted blamed the grid for the city's filth and crowds and its insular, corrupt ward politics. William Cullen Bryant, editor of the *New York Evening News*, declared that the city's street system encompassed the worst parts of London's, Boston's, and Philadelphia's street plans. Lacking formal plans, London developed along "the chance perpetuation of country roads," while "the winding ways of Boston" were but "old cow paths." William Penn designed Philadelphia's grid, but Bryant scorned it as "a sheet of paper ruled into rectangles, without any regard to the nature of the ground or the direction of traffic." Bryant expressed his contempt for these three approaches by deeming them "lack-thoughts," suggesting that they were intellectually indefensible. The problem was that the three "lack-thoughts combined [had] laid out New York."[7] In a lament that captured popular sentiment, Bryant concluded that the grid left New York without any useful instruments to monitor or adjust its growth.

Civic leaders, intellectuals, and elite reformers sought to govern, socially reform, and spatially reshape the increasingly dirty, ill-governed, and crowded city.[8] Steinway searched for a locale beyond Manhattan where he could employ a paternalistic corporatism that he believed would mitigate labor unrest and address the negative social effects produced by industrial development and capitalism. As part of the same reform impulse, and with similar strands of elitism, the search for a new urban form also reflected a search for an alternative urban experience that improved the city's physical conditions and its residents' way of life.

The influential New Yorkers who disapproved of the grid frequently looked to the trans-Harlem as a place to experiment in new urban forms. Such critics were frequently trans-Harlem boosters, including Green, Olmsted, and the engineer J. James R. Croes. Despite clear parallels, the grid's critics did not acknowledge the 1811 street plan as an antecedent to the new approach they proposed. Yet the grid and proposals for the trans-Harlem shared fundamental qualities: extensive public works and large-scale visions. Despite the frequency with which the grid was criticized in the 1860s, its creation had marked a new era in urban thought in New York: a commitment by the city to plan for future growth. Green and Olmsted looked to secure a similar commitment in the trans-Harlem. The grid had illustrated a new radical idea that municipal government could and should interfere with the private real estate market. In the 1860s, city officials eagerly anticipated such extramunicipal powers. That planning might enhance real estate values was attractive in both 1811 and the 1860s. And both the plan for Manhattan and the plan for the trans-Harlem merged public works boosterism with New Yorkers' optimistic faith in future growth.[9] Like the grid's authors half a century earlier, trans-Harlem boosters looked to chart a new course for urban growth.

Green anticipated that a comprehensive plan would help the city avoid useless expenditure and duplication of public works, increase real estate values, and rationalize growth. By the 1860s, the ramifications of uncoordinated development on the Manhattan-Westchester edge were apparent.

A BCCP survey of lower Westchester revealed a confusion of street plans. Green found at least seven independent commissions, all authorized by the state legislature, attempting to devise plans for lower Westchester. For example, in 1868 the state appointed an independent commission to lay out and construct streets in Morrisania. But individual town authorities exercised local control over the laying out and grading of nearby local streets without reference to these multiple plans.[10] Furthermore, feuds over costs between New York City and lower Westchester municipalities hindered the locating and construction of Harlem bridges and approaches. Green feared such piecemeal planning was stagnating trans-Harlem growth. Green's work predated zoning and the establishment of professional city planning by nearly five decades.[11] Yet in calling for a comprehensive plan, he anticipated the future of city planning: cohesive municipal infrastructure and the zoning and separation of upper-income residential districts, commercial and wholesale markets, and nuisance industries.

In the years 1865–1871, a series of state laws transformed New York City's park commission into a vanguard city planning authority and endowed Green with broad influence over the trans-Harlem. The trans-Harlem had no definite boundaries, but the term was used in reference to Mott Haven and Port Morris on the Harlem and Morrisania to the north, the district that eventually became the central Bronx. The BCCP's expanded powers first incorporated oversight of waterfront improvements. In 1865 and 1867 the state legislature empowered the BCCP to adopt new pier and bulkhead lines on Spuyten Duyvil Creek and along the majority of the Harlem River. Authority over streets and public works came next. In 1869 the board gained the authority to execute a street and parks system for lower Westchester west of the Bronx River as well as oversight over public works in the towns of Yonkers, West Farms, Morrisania, and East Chester.[12] Westchester and New York County were jointly to supply the necessary funds. While the home-rule charter of 1870 dissolved the BCCP and replaced it with the Department of Public Parks, the authority vested in the office continued to grow. The city's oversight eventually extended well beyond the area referred to as the trans-Harlem. By 1870 the department oversaw public works across roughly 8,400 acres of lower Westchester.[13]

The parks department functioned as a planning authority in the trans-Harlem and beyond. The term *comprehensive* was frequently employed in late nineteenth-century American cities to describe a range of special-purpose endeavors even if the projects were narrowly focused, such as a park plan that did not address streets or utilities. The department, in contrast, took a large-scale approach that integrated a truly comprehensive list of urban components before *comprehensive* routinely suggested systematic and professional projects.[14] State law authorized the department to regulate the Harlem riverfront and all bridges, tunnels, and rapid transit across the river; to make a topographical survey and formulate a street and drainage plan; to acquire land for associated grading and improvements; and to locate all public spaces, including cemeteries, fair grounds, race courses, and parks. The state also consolidated control of the water, sewerage, and gas systems of Yonkers, West Farms, Morrisania, and East Chester under the department.[15] In the process, the legislature empowered the department as a regional public works authority, albeit not yet headquartered beyond Manhattan.

Green's interest in the environmental characteristics of the trans-Harlem led him to advocate for broadly defined planning powers that might reduce the level of pollution in New York Harbor. He approached the waterways of greater New York as belonging to bordering municipalities in common. In 1869, the Bronx River became the eastern boundary of the BCCP's mainland jurisdiction.[16] Park commissioners declared the river a disadvantageous, "purely arbitrary" boundary.[17] Green predicted that urbanites would one day populate both sides of the river valley and that the common needs of a regional public would supersede this boundary. Infrastructure such as drainage and water supply required detail. The parks department's Bureau of Topographical and Civil Engineers inventoried rainfall, streams, elevations, and surface material. The bureau's stream gauges, which it installed on the area's larger rivers to estimate the aggregate amount of water in sixty square miles of lower Westchester, recorded an insufficient water flow. The bureau concluded that the territory could not provide an adequate supply of drinking water and suggested the department expand its jurisdiction east and incorporate the drainage of the twenty-five-square-mile territory between the Bronx and the Connecticut state line. An environmental frame of reference led Green and his fellow park commissioners to conclude that a regional approach would best provide clean drinking water and prevent ecological degradation from sewage and industrial waste. The department enjoyed broad powers in the trans-Harlem, yet in each annual report it asked the state for oversight over even greater territory.[18]

The threat of pollution further encouraged a regional approach to public works. The problems of water pollution and sewage systems derived from both sides of the Harlem and Bronx Rivers. Green worried that without regulation, "the amount of sewage and offal . . . [that] would be cast into the Harlem River from either, or both, shores would . . . be injurious to the healthfulness of both [New York and Westchester] and detrimental to navigation."[19] Nuisance industries on the headwaters of the Bronx River were an additional concern. Because of the topography and geography of the Bronx River Valley, the principal north–south depression in the area under city jurisdiction received drainage from both sides of the river: "the drainage from the westerly side obviously cannot be separated in any plans for its future discharge, from that received from the easterly side, and a mutual and common system of sewerage or outflow must therefore be devised and adopted."[20] Pollution proved to Green that local communities were financially, technically, and politically unable to keep the city's coastal hinterlands and the river systems that emptied into them clean. Green recommended the creation of a single system of sewerage and piped water across upper Manhattan and lower Westchester. Despite the latitude that the BCCP and then the Department of Public Parks received over the trans-Harlem between 1865 and 1871, Green sought even greater environmental oversight.[21]

In the same years that the city parks department formalized a regional approach to public works and environmental regulation, trans-Harlem residents, politicians, and business investors encouraged urbanization and looked to knit together city and periphery. Municipal infrastructure became a tool with which to physically integrate their communities with the metropolitan whole. The goals of lower Westchester communities often aligned with those of influential New York officials such as Green. Residents near Woodlawn Cemetery and the Jerome Park Racetrack looked forward to the day when the northward-expanding city would reach to and incorporate their subdivisions. New York City's expansion and its grid were not juggernauts. In 1868 Morrisania representatives secured an act in the state legislature to reproduce the city grid across two hundred acres of the town's farmland. For trans-Harlem residents, the grid guaranteed streets and boundaries across uneven, unorganized terrain and provided an important incentive for private investment. The grid embodied Manhattan's economic success, and Morrisanians dreamed of similar advances in real estate values.[22] The scale and orientation of Morrisania's grid aligned with New York's in anticipation that one day the city would extend across the trans-Harlem. This street building was proof, a resident claimed, of the "general belief . . . in the upward extension of the city."[23]

Trans-Harlem residents increasingly saw themselves as part of a regional city. In the late 1860s, Westchesterites envisioned annexation to New York as a way to streamline public works and encourage growth. Residents in search of improved services welcomed annexation. The fragmented local governments of the forty-four villages of lower Westchester could not institute large-scale improvements. For example, residents of West Farms and Kingsbridge wanted city water as well as financing for bridges and streets that only New York could provide. In 1868 Gouverneur Morris II sold his personal holdings in his Harlem River settlement to a group of investors who organized the Port Morris Land and Improvement Company, a group that subsequently advocated for annexation to New York City in the hopes of receiving better sewerage and rapid transit.[24] Annexation would also settle funding feuds between the city and Westchester municipalities that hindered the construction of Harlem River bridges and approaches.[25]

Combining jurisdictions through annexation would tie core and periphery together politically. In 1869 Mount Vernon resident Cornelius Corson partnered with William Tweed to present the first annexation bill in the state legislature. An infamous figure in the nineteenth-century city, "Boss" Tweed successfully consolidated Tammany Hall and expanded the influence of its machine politics and spoils system. Tweed enjoyed unmitigated power across New York's municipal government and used it to siphon stunning sums from the city's coffers. Notwithstanding corruption, Tweed's Tammany Hall provided a kind of positive government, constantly advocating for building projects, since public works contracts were a key source of voter loyalty and graft. Corson and Tweed, frequent coconspirators in Tammany business, had both invested in Mount Vernon real estate, and Tweed was a partner of Corson's Eastchester National Bank.[26] The two men hoped annexation would increase land values and deliver profits on their investments. Corson chose not to consult his fellow Westchester representatives, most likely to catch the senators off guard and minimize opposition. When Corson presented the bill in the state senate, the surprised but quick-thinking William Cauldwell rose from his chair to challenge it. A prominent citizen of Morrisania, Cauldwell sarcastically threatened to present a "bill to annex the City of New York to the town of Morrisania." The senator noted with satisfaction that his comment "hit the nail on the head," stopping the Corson bill until a truly regional set of stakeholders and representatives were involved in the process.[27] William Cauldwell and the representatives of West Farms and Kingsbridge demanded a say in the conditions of annexation and courted votes. In 1871 the election to the state legislature of proannexation Judge William H. Robertson signaled the extent of local support.[28]

Trans-Harlem voters overwhelmingly approved of annexation. In late 1873, property owners in Morrisania, West Farms, Kingsbridge, and the

surrounding unincorporated territory passed an annexation bill in a bind-
ing referendum. Ninety-six percent of residents within the territory approved the bill. In the rest of Westchester, 70 percent of voters approved
of annexation, as did 85 percent of New York City's voters.[29] On December 31, 1873, revelers crowded Morrisania's Town Hall to celebrate their
new status as citizens of New York City; at midnight, their cheers mixed
with booming guns to echo across the territory.[30] On New Year's Day 1874,
lower Westchester, the mainland half of the trans-Harlem, became the
city's Twenty-Third and Twenty-Fourth Wards. The Department of Public
Parks assumed control of the new wards—and a contentious decade of
planning debates followed.

*The Annexed District*

Annexation fundamentally shifted the meaning of New York City. For two
centuries New York was coextensive with Manhattan Island. Limited to
the island, the city encompassed 12,576 acres, or twenty-two square miles.
Annexation of the 12,317-acre new wards doubled the size of the city. The
ruminations of an *Architectural Review and American Builder's Journal* reporter, who in 1869 mused that Central Park might become "the South
Park" and "the Harlem river [would] yet be the centre of the ambitious
city," had proven correct.[31] Yet the shape the territory would take remained
to be seen. Between 1874 and 1879, debates over the appropriateness of the
street grid and parks as well as the utility of master plans characterized the
"Annexed District" era. The material nature of greater New York became a
framework with which to understand city boundaries.

Following annexation, the Department of Public Parks commissioned
five separate plans to lay out the wards in advance of urbanization. Despite

**2.2.** *Opposite:* The annexed district. Beyond the industrial riverfront of Mott Haven and
Port Morris, the new Twenty-Third and Twenty-Fourth Wards were sparsely populated. The
area's thirty-six thousand residents averaged roughly three people per acre. At annexation,
the typical village encompassed a few blocks, separated from neighboring villages by empty
lots, farmland and orchards, and meadows. *Source:* "Map of the 23rd and 24th Wards, New
York, Compiled for an Index to Volumes of Important Maps," in *Atlases of New York City Certified Copies of Important Maps Appertaining to the Twenty-Third and Twenty-Fourth Wards, City
of New York, Filed in the Register's Office at White Plains, County of Westchester, New York* (New
York: E. Robinson, 1888). Lionel Pincus and Princess Firyal Map Division, New York Public
Library.

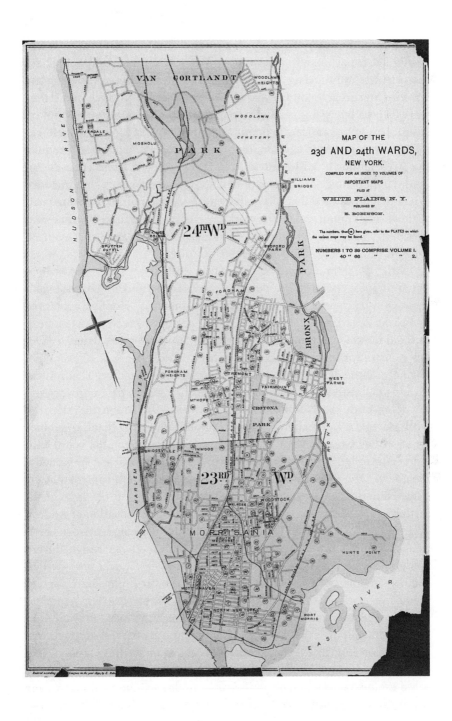

MAP OF THE

23d AND 24th WARDS,

NEW YORK.

COMPILED FOR AN INDEX TO VOLUMES OF

IMPORTANT MAPS

FILED AT

WHITE PLAINS, N. Y.

PUBLISHED BY

E. ROBINSON.

The numbers, thus (46) here given, refer to the PLATES on which the various maps may be found.

NUMBERS 1 TO 39 COMPRISE VOLUME I.
"   40 "  66    "     "    2.

the grid's recent notoriety, every plan extended it across the Annexed District.[32] The department's new president William R. Martin did not believe the grid was appropriate for the area. An uptown property owner and booster, Martin argued for expansion, comprehensive transit and public works, and a bourgeois aesthetic to ensure a certain class of cultural institutions and private development.[33] He led the West Side Association in lobbying for West Side development, which had included the extension of the BCCP authority to Manhattan Island north of 155th street a decade earlier. Martin belonged to a class of real estate entrepreneurs that included William Steinway and that successfully pushed personal development goals to the front of municipal public works agendas. Martin used his three-year presidency to encourage the city's regional expansion and experiment in urban form.

In 1875 Martin rejected all five Annexed District park proposals as unexecutable, claiming the designs encouraged "undecided, procrastinating, obstructive policy," not imaginative growth.[34] He then appointed two parks department members to replan the new wards: Olmsted, head of the department's Bureau of Design and Superintendence and nationally revered for his vanguard park designs, and J. James R. Croes, chief engineer of the department's civil and topographical office. That Martin turned to Olmsted, says historian David Scobey, was in part a provocation to Green. Little goodwill existed among these men. Green's relationship with Martin, a Tammany man, was strained. The 1870 home-rule charter abolished Green's BCCP and replaced the state-appointed board with a city-controlled Department of Public Parks. While Green remained a member of the department, Tammany Democrats took control and excluded him from all decision-making. Even after Green left the department for the office of the comptroller in 1871, his continued influence in the department put Martin on the defensive. Enmity also colored Green's relationship with Olmsted, and had since 1859. Olmsted twice resigned from work on Central Park because of conflicts with Green over park design, staffing, and management. After Green became city comptroller, disagreements over park appropriations kept tensions between the two men high. In May of 1876 Green even tried, without success, to oust Olmsted from the department after he accepted an unpaid post with the state government. Green and Olmsted recognized that the municipalities around New York were functionally one; both men also championed environmentally based regional development. Rather than finding commonalities in their approaches to pollution and water supply, the bad blood between them exacerbated their ideological differences.[35]

Despite the political tensions that threatened to derail park work, Olmsted accepted Martin's appointment as an opportunity to counter migration out of the city by building its first dedicated residential district. Olmsted reasoned that no city could "long exist without great suburbs."[36] Martin agreed, as did the *New York Times*, which avowed that New Yorkers would embrace suburban life in the new wards. Croes's public works experience made him an ideal partner for Olmsted. Croes began working on city engineering projects fifteen years earlier. He joined the Croton Aqueduct Department in 1860, working on the Central Park reservoir, and later took an appointment in the Department of Public Parks, overseeing surveys and plans for northern Manhattan. Between 1876 and 1877 Croes and Olmsted designed innovative street, park, and rapid-transit plans for the new wards.[37]

Olmsted devoted a substantial portion of his career to advocating for park planning in suburbs and theorizing the importance of suburban environmental aesthetics. For his first park commission in 1858, he designed Central Park with partner Calvert Vaux. It became a symbol as well as a practical blueprint for delineating natural environments in cities. Olmsted outlined his suburban planning theory in two late-1860s projects: the park and parkway plan for Brooklyn, completed with Vaux, and the elite suburban enclave of Riverside, Illinois.[38] In the late 1870s Olmsted took on the most significant large-scale public works of his career, including the Annexed District.[39] New York's new wards presented an exceptional opportunity for him to secure comprehensively designed residential districts as integral urban elements. Olmsted and Croes outlined a treatise on the importance of regional planning and suburbs for the growing city. They argued the Annexed District could be a model suburban district sensitive to natural boundaries of topography and hydrology and outfitted with comprehensive municipal infrastructure such as street and park systems and a steam railroad rapid-transit loop.

The Olmsted-Croes plan for the Annexed District made four important interventions in the debate on long-term planning for New York's hinterland: it eschewed the grid for a more flexible street plan, added suburbs to the cityscape, took a regional, environmentally sensitive approach to planning, and effectively wrote a master city plan. Since their plans diverged so sharply from the city's established grid, Croes and Olmsted presented their work as a general reflection in advance of formal plans to persuade the parks department of the merits of their theories.

Olmsted and Croes's first report on the Annexed District represents Olmsted's most systematic and persuasive attack on the grid. Olmsted objected that the grid's rigidity was monotonous. The pair declared that this monotony, part of "an attempt to make all parts of a great city equally convenient for all uses," disallowed functional differentiation and in fact made all districts "equally inconvenient."[40] The BCCP had created a street plan for the district south of Morrisania in 1868, and the parks department had commenced street planning in Riverdale in 1872. Olmsted and Croes believed a good street system might provide for rational development but thought that "many previous partial efforts" to open streets in the wards had failed in this regard. They claimed that "a large number of farms have been independently sub-divided . . . and rarely has any attention been paid in arranging the streets, to the manner in which the adjoining property was laid out . . . or to convenient connections and extensions."[41] The planners particularly disliked that blocks were laid out without regard to topography.[42] Since a grid already existed in Morrisania, it was designated as the wards' modest business center. They also reoriented Morrisania's new sections so that the long side of the street grid ran north–south in order to emphasize east–west exposures, maximizing the sunlight entering the homes facing these streets. Olmsted and Croes chose not to extend this grid and instead plotted a curvilinear street plan attuned to the region's topography.

The Olmsted-Croes reports offer the most complete articulation of Olmsted's theory on the differentiation of residential land use within the city's boundaries. Expanding New York could encompass mixed-use commercial and industrial nodes as well as suburban enclaves in a truly syncretic urbanism. Yet the new spatial reality of the sprawling city did not mean that all districts of the regional city could or should be uniform. The planners proceeded on the conviction that the differentiation between public and private spaces in the home could be applied to the city to rearrange metropolitan geography and improve the interrelation of its parts. "If a house to be used for many different purposes must have many rooms and passages of various dimensions and variously lighted and furnished," they urged, "not less must such a metropolis be specially adapted at different points to different ends."[43] Olmsted wrote earlier in his career that "constant increasing distinctness" between domestic and commercial spaces represented one of the laws of civilized progress. Innovative street design and housing controls ensured appropriate land-use divisions in the new wards.[44] While distinguishing between the core and periphery, the Olmsted-Croes plan ultimately presented the spaces as two halves of the same whole.

Olmsted and Croes experimented with the street layout of two Annexed District neighborhoods, West Farms and Riverdale. They platted moderate- and low-density suburbs for the northern sections around West Farms via a patchwork of variously orientated grids to create variety. River-dale was already a popular estate district. To preserve this character, they omitted shortcuts and "thoroughfares adapted . . . to heavy teaming or to rapid driving."[45] Croes and Olmsted approved of an exclusive villa district of large lots and curvilinear streets and preserved it in their design for River-dale. They also urged that streets be constructed selectively to leave room for future spatial innovation rather than repeating the same grid across the entire region. Rejecting inflexible grid plans and laissez-faire city building driven by real estate interests, the planners elevated a public agency, the Department of Public Parks, to institute active, long-range planning.

Olmsted and Croes urged municipal leaders to conceptualize New York City based on environmental and topographic features rather than corpo-rate limits. If the city were to be larger than Manhattan, what aspects of New York Harbor and its surroundings might it come to include? Croes and Olmsted envisioned an urban borderland that eschewed municipal ju-risdiction in favor of environmental characteristics common to the main-land wards and the surrounding Westchester territory. Just as the Croes-Olmsted plan reconfigured the relationship between core and periphery, it also reconfigured the traditional binary of city and environment. Rather than defining nature as something beyond the city and inherently antiur-ban, they defined urban limits by the natural environment, the system of ridges and river valleys that gave the mainland wards their character. Three long northeast-trending rocky ridges and three rivers—the Saw Mill, the Bronx, and the Hutchinson—corrugated the district from west to east. The territory's steep grades varied from 15 to 25 percent, and its ridges seldom dropped below an elevation of two hundred feet. The rugged ter-rain of broken ledges and steep hillsides was, the planners approvingly ob-served, "largely wooded and wild."[46] To preserve the natural environment of the mainland wards, the planners drafted, but never completed, a com-prehensive park plan. They proposed a chain of small linear parks along the Bronx River and Mill Brook valleys to preserve the scenic and recre-ational value of the rivers.[47]

To achieve regional environmental management, Olmsted and Croes recognized that the city needed extramunicipal powers.[48] While the city's jurisdiction ended at the Bronx River, Olmsted and Croes proposed bar-ring industrial plants along the streams and higher ground in Westchester that drained into the Bronx and East Rivers. Green argued in the 1860s that to mitigate water pollution the city needed to control the entire lower Westchester peninsula.[49] Olmsted and Croes did not acknowledge Green's earlier proposal, although their reports advanced an environmental framework for the regional city. The boundary waters around Manhattan had long demarcated New York's city limits. But the mainland wards shifted the city's frame of reference regarding its harbor system. Rather than places of bifurcation, the Harlem, the harbor, and its tributary rivers that ran through the mainland wards were reframed not as boundaries but as ligaments knitting together corners of the city.

The intense partisan politics of the 1870s ultimately derailed the Olmsted-Croes plan. Public works became a political battleground during the Tweed Ring's brief reign in municipal government (May 1870–September 1871).[50] The home-rule government spent profligately. During Tammany's reign, the parks department alone spent $2.25 million, a third of what the BCCP had spent in fourteen years. Tammany's spree of personally remunerative projects left the city with staggering debt and the looming threat of bankruptcy.[51] The Tweed Ring collapsed in 1871, although the struggle between the reform government members like Green and Tammany leaders like Martin continued. Under the reform city charter that followed the ring's collapse, Green was appointed city comptroller. In this role he allocated municipal finances and controlled public spending. While no longer a member of the parks department, Green's ability to withhold funding gave him nearly absolute power to veto projects he disliked.[52] Because of the financial fallout of the Tweed Ring and the depression that followed the economic panic of 1873, the comptroller championed fiscal conservatism and small projects funded via local assessments. Green and his allies in the parks department believed the city could not afford to finance experiments in suburban or urban form.

In the 1870s Green declared the city finished, and his office circumscribed large-scale plans for the rest of the century; his successor, John Kelly, continued Green's austerity policies and officially scuttled the Olmsted-Croes plan in 1877.[53] In a letter to the *Tribune*, Olmsted rebuked the decision as ideologically bankrupt and proof of the parks department's unwavering allegiance to the grid and private real estate interests. John C. Olmsted, his nephew, adopted son, and head of the family landscape design firm, who had drafted the proposed street plan for his stepfather, declared the move to be a "shocking waste" of the Annexed District's potential.[54]

Olmsted and Croes envisioned an upscale residential community for the Annexed District, a land use in which they saw value. In rejecting the Olmsted-Croes plan, municipal officials relegated the mainland wards to the laissez-faire development of the grid. Yet perhaps Green had also perceived the grid as a more suitable plan for the new wards, one that preserved a mix of commercial and residential land use in ways that were more democratic than Olmsted's enclave. In fact, landowners also rejected the Croes-Olmsted plan in favor of unrestricted private development and called for a straightforward grid rather than curvilinear streets.[55] That locals had little use for an elite suburb reveals how their goals for plans could diverge from orthodoxy. Olmsted and Croes's dismissal marked the end of an era in land-use planning in the trans-Harlem, an era of experimentation across the urban edge by Manhattan-based city officials. Yet the region continued to function as a laboratory for innovative city building. Planning for regional growth continued under a new set of urban design parameters. The ensuing era fostered a grassroots public works campaign and renegotiated the governmental relationship between city center and edge to create municipal oversight unique to the periphery.

*The North Side*

In the 1870s, Annexed District localism emerged as a grassroots response to perceived inferior treatment by the distant city government. While the citywide parks department had promised street construction and park development, residents complained that the government treated the mainland wards "as a mere suburban locality, more to be tolerated than recognized as a part of the City."[56] The Twenty-Third and Twenty-Fourth Wards were home to 36,194 residents by 1875, and nearly five thousand more arrived in the following five years.[57] These New Yorkers, boosters declared, deserved the same services as residents of the city center. In fact, residents north of the Harlem River came to eschew as a "misnomer" the moniker "Annexed District." To chart a new course of progress, boosters renamed their district the North Side and pushed for self-government over public works. The result was a local government reminiscent of home rule but without the characteristics of parochialism or a narrow focus on localized projects. Boosters and bureaucrats experimented with the levels of government oversight: North Siders advocated for, and achieved, local planning powers. This level of governance, unique in New York City at the time, led to three public works projects that encouraged growth: the creation of a North Side park system, a local street-building program, and Harlem River improvements. These projects marked the coming-of-age of the North Side from 1874 to 1898, reconfigured the relationship between core and periphery, and facilitated the consolidation of Greater New York.[58]

Annexation encouraged booster prophecies of a North Side boom under the parks department's supervision. Despite such predictions, delays and unfinished plans characterized the department's work in the North Side.[59] Following annexation, the department platted 450 acres of parks, but through the following decade the department acquired only a fraction of this land. Businessmen worried that this neglect drove potential new residents, and their tax dollars, to the more developed commuter regions of Brooklyn and northern New Jersey.

Resident John Mullaly founded the New York Park Association to address lagging park development. Mullaly surveyed the state of North Side parkland acquisition and reported the parks department's shortcomings to the state legislature. Legislature representatives found Mullaly's argument convincing and censured the department for failing to realize the "rural suburban parks" that it deemed "a metropolitan necessity."[60] The state and locals agreed that new parks were overdue.

To restart park development, the state authorized an exceptional park program for the Twenty-Third and Twenty-Fourth Wards. The New Parks Act of 1884 marked the first time the government—city or state— authorized park building in advance of settlement, in a unified system, for an entire geographic district. The law also marked a vanguard attempt at environmental preservation in greater New York. The Mosholu, Bronx and Pelham, and Crotona parkways linked the four largest parks, newly protected reserves of hilly woodland and marshy coastal plains. The system included seven parks comprising four thousand acres, including three massive parks that rivaled and exceeded Central Park's 843 acres: Bronx Park (635 acres), Van Cortlandt Park (1,069 acres), and Pelham Bay Park (1,740 acres).[61] Pelham Bay was located east of the Twenty-Fourth Ward's boundary, in Westchester. Although challenged, the city's constitutional right to annex parkland beyond its municipal boundaries was upheld in the State Court of Appeals.[62] In 1890, an enthusiastic Olmsted would write John C. Olmsted about the opportunity of the North Side parks. "Suppose we saw an opening to get the . . . new parks of New York!"[63] Over a decade earlier his plan with Croes had been scuttled, and the Tammany-influenced park board of commissioners had abolished his Office of Design and Superintendence, thus ending Olmsted senior's twenty-year influence over New York City's cityscape. Nevertheless, he still harbored hopes that he might finally get to design a comprehensive park system for the North Side.[64] While Olmsted did not secure the commission, the attention of the nation's leading park designer underscores the significant opportunity for design experiments that the North Side represented. The creation of the park system also proved the power of locals to demand urban improvements, a lesson that residents learned and used to secure street renovations.

North Siders and their representatives demanded improvements in not only the parks department's parkland acquisition but its work mapping and laying out streets as well, continuing the contest between ward-based and citywide oversight. The power vested in the parks department to plan streets in the territory dated to Andrew Haswell Green's work in the BCCP in the 1860s. But the department's street-building efforts were even less impressive than its park work. The local Taxpayer Alliance accused the parks department of ineptitude. The alliance declared that the few streets the department managed to complete were too narrow or ran without reference to main avenues. Concerned citizens claimed these roads benefited only interested politicians and powerful landowners.[65] Furthermore, the department had failed to create accurate maps. Accurate maps were powerful planning tools. The determination of street location, grades, and property lines depended on reliable mapping, not to mention tax and assessment collection. Existing maps were considered "so full of errors as to make them practically worthless, while the best" even included "very serious flaws of a dangerous and misleading character." For example, the twenty-foot-grade difference between Eagle Avenue and 161st Street had been ignored in municipal topographical surveys and the street plans based on them, resulting in an inaccessible intersection.[66] The challenge the city faced in mapping streets is a window into the difficulty New Yorkers faced in planning for expansion. If geographic spread and infrastructure could not be identified before it evolved, how could such growth be controlled? Good maps encouraged local development and projected a sense of what the city should look like. Bad maps hampered growth.[67]

In 1887, the Twenty-Third Ward Property Owners' Association turned to the state for support and convinced the senate to investigate negligence charges against the parks department.[68] Following a public hearing on the condition of North Side streets, committee members toured the district and received a "very practical introduction to the celebrated mud of the

2.3. *Overleaf:* To fill in the Harlem. This map of the Harlem River and Spuyten Duyvil Creek between the Hudson River (*top*) and the East River (*bottom*) illustrates Simon Stevens's proposed "Covered Water-Way." The proposed infill is shown in shading and labeled "covered waterway." Source: *Map of the Harlem River and Spuyten Duyvil Creek from Ward's Island to the Hudson River, Showing Project for a Covered Waterway Sixty Feet Wide to Be Built on the West Line of the Harlem River* [ . . . ]. Map, 1892. Lionel Pincus and Princess Firyal Map Division, New York Public Library Digital Collections.

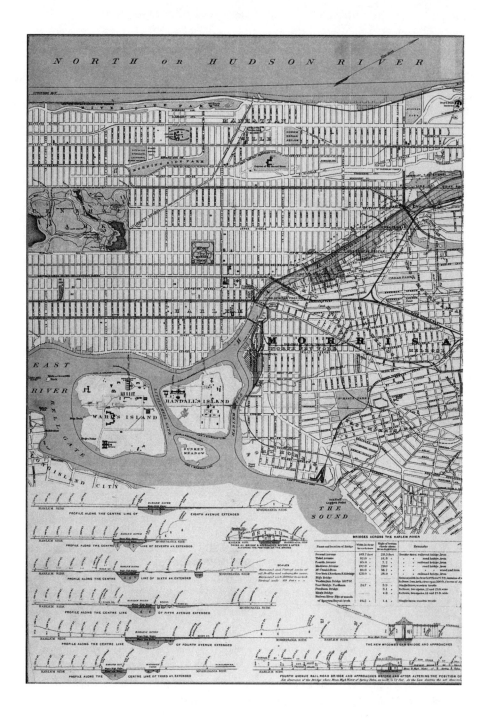

MAP OF THE

# HARLEM RIVER AND SPUYTEN DUYVIL CREEK

FROM WARD'S ISLAND TO THE HUDSON RIVER

SHOWING

PROJECT FOR A COVERED WATER-WAY 60 FEET WIDE, TO BE BUILT ON THE WESTERLY LINE OF THE
HARLEM RIVER FROM THE EASTERLY SIDE OF THIRD AVENUE TO 165TH STREET, NEW YORK CITY,
AND FILLING IN BETWEEN THE POINTS NAMED, SO THAT THE AVENUES AND STREETS OF HARLEM
MAY BE EXTENDED INTO MORRISANIA.

COMPILED FROM THE PUBLISHED

WAR DEPARTMENT MAP OF THE PIERHEAD AND BULKHEAD LINES OF
BOTH SHORES OF THE HARLEM RIVER & SPUYTEN DUYVIL CREEK, NEW YORK CITY

as recommended by the

NEW YORK HARBOR LINE BOARD

and approved by the SECRETARY OF WAR October 16th 1890.

FROM PLANS SUGGESTED BY SIMON STEVENS and G. THADDEUS STEVENS 61 BROADWAY, NEW YORK.

district by having their carriages break down and in having been compelled to wade ankle-deep in their shiny patent leathers to terra firma."[69] As a result committee members proposed a bill creating a North Side street authority. The proposal was so popular among North Siders that it earned the nickname "the People's Bill." On January 1, 1891, the legislature stripped the parks department of its power over mainland projects outside its park holdings and allocated such work to the new Department of Street Improvements of the Twenty-Third and Twenty-Fourth Wards.[70] While the parks department had functioned as a regional public works authority since the 1860s, it remained a Manhattan-based agency. In contrast, the Department of Street Improvements was the first regional public works authority actually stationed in the periphery zone of the mainland. Reflecting the neighborhood-focused nature of the street improvement movement, the People's Bill required that the head of the department reside in the North Side. Despite this insistence on local representation, locals hoped to overcome their perceived peripheral status by reducing any sense of difference between the city center and edge. The hope was that borough rule would achieve parity in street improvements between Manhattan and the mainland wards. Accurate maps and a street plan, like the new park system, would attract investment to the urban periphery and secure it as a principal, not marginal, district in the emerging metropolis.

Just as the New Parks Act created a vanguard home-rule parks system in the North Side, the establishment of the Department of Street Improvements marked the first time the state granted a district local control over municipal infrastructure. Reconfiguring the relationship between the city and its outlying districts, state officials tacitly acknowledged that the regional city required new governing approaches. Espousing a variation of home rule, new officials assumed public works orchestrated from the edge would better nurture development than the existing centralized and overextended authority. The new street commissioner had the authority to lay out all streets, devise plans for sewerage and drainage, locate all bridges and tunnels, and make contracts for all public improvements except those relating to parks and parkways. The commissioner also became a member of the Board of Street Opening and Improvement of the City of New York but could vote only on questions relating to his territory.[71] The department was truly local in the scope of its power. A nonpartisan Citizen's Local Improvement Party was created specifically to nominate for commissioner Louis J. Heintz, a well-respected engineer and businessman with property interests in the district. Tammany district leader Henry D. Purroy handpicked Heintz's opponent, but while the North Side was Democratic, Tammany never had a majority there.[72] Local interests prevailed, and Heintz won.

In his first year heading the Department of Street Improvements, Heintz supervised more new streets and sewer construction in the new wards than had been completed since annexation in 1874. In the seventeen years following annexation, the parks department had adopted 231 public improvement ordinances, an average of fourteen per year. From January 1, 1891, to September 30, 1897, Heintz adopted 471 ordinances, a yearly average of nearly sixty-eight projects. Increasing real estate values and building activity kept pace with improvements. The Twenty-Fourth Ward's assessed value increased by 69 percent, surpassing the increases in all but three city wards. The North Side Board of Trade pointed to rising real estate values as proof of the success, even during the depression of the 1890s, of Heintz's program.[73] In 1896 the department additionally completed a topographical survey and the first accurate mapping of the region. The official map represented an enormous expenditure, and at times more than forty teams of surveyors worked simultaneously to triangulate and map twenty square miles of territory. Neighborhood groups such as the North Tremont Association, the Twenty-Third Ward Association, and the Unionport Association celebrated the map's unveiling. With this record of local topography, planners could complete a public works program that made private investment profitable and attractive. The department's chief engineer, Louis A. Risse, enthused that topographical surveys, as the basis of street plans, could transform the North Side's "great area of farm into city lots and to make exact working plans upon which could be built the foundations of the great city of the future."[74]

Chief Engineer Risse and Street Commissioner Louis F. Haffen, who replaced Heintz after Heintz's death in 1893, predicted official maps, when combined with a Harlem River Canal, would transform the North Side. Like the North Side park system, maps, and street plan, Harlem River improvements helped define the North Side as a region and refocus the city-hinterland relationship. Canal construction might make the North Side not just a garden spot of the city but a vigorous commercial rival and transportation hub.[75] Risse's confidence in the canal's importance reflected three decades of advocacy on the part of North Side industrial and commercial boosters. The development of parks and streets for the North Side had primed the urban edge for investment and urbanization; the channelization of the Harlem River, an additional segment of modern metropolitan infrastructure, was designed to expand this work to the riverfront. The mid-decade opening of the Harlem River Ship Canal crowned nineteenth-century city building across Manhattan's developing hinterland.

As early as the 1860s, Green had argued that the Harlem could easily be "made available for the purposes of commerce."[76] Compared with London's

Thames, the Harlem featured a longer waterfront and far more moderate average tidal changes, making it more conducive than the Thames for trade. Green declared the Harlem ran "through the heart of New York" and that "something more should be made of it."[77] Yet the environment of the Harlem prevented intensive commercial use. Unlike its mighty neighbor the Hudson, the Harlem was narrow, shallow, and crooked. At its widest the Harlem reached only 450 feet across; its narrowest sections measured a meager 200 feet.[78] A mixture of sand and "vegetable matter," decomposing organic material resembling peat, made deep, soft layers on much of the river's banks and bed. The littoral's nature, oozing mud, and shallow estuaries made riverfront construction challenging and docking impossible for deep-keeled vessels.[79] North Side investor and booster Fordham Morris irritably noted that the river was ridiculed as "a *ditch* and *mud hole*" and had garnered the name "Harlem sewer" from the *Evening Sun*.[80] The river could not become a commercial artery to rival the Thames in its natural state, and the economic potential of Mott Haven and Port Morris remained untapped.

Like William Steinway in Long Island City and P. T. Barnum in Bridgeport, Henry Lewis Morris, the founder of Port Morris, typified the entrepreneurs who anticipated regional expansion and courted growth on the urban edge of the Harlem. If larger trade vessels could navigate the Harlem, it would eliminate a twenty-mile trip around the tip of the island from the Hudson to the piers of the East River and, Morris hoped, increase commerce in his settlement. General John Newton similarly expected that deepening the Harlem would attract the grain trade and spur the construction of new waterfront warehouses.[81] Chief engineer and commander of the Army Corps of Engineers (1884–1886) and New York City's Commissioner of Public Works (1886–1888), Newton spent a significant portion of his career overseeing coastal engineering projects in greater New York, including the eventual reconfiguration of the Harlem he and locals such as Morris hoped for.

Government bodies like the Army Corps and locals like Morris advocated for coastal improvements, but the lure of new, modern dock space also appealed to regional investors who worried over the state of Manhattan's waterfront. Steep grades made the Hudson waterfront of uptown Manhattan "practically inaccessible" and curtailed commerce there.[82] Squalor-inducing nuisance industries such as breweries, coal yards, and night-soil shippers competed with trade interests over downtown Manhattan's deteriorating commercial frontage. By 1870, because of dilapidated wooden wharves, decaying floats, and abandoned tugs, less than 10 percent of the island's docks were in good or fair condition.[83] In contrast, development on the Harlem was still in its infancy. Perhaps there, boosters urged, a modern waterscape could be built.

Real estate speculators, shipping interests, and business interests could not agree on the best way to utilize the Harlem, debating whether public works should facilitate traffic across the river or shipping along the river. Coastal trade interests advocated for a Harlem River Canal, but their opponents argued that transportation across the river was more important. In 1890 four Connecticut members of the House of Representatives demanded oversight powers to veto or approve all Harlem riverfront proposals to ensure they did not hamper transit between Manhattan and its developing hinterland.[84] Among those who favored transportation across the river was New Yorker Simon Stevens. In 1891, Stevens proposed filling in the Harlem. According to Stevens, the river's slow-moving drawbridges divided New York "just where it ought to be bound as closely together as possible."[85] Stevens lobbied the state legislature to connect Manhattan with the mainland between Third and Eighth Avenues and turn the Harlem into a subterranean river.[86] He argued that geographically unifying the island and mainland would facilitate trade, save more than eight million dollars in bridge and dock improvements, and create 235 new acres of land worth at least ten million dollars.[87]

Stevens's call to fill the river had precedent. Green once called for increased connectivity across the Harlem, although he suggested the city construct more than a dozen bridges along the eight-mile river.[88] While Stevens proposed a more extreme plan, he nevertheless enjoyed a certain level of support in Congress. In a federal River and Harbor Committee hearing on the proposal, Senator Joseph Roswell Hawley from Connecticut claimed that "what New York needed was standing room. Wherever the mud flats had been filled up they had been built upon, and if the Harlem River and flats were filled up they would be built upon."[89] When Stevens sent letters to Washington, DC, to request the federal government oversee his land-making plan, however, the House of Representatives and Board of Engineers of the War Department rejected the proposal. While Stevens's campaign failed, a number of officials and reporters embraced the notion of regional growth with seriousness and enthusiasm.

Shipping interests challenged Stevens's plan and proposed improving the Harlem's navigability. Coastwise traders advocated for a dredged channel down the center of the Harlem and the construction of docks and piers. Shipping boosters hoped that channel improvement would make the Bronx a steamship terminus and commercial port. The state Legislative

Committee of the Real Estate Exchange endorsed a similar plan.[90] This required reshaping the natural waterscape. Larger boats could traverse the southeastern portion of the river's ten-foot-deep channel, but only rowboats could navigate the entire river. The channel through Dyckman Meadows and Spuyten Duyvil Creek, the link between the Hudson and the Harlem, was particularly shallow and serpentine. The creek wound inefficiently through nearly twenty acres of unimproved salt meadows.[91] The six-foot drop between high and low tide exposed expansive mudflats. In 1885 General John Newton's Army Corps famously dynamited Flood Rock, the culmination of sixteen years of work to clear the dangerous reefs of Hell Gate in the East River. Green admired Newton's work and believed such improvements were overdue on the Harlem.[92]

To remake the Harlem, civil engineers had to first decipher its tidal mechanics to judge what would improve navigation. The river featured particularly vigorous tidal movement, but channelization might inadvertently impair the river's flow, and the movement of tides could either facilitate or hinder navigation and trade. Monitoring tidal movement and tide lines, civil engineers discovered that several factors combined to make the Harlem run—and make it conducive to shipping. First, high and low tide occurred at different times and for different durations in the Hudson, Harlem, and East Rivers, with high tide occurring earlier on the Hudson than the Harlem. Second, the volume of the tides in each river differed: the mean rise and fall of the tide in the Harlem was roughly two feet more than in the Hudson. Third, the Harlem experienced a higher mean high tide. These conditions created a strong current that flowed preponderantly toward the Hudson and made it possible to keep an open channel in the Harlem no matter the tide. Channelization would amplify these natural patterns.

Nineteenth-century coastal engineers focused on preserving navigability and preventing stagnation of harbor waters. General Newton explained that a channel 350-feet wide and fifteen-feet deep at mean low water would maintain a strong current and prevent the river from becoming a "cesspool."[93] In the 1860s Green worried that the unregulated release of sewage and offal into the Harlem would be "detrimental to navigation" and "injurious to the [river's] healthfulness."[94] Time and General Newton's attention to pollution proved Green's worry well founded. By the end of the century, six of the city's twenty-nine largest sewer outlets discharged raw sewage into the Harlem. The East 110th Street sewer, a drain for seven hundred acres and Manhattan's second-largest sewer, emptied into the Harlem and made piling and bulkhead work "difficult and slow"—and malodorous.[95]

Coastal engineering would alter the Harlem's tidal, maritime, and adjacent upland ecosystems and landscapes to turn the river into a nautical highway for commerce.

On the Harlem, regional management reshaped two of the city's edges at once: the metropolitan periphery and the waterfront. In 1876, the state turned the project over to the Army Corps. The state had authorized the city parks department to develop harbor lines and create an improvement plan for the riverfront, but for five years, the Department of Docks fought for control of the project.[96] Once in control of the Harlem, the Army Corps slogged through a twelve-year legal process to acquire both land and underwater property rights.[97] Construction finally began in 1889. Yet even through the creation of environmental infrastructure on the Harlem, the river's natural features continued to shape the waterscape.

Strong tides proved a challenge to this intervention in the harbor's hydrology. To channelize the Harlem, workers erected dams at both ends of Spuyten Duyvil, 1,200 feet apart, pumped the river dry, and dug and blasted a new riverbed 375-feet wide and twenty-eight-feet deep. But reshaping the river gave workers trouble. A strong gale or nor'easter could transform the Hudson into roiling slush and swamp the dams. In late April 1893, a gale from the northeast impeded the ebb of tides, raising flood tides higher than usual. The dams collapsed from the enormous pressure. The *Times* estimated that in two minutes, fifteen million gallons of water entered the cut and "mingled in a mighty rush."[98]

Even when the dams held the tide, the soil characteristics of the Harlem shoreline, its layers of peat and mud, made construction difficult. As an author in the *Engineering Record* reported, "the cut through the soft material of the marsh . . . caused considerable anxiety and difficulty." The Army Corps studied the nature of the riverbed. One hundred forty-one borings revealed the riverbed was semifluid mud 60–100-feet deep.[99] The river's soft mud oozed its way around the dam at the Hudson; along Spuyten Duyvil, it was sufficiently fluid to require pumping rather than excavation. The corps built a seven-mile-long retaining wall to keep the mudflats from sliding into the new channel. The project was typical of landfill operations of the era in that it produced straight, deep bulkheads to facilitate navigation and the docking of commercial vessels. Over seven years, workers eventually removed 550,000 tons of rock, dredged one million cubic yards of earth, and deepened and straightened northern segments of the river.[100] The canal cut a new course for Spuyten Duyvil Creek and, astonishingly, turned the fifty-two-acre promontory of Marble Hill into an island to allow for a straight Harlem–Hudson channel—in 1914 the former bit of

Manhattan would be attached to the Bronx through fill, although it legally remained part of New York County.[101]

The Harlem River Ship Canal officially opened on June 17, 1895. The canal would not be complete until well into the twentieth century, and complaints over mud filling the channel were not uncommon in subsequent years.[102] Nevertheless, the North Side Board of Trade sponsored a lavish opening festival. Steam launches, tugs, rowboats, scows of all sizes, and military floats traversed the canal under blue skies. Warship guns saluted in celebration. A land parade of ornate floats celebrating North Side businesses—brewery exhibits, ice making, piano making, and even house construction—simultaneously snaked through adjacent streets. William Steinway noted in his diary, "The Harlem Ship Canal is being opened today[,] my Theodore being present in a Daimler Motor boat[.] Both the land and water parade pass off without a hitch and the whole thing is a fine success."[103]

Mayor William Strong led the ceremonial opening of the canal with a narrative of natural environments giving way to development, a symbolic step of progress. The mayor told the crowd that the canal had spoiled his favorite fishing spot.[104] At midcentury, the Bronx Kills, connecting the East and Harlem Rivers, was the location of the Willows, a well-known fishing resort.[105] Through the 1880s anglers frequented Morris Dock, King's Bridge, and Spuyten Duyvil. Striped and banded killifish, Atlantic silversides, and winter flounder once lured both anglers like Strong and marsh birds such as snowy egrets.[106] But the river Strong recollected was gone by 1895. New York and Putnam Railroad tracks lined the North Side riverfront between Macomb's Dam Bridge and Marble Hill. In Port Morris at the confluence of the East and Harlem Rivers, the train yards of the New York, New Haven and Hartford Railroad dominated the shore.[107] The river's environment never played a prominent role in the two-pronged project of improving navigability and preserving tidal patterns. Canal boosters understood the material nature of the river through the entwined problems of pollution and insufficient shipping channels. Strong declared that the promise of commercial productivity and increased awareness of the district justified this lost contact with nature.[108]

For boosters, the waterfront as commercial infrastructure was the most valuable iteration of the river, for it promised regional growth. Municipal systems such as streets and the environmental infrastructure of parks and the canal reveal that nineteenth-century growth was more than dense streetscapes and ever-taller buildings. In the city's borderlands, expansion incorporated urban, suburban, and rural land uses. Rather than diffuse the

influence of urban government or balkanize locals to view downtown bureaucrats as foreign invaders, expansion knit core and periphery together
to form a vision of a New York City far larger than Manhattan.


*Conclusion: The Birth of the Bronx*

Home-rule politicians, Manhattan-based officials, and state-appointed
planners all promoted a consistent vision of expansion, long-range planning, and regulation of urban form for the trans-Harlem. Residents of the
city's northern periphery embraced large-scale planning, but locals' goals
did not always align with those of bureaucrats headquartered in the city
center. The rise of the regional city was a give and take between the interest groups of the urban core and those of the borderland district undergo-
ing urbanization. City-centered visions of the region did not always align
with the goals forwarded by actors hailing from the periphery. The work of
the Department of Street Improvements offers a case in point. The department marked the first time that the city granted neighborhood control over
public works. The department embraced the grid despite the decades of
withering criticism leveled against it by Manhattan-based bureaucrats. So
central was the grid to locals' aspirations for economic growth, Street Commissioner Heintz frequently had to assure worried real estate investors of
his commitment to the design.[109] Heintz's proto-borough-rule department
implemented a regional vision defined in large part by North Siders.

Developers in the trans-Harlem fostered the perspective that the modern metropolis could and should be planned, and they identified comprehensive planning, expanded administrative limits, and annexation as
central elements of the modern, enlarged city. Andrew Haswell Green
first publicly proposed consolidation in an 1868 BCCP report. The state
Chamber of Commerce and the city's mayor Abram S. Hewitt endorsed
municipal expansion in 1888. Between these two events, the city annexed
the Twenty-Third and Twenty-Fourth Wards, and the state authorized the
North Side Park system. Local drives for better services had contributed to
the first round of annexation in 1874. The North Side's home-rule park and
street commissions strengthened this tie, encouraging local development
with an eye to expansion. Parks linked the city and its northern hinterland,
a stepping stone to the creation of a river- and island-spanning regional
city.

The progrowth political agenda that drove public works also informed
planners' approach to the natural environment of greater New York. In

advocating for the consolidation of Greater New York in the 1880s and 1890s, Green continued to rationalize a regional city through an environmental lens. His expansionist agenda informed when and where underdeveloped land was set aside in parks. The creation of Pelham Bay Park was a step toward an enlarged New York. The 1895 annexation of the East Bronx, the 26,620-acre territory around the park, was the first successful vote in the four-year process of municipal consolidation. The *Times* judged annexation of the East Bronx to be "the natural and logical outcome" of the parks department's work there.[110] Green advocated for planning and expanded political boundaries that took account of the environment of the edge: its tidal patterns, drainage systems, and topography. He foretold "serious embarrassments" concerning sewage and water supply because of artificial political boundaries.[111] Green included all of New York Harbor, a vast complex of rivers, estuaries, wetlands, and more than 585 miles of waterfront and adjacent territories in his definition of the city's environmental systems. Fragmented jurisdiction meant that the protection of the area's single greatest asset, the navigable water system—the concern of all—had become the duty of none. Green charged that Brooklyn and New York were each "injecting smoke, stenches and sewage" into neighboring jurisdictions. "The waters and atmosphere which penetrated and surrounded the metropolitan district," Green said, required comprehensive regulation.[112] Rather than seeing the harbor's waterways as barriers or division lines, Green urged city leaders to consider them to be "the means by which communities meet and mingle." The Harlem River, for example, was not a boundary line but a segment of an indissoluble natural system. Nature set the stage for a unified city spanning the islands and peninsulas of the harbor.[113]

With the development and then annexation of the mainland wards and western Long Island, the meaning of "regional" shifted in New York. The geographic area referenced by the term was larger at consolidation than it was in the 1860s when Green began work in the trans-Harlem. Consolidation's supporters embraced the enlarged city as the outcome of peripheral development and farsighted investment on the urban edge. Greater New York officially came into existence on January 1, 1898, a municipality of unprecedented scale. Consolidation merged forty municipalities, including the nation's first and fourth largest cities, New York and Brooklyn. Overnight its population increased from two million to 3.4 million persons. The behemoth city ranked first in the world in geographic size, covering over three hundred square miles, and second in terms of population.[114] The *New York Times* recognized the role of the urban periphery in the establishment of the regional city. Home-rule boosters who reconfigured the city's edges had nurtured the idea of a greater New York. The *Times* praised William Steinway as anticipating the consolidation of New York City and Long Island City, declaring "what may have been considered by some of his contemporaries a visionary scheme has now come to pass." The North Side Board of Trade issued a similar celebration of the new borough of the Bronx, on the city's mainland territory, heralding annexation as "the manifest destiny of the district . . . the Greater New York beyond the Harlem."[115]

By 1898 a new geographic scale was needed to define the city; Manhattan, once the extent of New York City, was but one small slice of the metropolis. The majority of the city now lay in the new boroughs. The new boroughs made Greater New York a city of dramatic contrasts. In the crowded older wards on Manhattan, more than 220,000 people could be found living in a square mile district. But as a federal census worker observed, the eastern Bronx was "an almost purely country district, with hill and dale, upland and meadow, forest and open."[116] Northeast of Manhattan and the trans-Harlem, along the upper East River, the outer boroughs of Queens and the East Bronx remained largely undeveloped natural areas. This open territory proved conducive to suburban and leisure-based development. In the coming decades, the city building and reconfiguration of the urban-suburban-rural threshold continued unabated in the outer reaches of the sprawling metropolis.

# 3

# Working-Class Leisure on
# the Upper East River and Sound

Perched at the tip of Rye's Milton Point in Westchester County, the ornate shore station of the American Yacht Club offered its exclusive membership luxury and a panoramic vista of southwestern Long Island Sound. Financier Jay Gould founded the club in 1883 after being scandalously refused membership to the New York Yacht Club, the apogee of Gilded Age sporting clubs. Built in 1887–1888, American sponsored sailing races complete with naval welcomes and galas catered by Delmonico's of New York City.[1] Selective nominations, initiation fees, and annual dues kept clubs like American exclusive. The club's setting added to its cachet. The adjoining estate community of Milton Point attracted wealthy New Yorkers such as Simeon Ford, co-owner of the Grand Union Hotel, who summered there in a forty-eight-room manor. Rye boasted some of the grandest estates in Westchester. Journalists waxed poetic that "the Sound shore, between New Rochelle and Port Chester," which included Rye, was "almost an unbroken line of beautiful private estates."[2]

Country manors and private clubs, markers of genteel social life, constituted the archetypical landscape of Westchester's Gilded Age coast. But at the turn of the twentieth century, the Sound's waterfront was actually a mosaic of elite and working-class resorts such as William Steinway's Bowery Bay Beach. In 1908 the *New York Tribune* matter-of-factly recognized coastal Westchester as a "suburban zone of the greater city" and the Sound as a "summer playground district of the great metropolitan zone."[3] In fact east of Milton Point, mile-long Rye Beach lured working-class excursionists. In 1872 its first commercial resort opened. Hotels, restaurants, bathhouses, shooting galleries, and mechanical amusements such as roller coasters followed. Less than a mile from American Yacht Club, Rye Beach was nevertheless far removed from the social prestige of Jay Gould's world.

Rye Beach is a window into the wide spectrum of leisure resorts along greater New York's coastal hinterlands. In 1860 a reporter remarked, "It is strange how few New Yorkers appear to be aware of the charming little villages that dot the shores of Long Island Sound between New York and New Haven."[4] This comment reflected the preferences of the city's wealthy, who sought the status of enclaves such as Milton Point, Saratoga Springs in upstate New York, or Newport, Rhode Island.[5] In reality, a working-class leisure market flourished along these shores, attracting blue-collar clientele who could not afford a full summer season away. Advances in mass transit increased the availability, speed, and comfort of regional travel between Manhattan and Queens, the Bronx, and southern Westchester. Steamboat routes began in the 1820s and railroads in the 1840s. A web of trolley lines sprawled across Queens and between the Bronx, Westchester, and Connecticut from the 1870s on, and high-speed, high-capacity subways were built in Queens and the south and central Bronx between 1905 and 1920.[6] The expanding rapid-transit network brought new coastal places into the range of urban and suburban life. Leisure activities and residential development produced regional growth along vernacular and working-class aesthetic lines very different from those envisioned by Andrew Haswell Green and Frederick Law Olmsted. Spillover development, blue-collar suburbs, and commercial resorts were major factors in the movement of people to the coastal borderlands of the regional city.[7]

The search for nature took urbanites not just to national and city parks but to the undeveloped spaces of the urban periphery. The natural world was an important part of an excursionist's experience on the urban edge. Traveling by steamer up the East River, city dwellers left behind the urbanized port for the "many pretty islands that lie in the lower part of the Sound" and "the phenomenally green meadows that run down to the water's edge."[8] While pleasantly scenic from a passing steamer, the intertidal marshes that dominated this coast also had less attractive environmental features. The shores lacked surf and were generally narrow, rocky, and reedy. Waters were brackish and dark colored. Pungent black mudflats were exposed at low tide. The bathing spots paled in comparison to southern Long Island's wide, sandy Atlantic beaches. But appreciation of this environment was not wholly dependent on beauty. The shore took on new value as a refuge from the pollution of Manhattan's waterfront. In a typical report on Bowery Bay Beach's setting, a New York Times journalist extolled the wooded shores and natural bluffs that framed Bowery Bay. The author also emphasized the bay's water quality: an "advantage . . . which should not be lost sight of is . . . that no sewage whatsoever enters the beautiful sheet of

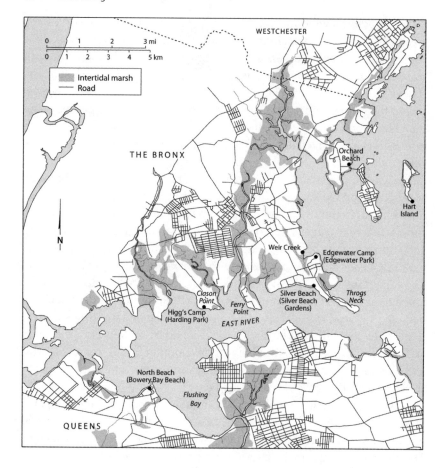

**3.1.** The informal leisure corridor of the Upper East River. The dark gray–shaded areas represent intertidal marsh that originally covered a substantial portion of the shoreline at the turn of the century. *Source*: Map by Bill Nelson.

water at this point, and, therefore, the bathing here is absolutely healthful and pleasant."[9] By 1896 the upper East River still remained free from the severe pollution of its lower shores at Long Island City and Brooklyn Heights. Water quality improved the farther one traveled from Manhattan up the Sound. Prudent blue-collar leisure seekers assigned cultural values to this shore, making the most of its recreational potential, since it was comparatively "healthful and picturesque" and within "easy commuting distance" of New York.[10]

This history of the working-class leisure spaces of the upper East River and Sound can be understood in three parts: the rise of amusement park districts, the opening of summer colonies and their evolution into permanent neighborhoods, and finally the limits of the corridor's environment and public spaces. Leisure was a space-structuring tool on the city's coastal hinterland. The amusement parks of Bowery Bay, Queens, Clason Point in the Bronx, and Rye Beach launched the leisure corridor in the 1880s. The corridor stretched northeast from the upper East River to the Sound, but its edges were continually shifting as improved rapid transit brought new areas into orbit. Dozens of shore spots attracted excursionists. The six locales and nine resorts highlighted underscore the diversity of the corridor's camps, amusement districts, and beaches. Closest to Manhattan, North Beach and Clason Point marked the corridor's western end. The resorts lay kitty-corner across a three-mile-wide section of the upper East River. Another three miles up the river from Clason Point, Throgs Neck was home to the summer colonies of Silver Beach and Edgewater Park, the latter of which was adjacent to Solomon Riley's African American beach club on the neck's northwestern corner. Across the bay in Pelham Bay Park was Orchard Beach, a tent colony and bathing spot. Orchard Beach looked out on the Sound and its numerous islands, including City and Hart Islands. Riley built a second resort at the southern tip of Hart Island, two miles from the park. Rye Beach and Milton Point were less than ten miles up the Sound. While noncontiguous, the coastal resorts of the Bronx, Queens, and Westchester shared development patterns that in aggregate overrode official plans or civic elites' development goals. The shared history of these leisure spaces challenges the stereotype that confines laborers to cities and defines suburbs as preserves of the affluent.[11]

## Amusements on the Urban Periphery

In the last decades of the nineteenth century, urban laborers experienced economic and workday changes that contributed to the rise of a diverse working-class leisure culture. The turn of the century was a heyday for amusement parks. Seaside resorts became a quintessential feature of East Coast metropolitan development.[12] Amusement districts followed similar trajectories, maturing from individual concessions into resort districts with comparable rides, games, and restaurants that hosted a variety of gymnastic expositions, political rallies and parades, and singing festivals. Ad hoc development created distinctive leisure landscapes that nurtured activities discouraged in the city center, such as enjoyment of a beer garden or contact with nature on a beach. In June 1886 William Steinway and George Ehret opened Bowery Bay Beach. On a typical summer day, half a million New Yorkers bathed, drank, and danced there.[13] Clason Point's first amusement park opened the following year. Established in the 1870s, by the early 1900s the hotels at Rye Beach, such as Edward's Beach Hill, catered immensely popular "Rhode Island" clambakes that served up to six hundred pounds of bluefish, four thousand ears of corn, and one thousand chickens to two thousand revelers. The district attracted trolley-riding excursionists from the Bronx and southern Westchester's small industrial centers of Port Chester and White Plains.[14] Solomon Riley, a wealthy black real estate investor, built the last of the upper East River resorts in the 1920s replete with a pier, boardwalk, and bathing facilities.

At the turn of the twentieth century, nonfarm employees nationwide witnessed an increase in wages and a decrease in the cost of living that made their dollars go farther. The average manufacturing worker also enjoyed a decline in hours worked, working 3.5 hours fewer a week in 1910 than in 1890. Workers in unions and blue- and white-collar jobs enjoyed an even greater decrease in hours. In addition, a growing number of businesses began closing early on Saturday, particularly in the slow summer months, giving rise to the Saturday half holiday.[15] Because of these changes, large numbers of urban laborers, immigrant and native born, acquired the time and money to participate in leisure activities.

Steinway's Bowery Bay Beach exemplified the growing popularity of commercial amusements and the investment strategies that gave rise to the leisure corridor. In the 1870s and 1880s, Bowery Bay was a popular destination for bathing and picnicking parties, but the undeveloped waterfront lacked bathhouses and a convenient trolley stop. To commercialize the shore and turn a profit, Steinway extended his trolley line to the shore. The Astoria ferry, in which he owned a controlling interest, also extended service there from Manhattan and nearby College Point, Queens. Like Bowery Bay, Clason Point and Rye Beach centered on grand steamboat piers. Rye welcomed steamers from Long Island, nearby ports, and New York City. Clason Point excursion boats arrived from Manhattan, Long Island, and the western Bronx. Until 1910, when a trolley line reached the point, most visitors arrived via steamer, since previously the closest trolley stop had been nearly a mile-long walk or ride on an open bench wagon. Trolley lines proliferated at the turn of the century. It became possible to "trolley" along the entire shore of the Bronx, Westchester, and Fairfield counties, and formerly sleepy agricultural and fishing villages came into the purview of day-trippers. By the early 1900s the *Brooklyn Eagle* approvingly declared that "the democracy of the trolley car" was transforming the rolling hills, fields, and beaches of greater New York into "everybody's" playground.[16]

Since the leisure corridor's modest amusement parks were constructed piecemeal by individual concessionaires over many seasons, they exhibited neither formal spatial organization nor aesthetic unity. At Bowery Bay, Steinway & Sons never monopolized real estate or attractions as a traditional company town might. William Steinway and George Ehret's improvement company leased the empty blocks south of Grand Boulevard to independent concessionaires. The boulevard, which ran along the shore, became the resort's unofficial midway. Each amusement district featured a similar main thoroughfare lined with a typical assortment of attractions: independently run bandstands, pools and bathhouses, ice cream booths and restaurants, and varied mechanical amusements such as Ferris wheels, gyroplanes, and roller coasters.[17] Located as they were along the shores of the upper East River and Sound, the resorts all had bathing facilities like the pavilion Solomon Riley built at his Hart Island resort, which capitalized on the site's views of Pelham Bay.[18] These districts were precursors, albeit in more modest forms, to the grand amusement epoch ushered in by the Midway of the 1893 World's Columbian Exposition in Chicago and most famously replicated at Coney Island in Brooklyn.

New York City's recreational hinterland fostered a culture often at odds with reform-minded cultural institutions such as Central Park and the work of genteel reformers intent on disciplining, refining, and instructing the cultural expressions of the worker in urban-industrial society. The region's beer gardens and bathing beaches offered alternative social and cultural values and welcomed a wider range of behavior. As historian John Kasson argues, bathing beaches in particular encouraged the relaxation of proprieties and a loosening of social rigidity signaled by bathing costumes, while their mechanical rides further eroded behavioral expectations through forced contact and physical exhilaration.[19]

German social customs and values were particularly evident at Bowery Bay Beach. The resort was after all built by two German immigrants, one of whom was a major beer producer. The centrality of the beer garden tradition at Bowery Bay distinguished Steinway's company-built settlement from others such as Pullman, where aspects of working-class culture, such as drinking, were barred in the hope that temperance would promote industriousness and order. Ehret, who owned Hell Gate Brewery and forty-two saloons across the city, established additional sales outposts at Bowery Bay. He personally leased land from the corporation, built wooden concession stands, and re-leased the stands to small-time theater and restaurant

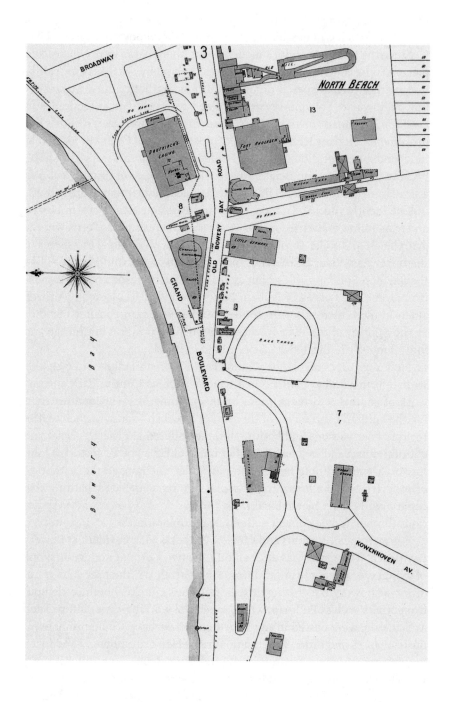

**3.2.** *Opposite*: North Beach in 1903. The amusement park's built environment had matured by the turn of the century. Between 1903 and 1914, private concessionaires filled in some surrounding lots, and a general reshuffling of rides occurred, but the density of the built environment stayed the same. For example, the fields around Kouwenhoven Avenue, in the southwest corner, remained open in 1914. *Source*: Queens V. 5, Plate No. 2 (map bounded by Broadway, Jackson Blvd., Bowery Bay). Lionel Pincus and Princess Firyal Map Division, New York Public Library Digital Collections.

**3.3.** The Grand Pier and shores of Bowery Bay in 1892. *Source*: BPQ-000341 "Grand Pier at North Beach, 1892." Courtesy of the Queens Borough Public Library, Archives, Borough President of Queens Collection and the NYC Municipal Archives.

managers, sometimes on condition that they sell his beer exclusively.[20] Bowery Bay had a distinctive Teutonic character. The names of its establishments and concessionaires spoke to its ethnic base: George W. Kremer ran a carousel; Paul Steinhagen and Julius Kelterborn operated Villa Steinhagen across the street from Heicht's picnic grove. As a symbol of the resort's German culture, Friday firework displays often ended with a sparkling beer stein topped by a twinkling cascade of beer foam.[21]

In the 1850s German immigrants in New York faced intolerance for their beer-drinking culture. From midcentury on, foreign-born residents consistently made up nearly half the city's population. In this first wave of nineteenth-century arrivals, the Irish, the Germans, and the British were the city's three largest immigrant groups. The city's Protestant elites expressed anxiety that immigrant drinking culture, particularly that of the Germans and Irish, would overrun and degrade society. German beer gardens were controversial with some native-born New Yorkers. Influential citizens who advocated temperance, such as Henry Ward Beecher and P. T. Barnum, equated alcohol consumption with the ruined man. Others, such as James Gordon Bennett, the founder, editor, and publisher of the *New York Herald*, equated German drinking culture with social disorder and poverty.[22] The city's elites stereotyped beer drinking as the pastime of poverty-stricken immigrants in the dives, opium dens, and deteriorating theaters of the Bowery. By the 1870s, the Bowery and neighboring Broadway offered a variety of commercialized sexual activities to a mass working-class audience. Outraged by activities that diverged from their own values, the city's native-born elites pejoratively decried the area as a debauched "vice" district.[23] That German newcomers settled on the east side of the Bowery in lower Manhattan exacerbated prejudiced stereotypes. Critics claimed that "good men do not grow out of the boys who spend their Sundays at Volks Gartens and Volks Theatres."[24] It was outrageous, one reverend concluded, "to sacrifice the authority of God's Day and the moral interests of multitudes simply that a German can have his lager fresh."[25] Germans in Chicago faced similar discrimination from conservative native residents; the city weathered the so-called Lager Beer Riot in 1855 after its anti-immigrant Know-Nothing mayor Levi Boone ordered taverns in immigrant neighborhoods closed on Sundays.[26]

Germans saw their distinctive leisure culture and beer gardens as reputable community spaces where families, working men, and social groups such as singing societies gathered. "The common German . . . laborer," a sympathizer explained to the *Times*, "wants his glass of beer on Sundays . . . because that glass of beer symbolizes 'Gemuethlichkelt,' [Gemütlichkeit] good fellowship," and a release from "the dull grind of the week's work."[27] Ehret and Steinway's hinterland resort welcomed German beer-drinking culture and promoted this *Gemütlichkeit*, a word that evokes feelings of unhurried cheerfulness, peace of mind, and social acceptance. On the 1886 opening of Bowery Bay Beach, Steinway happily observed that while decorous, the immense crowd drank the resort dry: "At 5 P.M. all the beer is gone, and people overflow Steinway village and drink all the beer there."[28]

In the 1890s and 1900s, more brewers opened beer gardens and saloons along the bay. The resort's well-patronized beer gardens sold over one hundred half barrels of beer on a typical weekend.[29] Nevertheless, to bolster its status, and to distance it from the debauched reputation of the Bowery, Steinway and his fellow investors renamed it North Beach in early 1891. Steinway observed, "It certainly was a happy idea to change the name to North Beach, as people would persist in connecting Bowery Bay Beach to the 'Bowery' in New York."[30] The German benefactor-cum–city builder was determined to market the resort as a reputable, if also lager-filled, destination.

Steinway and Ehret profited from ethnic working-class leisure culture, but they also created social spaces at the resort for their elite business colleagues. The investors belonged to a well-off German community that frequented the higher-end establishments of North Beach. Steinway often joined Ehret and other acquaintances at Astoria Schuetzen Park or at Steinhagen's, where Steinway celebrated the wedding of close friends in the summer of 1891.[31] Middle-class patrons such as the Geipels from Astoria, a family with the time and money to frequent the resort, purchased a four-dollar season pass to Frederick Deutschmann's Silver Spring Bathing Pavilion in 1897.[32]

Elites enjoyed North Beach's high-end destinations, but working-class excursionists joined them in enjoying the district, if not the same exact venues. The *New York Times* focused on the economic diversity of greater New York's resorts and praised North Beach for including diversions to suit all "palates and purses." The paper similarly deemed Rye Beach "one of the best beaches along the Sound," because "from expensive restaurant service to hot roasted peanuts from a plebeian stand, you [could] find whatever refreshment suits your taste" and budget.[33] In a 1907 standard of living survey, political scientist Robert Coit Chapin reported that the city's leisure seekers included "hundreds, if not thousands of wage-earners and small tradesmen of the middle-class who are self-respecting and self-supporting." Chapin estimated that workers needed an annual income of at least eight hundred to nine hundred dollars to pursue commercial leisure.[34] He also recognized, however, that the city's "miscellaneous" class of recreationalists earned a wide spectrum of wages and exhibited "considerable latitude" in recreation spending.[35] Chapin used the term *middle class* to describe the city's workers engaging in commercial leisure, but no consensus existed as to whether nonmanual workers such as clerical and sales employees counted as white-collar or middle-class wage earners.

Trolley and steamer destinations catered to a spectrum of excursionists by the early 1900s, but social scientists could only estimate the general economic threshold at which workers could afford commercial leisure and the frequency with which they did so. Funds spent on leisure travel can only be estimated from scattered data on total travel outlays, since individuals frequently paid for recreation with "spending money" not itemized as recreation outlays.[36] Discount tickets for outings sponsored by voluntary societies, religious or fraternal organizations, or unions create similar problems in tracking leisure spending, since such spending was not always differentiated from society payments.[37] Tickets such as those for a 1901 Retail Butchers Association outing at North Beach, which cost a dollar, offered union members substantial savings. Such events routinely included transportation costs, such as a twenty-cent steamer fare and a five- or ten-cent trolley trip as well as nickel charges for individual amusements and beers.[38]

The cheap amusements and nominal-fee venues that characterized the working-class leisure corridor opened participation to laborers who could afford minimal or no recreation expenditures.[39] Many families made no recreation expenditures; of families with annual incomes of eight hundred dollars, 15 percent spent nothing on recreation while 28 percent of families with incomes of six hundred dollars refrained.[40] Even within these groups, a prudent worker might walk to work and save the carfare, as one young woman explained, for an excursion "'away out in the Bronx.'"[41] A North Beach patron expressed that she found a trip to amusements on the urban edge affordable, remarking, "I want some fun, and it only costs ten cents."[42] William Kells's father ran a dance hall at North Beach that he described as "just a little thing." His father eventually converted the space into a barebones theater. Admission was free. The Kells sold whiskey, Ehret's beer, and lemon or sarsaparilla soda to realize a profit, but a purchase was not a requirement for entry. "We only had four reels, and by the time a half hour would go by, a nice place to sit down . . . whole families would just walk in."[43] At Rye Beach, excursionists who wanted to avoid bathhouse fees changed into their swimsuits in the back of cars parked on adjacent roads.[44] A number of North Beach venues were simply fenced-off grassy spaces, sometimes tented, with "just tables and chairs, and . . . a sign that'd say 'Picnic Parties Welcome.'" North Beach's German Castle, Kells recalled, "let people buy half a keg, [and] would take it outside to open land and tap it for you."[45]

Leisure-corridor resorts amused the masses, but their location on the coastal periphery provided more than opportunities to play. A formal gate gaily decorated with flags welcomed revelers to Kane's Amusement Park on Clason Point, but beyond it high grass covered the open peninsula. Open land, lacking surveyed streets, backed the amusements of the East Bronx and Queens. Upland forests of oak, chestnut, hickory, and walnut

**3.4.** The undeveloped East Bronx. This July 1928 image shows the roller coasters at Clason Point Amusement Park and the dancing pavilion to the far right. The upper East River leisure corridor included both shaped space, such as fenced-in amusement districts and the unshaped space of fields and beaches. The latter appealed to excursionists with small recreation budgets. *Source*: Photo by Percy Loomis Sperr. ©Milstein Division, New York Public Library.

gave way to more open woodlots containing sugar maple, crab apple, and cherry trees and a multitude of wild berry bushes. Close to the water's edge, woodlots gave way to extensive marshland of rose hips, fiddlehead ferns, milkweed, and an abundance of wetland plants such as cattails.[46] New Yorkers embraced the borderlands of the upper East River as common land, informal extensions of resorts where they could enjoy the nature of the beach and bay, upland fields and woodlots, with little or no charge. Across the open points of the East Bronx, summer camps sprouted with the hopes of attracting not just day trippers but summer renters. While leisure first tied the city core and edge together, over time, these modest resort communities developed into enduring blue-collar suburbs.

*Summer Shore Colonies*

The coastal leisure corridor's amusement parks attracted day-trippers as well as summer campers on longer stays. As a vernacular landscape, the coastal corridor both grew from and fostered unstructured activities and informal spaces that serviced working-class leisure needs, particularly affordable homes and clean, healthful environments for recreation and residency.

Entrepreneurs such as Thomas Higgs on Clason Point and Augustus Halsted in Rye opened campgrounds adjacent to amusement districts. Camp colonies developed vernacular street systems shaped not by city officials but ad hoc by the ways people used camps. The outer districts of the regional city remained informal spaces, not yet laid out under the comprehensive plans advocated by public works officials who equated formal planning and uniform development with modern urbanism. Over time renters at East Bronx summer colonies winterized their tents. The distinctive property regimes of these owner-built suburbs secured community permanence and dictated the first wave of residential development in this segment of the city's coastal hinterlands.

In the late 1800s and early 1900s, countless working- and middle-class summer colonies flourished on the eastern seaboard from Martha's Vineyard across New England. On Connecticut's beaches the communities of Little Danbury and Little Bridgeport, named for their inhabitants' permanent homes, spoke to the urban populations who recreated on the Sound. The Hudson's wooded riverfront and New Jersey's sandy Atlantic beaches, including the famous resort of Atlantic City, also offered summer attractions. Vacationers flocked to New England and New York's mountain destinations as well. New York's Catskill Mountains became known as the borscht belt, a Jewish resort destination that annually attracted a million summer renters to five hundred different bungalow communities at its peak in the mid-1900s.[47]

Rye Beach was typical of the Sound's summer colonies. By the early 1900s, half a dozen bungalow courts and campgrounds offered a range of affordable housing options. August Halstead, a major Rye landowner and businessman, largely founded Rye's bungalow rental market in the early 1900s. Halsted converted his waterfront property, in the family since the 1740s, into a summer camp. Renters chose from nearly two-hundred tightly packed wood camps fancifully named Walk-In Cottage, Camp Washout, and Dreamland Cottage.[48] When the town of Rye condemned his property to build Rye Town Park and Oakland Beach in 1909, Halsted simply relocated some of his condemned bungalows to nearby streets.[49] In the early years, canvas tents costing around sixty dollars for the summer dominated the shore, but additional investors followed Halsted's lead, and bungalow courts eventually came to outnumber tents. Substantial six-or-seven-room furnished rentals such as Bellchamber Bungalows, which rented for two hundred dollars a month, were built, and Rye Beach came to welcome campers and renters of varying means.[50]

Workers weary of the heat and crowds of tenement districts could find relief within city limits. Coastal colonies developed in the East Bronx as well as in Brooklyn, at Coney Island and Gravesend Bay, and the Rockaways in Queens. Working fathers often remained in the city during the week, traveling by train, trolley, el, and steamer to join their wives and children alongshore for the weekend. Margaret O'Shaughnessy recalled that as a girl summering in the East Bronx in the early 1900s, "we spoke of our homes as being 'in the city,' of our fathers as working 'in the city,' although, Orchard Beach was, of course, itself part of the city."[51] The parks department ran a public tent colony at Orchard Beach in Pelham Bay Park. Campers enjoyed tents in grassy fields fringed with apple trees and blackberry bushes and framed by Pelham Bay.[52] But campers like the O'Shaughnessys could choose from a number of summer colonies on the East Bronx's peninsulas. Higgs Beach campers rented cool cotton tents in a wooded campground adjacent to Clason Point's amusements. At Silver Beach on Throgs Neck, bungalow porches cantilevered over a nearly thirty-foot bluff. From this vantage point, residents enjoyed breathtaking panoramic views of Manhattan and Queens to the southwest. At both Sound and East River camps, rocky beaches for bathing and clamming lay just steps away.

Camps flourished on the East Bronx waterfront because nature, the coastal morphology, intertidal landscape, and geographic distance discouraged urbanization. The marshy, marginal lands of its estuarine shore were of little value for agriculture. The shallowness of the bays and the lack of rail infrastructure discouraged manufacturing and transportation.[53] Industry concentrated in the western Bronx. The north shore of the Harlem River became the manufacturing district of the Third Avenue corridor, where industrialists had secured channelization to facilitate trade. Gouverneur Morris II and Jordan L. Mott were two of the hundreds of entrepreneurs who, like Steinway, built industrial nodes around Manhattan. Mott Haven and Port Morris on the Harlem initiated urbanization in the South Bronx.[54] The elevated railroads, trolleys, and subways built between 1886 and 1905 connected these Harlem ports to Manhattan's north–south transit orientation. Rapid transit fostered a linear urbanized corridor around Third Avenue. In 1900, 58 percent of the Bronx's 200,500 residents lived west of the Bronx River in this corridor. The borough's population rose to 430,980 in 1910 and 732,016 in 1920, yet even by the latter decade, 90 percent of residents remained concentrated in the Twenty-Third and

Twenty-Fourth Wards west of the river.[55] Three-fourths of the borough, including nearly the entirety of the East Bronx, remained parkland, under agricultural use, or else open fields and marshland in 1910. While the *Bronx Home News* decried the "Unnecessary Isolation of [the] East Bronx," this characteristic was precisely what allowed the leisure corridor to flourish.[56]

The dominant land use of East Bronx peninsulas through the nineteenth century mirrored that of the Sound, not the urban-industrial landscape of Manhattan. The East Bronx's topography and environment were also more similar to the Sound shore than the precipitous elevations of the Twenty-Third and Twenty-fourth Ward's ridges. Rolling hills and low-lying coastal marshes characterized the coastal plain of the East Bronx, Westchester, and Connecticut. Estates and pastures dominated this shoreline. In 1823 the former Yale College president Timothy Dwight observed that from Throgs Neck to New York, "a succession of handsome villas is seen at little distances on both shores . . . embellishing the landscape, and exhibiting decisive proofs of opulence in their proprietors."[57] By the end of the 1800s, however, the rich had abandoned this stretch of coast for country places farther afield on Long Island and in Westchester's elite enclaves such as Milton Point. High-end summer retreats and early elite suburbs encompassed most of New Rochelle's shoreline, including Davenport Neck and Premium Point, and the numerous islands that clustered near the coast, such as the New York Athletic Club's Travers Island. Westchester's population grew slowly in the early 1900s. By the 1920s planners summarized the county as "comparatively rural with some closely developed urban communities," the satellite urban nodes of Yonkers, New Rochelle and Mt. Vernon. Of the three urbanized centers of lower Westchester, only the last was on the Sound, and it remained the smallest, with just thirty-six thousand residents by 1920.[58]

At the start of the twentieth century, upper East River and Westchester Sound communities had the same population density relative to the population concentrated around New York Harbor. Parts of Manhattan, Long Island City, and Brooklyn reached 75,000–100,000 people per square mile (and in places up to 300,000), while the density of the East River–Sound corridor remained less than five thousand.[59] The city continued to grow, but into the mid-1920s, 47.5 percent of the gross area of Queens, the Bronx, and Brooklyn remained open, and development concentrated in areas fronting Manhattan.[60] The city's coastal hinterland remained undeveloped.

In the East Bronx, mansions crumbled and once-landscaped grounds reverted, in the words of a local, to "just dirt pathways and some tar rounds, and swamps. . . . It was like going through the central part of the jungles just to find [Clason Point]."[61] Grazing cows remained a common sight along Soundview Avenue, the point's main road, into the 1920s. Throgs Neck, the easternmost Bronx peninsula, remained positively remote. While the city approved a street plan for the East Bronx in 1898 and recorded it on official city maps, most of the proposed streets were never surveyed or opened.[62] In fact very few official roads cut through the farms, marsh, and old estate grounds. Other than Fort Schuyler, built on the neck during the War of 1812, no municipal infrastructure existed to remind visitors that this territory was part of the nation's largest city.[63]

Boosters had long celebrated the environment of the undeveloped upper East River waterfront. In the 1880s the commission to select parks for the Twenty-Third and Twenty-Fourth Wards had declared "the tens of thousands of dwellers on our water-front . . . who breathe the air which sweeps over the fetid outpour of sewers and the poisonous refuse of factories and gas houses, filth, and abominations that are ever on the increase would, if consulted on the subject, soon dispel" any illusion that Manhattan might still provide fresh air and bathing beaches.[64] Industry made the lower Hudson and East River unlikely sites for waterfront parks. While city guidebooks declared the charms of rowing on the Harlem, in reality the success of trans-Harlem boosterism meant that foundries, textile mills, and dye works had taken over its shores. The commission concluded, "any park situated there would soon be environed with the smoke of furnaces and forges, and the noise of the triphammer."[65] When New Yorkers praised the clean environment of the upper East River and Sound they drew on long-standing ideas of the healthiness of the countryside. Health officials and the popular press that characterized suburban nature as healthier than the city understood the landscape in terms of its benefits to human health even in light of the rise of germ theory.[66] In 1909, the city Board of Heath avowed "there can be no question that these summer camps are of great value" to urbanites looking to escape crowded tenements and "liv[e] in the open air during the summer months."[67]

The benefits of the coastal environment included clean water and a cooler climate due to breeze patterns and lower temperatures. Large cities like New York create a climatic effect known as urban heat islands wherein the densest neighborhoods with the largest buildings maintain the highest temperatures. On sunny days, paved surfaces and stone and brick buildings heat more quickly and store solar radiation as heat in greater quantities than plants, soil, and water.[68] Rural and suburban environs lose a day's heat and provide cooler evenings, but built environments absorb more heat and cool at a slower rate, even when temperatures drop overnight. The temperature difference between the urban heat island and a leafy suburb during a clear summer evening averages around ten degrees Fahrenheit, but it can reach a twenty-degree differential.[69]

Not only did less developed landscapes remain cooler than dense urban centers, the open spaces of a coast offered additional climate benefits: sea breezes. "There is never a day when the East River has not its breeze. The great wind areas of Long Island Sound and the Lower Bay are connected by this strait; and the air, like the water, draws through from one to the other," claimed writer John Van Dyke. Coastal spots were cool even in warm weather "when the Central Park [was] like an oven."[70] In the 1880s, park advocate John Mullaly championed the East Bronx climate, calling it a "great reservoir" where the working class might "drink in new life and health in its refreshing, invigorating breezes." Mullaly even claimed that Pelham Bay Park had a greater abundance of fresh air due to the winds that blew from the Sound.[71] In reality, the entire upper East River–Sound corridor benefited from sea breezes.

Coasts are subject to a daily pattern of breezes that cool adjacent land. Sea breezes, also known as onshore breezes, exist when winds blow from the water toward the shore. They develop when the land becomes warmer than an adjacent body of water. As warm air rises on shore, the relatively cool air over the water surface replaces it, lowering temperatures. At sunset, the breeze reverses, since land cools more rapidly than water. At night, the warmer air over the water rises and cooler air from the land moves to replace it, initiating a breeze even in calm weather.[72] Onshore breezes bring cooler air during the day, and nighttime offshore breezes keep air circulating and cooling winds moving, offering constant relief. Since suburbs and rural areas cool off faster than cities, the pattern is more noticeable outside of urban centers.

These patterns appeared on the Sound. The proclamation by a Connecticut newspaper that "every body [sic] pants for the free air, the bracing breeze, and the out-door life" of the region's shores in fact captured this environmental attribute of the coastal climate.[73] Shore breezes had in fact convinced P. T. Barnum in 1867 that he should leave his inland Golden Hill estate Iranistan. A summer's sojourn to a farmhouse adjoining Seaside Park, "with its delightful sea breeze, both bracing and refreshing," led Barnum to conclude "future summers must be spent on the Sound."[74] The same desire brought campers to the East Bronx peninsulas. While their cotton tents were a far cry from Barnum's Waldemere, there was room for both on the coastal corridor, and breezes did not discriminate.

Campers first attracted to the cooler climates of summer camps on East Bronx peninsulas and Sound shores established colonies that became

3.5. Alongshore at Edgewater Camp, ca. 1915. This image captures the transition from canvas tents to more permanent bungalows and the proximity of camp lots. *Source*: X2011.34.2566. Courtesy of the Museum of the City of New York.

permanent features of the built environment. Bungalows came to constitute a considerable swath of Rye's five-and-a-half-mile coast, forming an unplanned blue-collar buffer district between the resort district and the rest of the town's waterfront.[75] They also fostered a unique property regime on Clason Point and Throgs Neck. East Bronx renters formalized their communities by converting tents into permanent residences. In the process they produced a distinctive blue-collar, owner-built suburb that diverged from the nineteenth-century civic elites' vision for the mainland wards and both Olmsted's vision of bourgeois suburbs and the speculative real estate market encouraged by the Department of Street Improvement's grid.

Unlike the politicians, boosters, and planners who sought to knit the urban core and outlying districts closer together through public works infrastructure, the individuals who populated the leisure corridor did so because they found its distance and differences valuable. The regional city developed through projects designed to unify districts but also at times through projects that marked the edge as distinct. Accessible enough that tenants were still, in the words of one camper, "within a short distance of everything that the City provide[d]," the distance between camps and the urban core was fundamental to their appeal as summer retreats. Summer camp colonies were an act of local redefinition of the city as a place for leisure and suburban living. The liminal quality of outlying districts also cultivated activities discouraged in the city center, such as owner-built housing in ruralesque settings.

While park planners and government officials from the Department of Street Improvements had substantial power to shape the cityscape of the Bronx, local development overrode official plans to establish a different trajectory and vernacular urbanism. At Edgewater Park, Higgs Beach, and Silver Beach, many streets depicted on the official borough map were in fact filled with bungalow lots and bathhouses by the 1920s.[76] The actual street and block patterns also created a more varied, smaller-scaled, and compact built environment than official maps. A typical camp lot measured from twenty-five by forty feet to thirty by sixty feet, although each colony had its own unique lot arrangement and street system. Olmsted and J. James R. Croes proposed curvilinear streets for the hilly western Bronx, but for high-end villa districts. No evidence indicates that they considered such a format might be reimaged in a laborers' district in their plans. But both Edgewater Park and Higgs Beach streets followed curvilinear paths. Higgs Beach included a mix of narrow streets and even narrower alley-like lanes. The roads of Silver Beach tapered into footpaths. At Orchard Beach they remained grass trails. Since traffic consisted of pedestrians and horse-drawn wagons, actual street construction was minimal. Edgewater lacked even an unofficial road system, and its curving lanes remained unnamed. Residents identified their bungalows by neighborhood section and number, such as "B-6."[77] While federal census workers capitulated to local reality and bypassed official addresses for the camps, the city ignored the East Bronx's vernacular streetscapes. Summer colony streets did not appear on the official borough president's map until after World War II.[78]

Although a gradual process, camp winterization became an increasingly common matter of economy during the housing shortage of World War I. Most of greater New York's workers could not afford summer rentals. Those with camps "were considered a little better than ordinary because you had a summer place. . . . You could get away from the city."[79] But even those who could afford to spend the summer in the East Bronx were not particularly wealthy. For example, Benjamin P. Waring, a laborer, moved his family from their central Bronx apartment to Higgs Beach Tent 82 in 1917. A tent cost less than a more centrally located apartment, and with two sons serving in the war, the family needed to conserve funds.[80] Camp winterization was a type of owner building, a process by which working people pursued economic security across the nation in the first half of the twentieth century. A journalist captured the process common to New England and New Jersey's summer colonies: "Summer after summer witnesses extensions and improvements. Season after season the family stays later and later . . . before the return to the city flat takes place. To pay rent for an apartment occupied only four or five months appears foolish. The family, perhaps in a season when the income has been reduced, resolves to remain all winter."[81] Self-provisioning eliminated rent as well as mortgage payments, since improvements were made only when materials or moneys were available, a construction approach that did not require lump sums or long-term financing.[82]

Owner building was characteristic of unplanned and unregulated blue-collar suburbanization nationwide. Here the power of localized, private-sector investments shaped the urban form even in America's most important city. When Waring winterized his camp, he replaced canvas walls with wooden boards and lattice panels and added an overhanging roof on one side to form a protected walkway. Self-provisioning resulted in jerry-built yet creative homes that met the needs of coastal living, such as wooden back annexes to serve as postswim changing rooms.[83] The clotheslines and chicken coops that bungalow residents added combined with the apple and peach trees to make a country village aesthetic that differed markedly from the urban core's tenement districts.[84] Over time camp proprietors installed electricity to replace kerosene lamps and gas, and they replaced the old communal water spigots with piped water. Camps became small homes on tiny lots—on the tightly packed lots of Higgs Beach, structures stood just twenty feet apart—but precedent existed for this very modest type of suburban housing.[85] As historian Elaine Lewinnek argues, the basic model in Andrew Jackson Downing's *The Architecture of Country Houses* (1850), the foundational design book of American domestic architecture, was an eighteen-by-twenty-six-foot bungalow. As late as 1947 developer William Levitt built similarly small models.[86]

The winterization of bungalow colonies in the late 1910s through the 1930s gave rise to bungalow communities' two most distinctive features: owner building and owner-owned homes on ground-lease lots. That the peninsulas of the upper East River remained outside of the city's real estate markets made it possible for locals to nurture a unique property regime that combined ground leases with homeownership. Although rental bungalows remained available, many families owned their homes but paid ground rent for the land.[87] For example, when Bessie and Charles Miller moved to Higgs Beach around 1909, they paid a monthly ground lease of five dollars (rents increased to around thirty dollars by the midtwenties). Like the Millers, the Warings owned their winterized bungalow but rented their lot.[88] Building a house on leased land had inherent risks, since the security of the investment rested on the length of the lease. Per the original written leases for the lots, tenants could sell or assign rights to the balance of a lease's term only with the landlord's consent. But these restrictions on the transfer of homes and leases were not generally enforced. Informal oral agreements granted residents longer tenancies and encouraged renters to treat bungalows as permanent homes.[89] Despite the availability of land in the East Bronx, real estate was still prohibitively expensive for many aspiring homeowners. Ground leases created new possibilities for working-class suburbs by removing land acquisition expenses from the cost of homeownership.

The transformation of the leisure corridor's summer camps into blue-collar suburbs was largely complete by 1930. Only Orchard Beach remained a seasonal colony. The city barred campers from building permanent structures, and tents were dismantled and packed away each Labor Day. The rest of the East Bronx colonies developed a distinctive vernacular streetscape. Gradual and informal building resulted in an ever-shifting mix of canvas tents, homes under renovation, and fully winterized bungalows. At Silver Beach in 1930, only forty-six of the available 284 residential lots, all interior lots, remained unoccupied. Higgs Beach, renamed Harding Park in 1924 in honor of the recently deceased president, included 250 residences; Edgewater Park included more than 650.[90] Winterized bungalows ranged in value from $200 to $700 and from $1,200 to $1,500, reflecting the gradual process of owner building. Having built homes on affordable but leased lots, residents at Edgewater formalized their unique property regime. In the 1920s Edgewater residents had incorporated the stock corporation Park of Edgewater, Inc., to collectively purchase their forty-acre campground and organize a cooperative. This ownership scheme was unique, since cooperatives were generally limited to apartment buildings and middle- and upper-income communities. Through cooperative organization residents protected their neighborhood's rare combination of ground leases and owner-built single-family homes. Silver Beach and Harding Park residents would eventually also organize as neighborhood cooperatives, but not until the 1970s and 1980s.[91] The ownership–ground-lease system established by the end of the 1920s guaranteed that all three communities would remain inexpensive to live in for decades to come, since property taxes continued to reflect only a lot's value, not a bungalow's worth.

*Working-Class Permanence*

East Bronx summer colonies developed and winterized without competition from estate owners since the wealthy abandoned the waterfront in the

**3.6.** 1925 aerial view of Rye Beach. This aerial looks southwest along Long Island Sound. Manursing Island, with its private country clubs and estates, dominates the foreground. South across the marsh in the center of the image is the location of the curving beachfront of the amusement district. In the background is Milton Point. *Source: Manursing Island,* photograph, October 15, 1925. Courtesy of the Westchester County Archives.

nineteenth century. In Rye, however, wealthy commuters and estate owners attempted to erase the working-class leisure landscape of Rye Beach.
And unlike the East Bronx, Rye's bungalow district did not develop on a
territory owned by a single landlord but by a number of individual proprietors, and large-scale cooperative purchase of the community was not
possible. Residents had to fight to preserve their neighborhood in local
politics and the courts. The 1920s fight over Rye Beach's amusement parks
and summer colony illustrate the divergent class-based opinions on waterfront land use. The working-class leisure corridor thrived, but not in a
vacuum and not without challenges. Control over property development
was central to defining the use and aesthetics of greater New York's suburban districts. Camps began as investment tools for local property owners
Thomas Higgs on Clason Point and Augustus Halsted in Rye. Twentieth-
century suburbanization and its resultant real estate price hikes brought

new middle- and upper-class commuters to greater New York, but these trends did not push everyone of modest means from the shore. The staying power of the amusement district challenged assumptions that the Sound and its adjacent suburbs should be exclusive playgrounds for the wealthy.

Rye's bungalow and amusement district sat squarely between two more genteel locales. The exclusive Westchester Country Club owned the waterfront to the east, while Milton Point stretched beyond Dearborn Avenue, the western limit of the district. Rye's amusement parks shaped the development of Milton Point. Between the shore and these neighborhoods, an intermediary zone developed of modest bungalow blocks mixed with commercial establishments. Across the street from Pleasure Park, rentals that cost seventy-five to one hundred dollars sat among privately owned residences valued at fifteen thousand to thirty thousand dollars.[92] As one traveled west from Rye Beach, inexpensive rentals gave way to increasingly substantial homes and eventually to estates. Milton Point's expensive homes were located farthest west from Rye Beach. John Howard Wainwright owned a hundred-acre family compound on the point. In the 1920s his son J. Mayhew, a senator and Assistant Secretary of War, built a French chateau inspired by his World War I headquarters. A second son, Stuyvesant, lived nearby in a hundred-thousand-dollar mansion; a third son Richard, Commodore of American Yacht Club, owned the eighty-thousand-dollar mansion Coveleigh near the club. Two socioeconomic groups populated the elite enclave: estate owners and their immigrant gardeners, servants, and chauffeurs.[93] Local property patterns fostered a sense of elitism and community identity for Milton Point residents that informed an anti–Rye Beach campaign.

In the 1920s the trickle of elite summer estate owners into Rye became a tide of wealthy commuters. Rye speculators lured upper-class buyers who could not afford Milton Point to the intermediary districts between and around the estate enclave and the bungalow district. Wealthy newcomers moved to Soundview Park (1892), Lounsbury Park (1901), and Ryan Park (1910). Their two-to-three-acre lots came with deed restrictions that ensured the subdivisions would be noncommercial and expensive: restrictions in Soundview Park required houses cost no less than four thousand dollars.[94] Like Milton Point elites, newly arrived well-to-do commuters held expectations of suburban living that did not include dense bungalow colonies and amusement parks.

Milton Point estate owners and their new neighbors in well-to-do subdivisions joined the Rye Citizen's Committee and Rye Welfare League to protest the leisure district. They complained to the Rye Board of Health

**3.7.** Rye Beach. Amusement parks and restaurants line the narrow, rocky beach. *Source: Paradise Park—Rye Beach*, photograph, July 4, 1924. Courtesy of the Westchester County Archives.

of "impending nervous prostration" due to "the strain of listening" to mechanical rides.[95] "Our homes are getting hateful to us," exclaimed a witness in a 1920s State Supreme Court investigation of crowds. "We no longer own our souls, not to speak of our front lawns—both belong to the visitors that come by thousands every Saturday and Sunday."[96] So-called lowlifes attracted to Rye Beach allegedly infringed on the private domesticity of gracious Milton Point property. One resident complained that swimmers disrobing in cars in front of his home offended his wife.[97] While the *Bronx Record Times* complimented the ten-piece jazz band and popular chicken dinners at Beach Hill Inn, critics deemed the area's bungalows "shacks" and the strip a hodgepodge of "cheap ramshackle hotels, shanties, and cheap, rundown bath houses very little above the level of city slums."[98] Rye's elites looked to exert control over the beach by criminalizing nonresidential practices and closing the amusements.

Having gained a critical mass by the mid-1920s, estate owners and the new commuter class joined together and entered local politics to wage war against Rye Beach. By this decade the village of Rye was home to 5,308 people unevenly distributed among low-density estate districts, tightly packed bungalows, and the village's small downtown.[99] In March 1925 Livingston Platt, of the New York City law firm Platt, Field & Taylor, and John M. Morehead, a member of the Milton Point elite, formed the Village Welfare Party and entered local politics. Platt and Morehead ran for positions on Rye's board of trustees. To earn the votes of the Rye Citizen's Committee and the Rye Welfare League, Platt and Morehead pledged to abolish amusements and "purify" the beach.[100] Local businessmen dubbed themselves the Poor Man's Party and ran in defense of the district. Ernest W. Elsworth, a contractor, and John H. Halsted, a lifelong resident who owned substantial property and several garages in town, headed the party. The Poor Man's Party characterized Platt and Morehead as interloping commuters unaware of and uninterested in the health of Rye's economy. The *New York Times* observed the battle to be "of 'the rich against the poor'"; the *Rye Tribune* deemed it "a bitter fight between commuters and townspeople," highlighting class antagonisms.[101] The Poor Man's Party was also dubbed the "Old-Timer Party," differentiating between longtime locals and interloping commuters, and drew support from amusement and bungalow proprietors and the district's grocers and druggists who catered to bungalow residents. The Poor Man's Party declared the Village Welfare ticket represented a minority of outsiders uninterested "in the welfare of the village, being in it practically only long enough to sleep."[102]

Despite the Poor Man's Party's campaign, the Village Welfare ticket won the spring 1925 board of trustees election.[103] Platt and Morehead immediately commenced their crusade to make Rye's shoreline uniformly genteel and noncommercial. That spring they rejected three applications for licenses to operate bus lines to the shore. The new board aimed to keep urban crowds out of Rye; it did not object to ferries from Long Island but specifically legislated against boats from New York City.[104] The board also prohibited parking along Forest Avenue on Milton Point and Rye Beach Avenue fronting the parks, a move directed at day-trippers who arrived by automobile. In a final blow, the new board instituted Sabbath laws that shuttered amusements on Sundays.[105]

The proprietors of Rye Beach's leading attractions fought to preserve the mixed-use shore district. Fred H. Ponty, owner of Paradise Park, and Colonel Austin I. Kelly, owner of Rye Beach Pleasure Park, sued the village board over the parking bans and Sabbath laws.[106] Kelly was doubly invested, as a resident of the beach community and a businessman, to fight Platt and Livingston. In May of 1925 the county court issued a temporary injunction restraining the board of trustees from arresting persons who operated amusements on Sundays. Two hundred residents, largely shop-keepers and business owners who opposed blue laws for fear that the tourism trade would be injured, filled the courtroom in support of the injunction.[107] Ponty and Kelly additionally brought a conspiracy suit against the village president and the board of trustees for applying Sabbath laws discriminatorily to amusements while nearby clubs such as the Apawamis and the Manursing Island Beach Club remained open. The amusement park owners threatened to turn the Sabbath laws on golfing at private clubs. The court ruled that the town's behavior toward the parks was oppressive and that the 1877 Sabbath law did not reflect modern modes of entertainment.[108] Rye Beach remained a resort destination for the masses.

The failure to close Rye Beach's amusement district reveals the success with which amusement parks cleared geographic space for blue-collar suburbs in elite coastal enclaves. The preservation of the buffer zone of bungalows paralleled the evolution of East Bronx summer colonies. Private-sector property influenced regional development patterns. In aggregate, summer rentals established local property regimes that gave rise and form to the coastal corridor. The rental configurations of East Bronx bungalows and the patterns of Rye Beach bungalow courts underscore the complex mosaic of land use that challenged narrow definitions of suburbia. West-chester became a national prototype of elite suburbia. By the 1930s three-quarters of the county remained nonurban, and estates covered 40 percent of this territory.[109] Yet the county's reputation as a place of gracious suburbs and estates, of places like Milton Point, overshadowed the diversity of New York City's suburbs.

Despite the staying power of the leisure corridor's bungalow districts, real estate investors, like municipal mapmakers, considered blue-collar suburbs marginal real estate and thus unimportant. One leading Bronx real estate investor dismissed Silver Beach because of its self-built aesthetic and its minor role in the borough's real estate market: "I have heard of [Silver Beach] very often, but personally I am not interested in it—in these community centers, because I am not in the—well, I don't know how to phrase it . . ."[110] The investor stumbled as he tried to avoid admitting that the aesthetically jumbled architecture of winterized camps did not meet his profession's expectations for "appropriate" suburban design. What this investor failed to see is that self-built bungalows made possible suburban living on an extremely modest economic scale.

More egalitarian observers recognized and championed the suburbs that grew from summer colonies. Winterized bungalows, the *Rye Chronicle* similarly assured readers, were an ideal, if somewhat unfashionable, way for people of modest means to enjoy coastal living. Renters and "those who build bungalows for permanency need not, in our opinion, fear that they will ever be ashamed to live in it."[111] In the early 1900s the borderlands of greater New York's leisure spaces remained liminal to the region's real estate markets. But throughout the century the threshold of the city's urban and suburban real estate markets expanded: as one approving journalist explained, resorts characterized by ramshackle dance halls and scattered rides had matured into good-sized suburban communities.[112] The spatial extent of the leisure corridor did not shift, but the land use that characterized it did as suburbanization spread. In 1925 a Westchester official observed, "the sparse shore-line settlements of only a few years ago are spreading rapidly to form a continuously built up section and there is a strongly moving trend toward more intensive use of land for residential, commercial, industrial and public utility purposes."[113] Through this evolution, the small-scale nature of growth and the fact that nonelites were part of the first wave of development meant the region remained accessible to social groups often overlooked in characterizations of suburban growth.

The leisure corridor's blue-collar bungalows speak to the diversity of the regional city's suburban districts. The *Rye Chronicle*'s editor declared that even Rye's least expensive bungalow could be a stepping stone to suburban living in the Sound's sought-after shore towns. The waterfronts of Milton Point and Clason Point both offered panoramic vistas. The cultural values that working-class urbanites assigned to the upper East River–Sound environment reflected its early excursionist uses. The natural world was valued for recreation, as a place to boat, bathe, and enjoy the breeze. Even though the shore lacked the aesthetics and pleasure of surf and broad sandy beaches, prudent leisure seekers found value in its environment of fields and marshes, mud and rocks. Over time, as summer colonists became permanent bungalow residents, they embraced new values associated with ruralesque environments, building a working-class suburb of gardens and fruit trees. Unlike Milton Point, Harding Park's shores and groves were home to policemen, bookkeepers, telephone operators, and a range of laborers of Irish, Scandinavian, German, and Italian descent, nobody, one resident explained, "that you could consider rich." By the late 1920s, Clason Point even boasted four modest yacht clubs, including Point Yacht Club, which a member proudly declared "was never a fancy [club]. . . . It was always a blue-collar thing."[114] East Bronx and Rye bungalow residents carved out a modest version of the waterfront living and country clubs more often associated with the likes of Gould's American Yacht Club than with urban workers. And like Rye's elites, they also tried to limit access to the shore. On the islands and peninsulas of the upper East River, battle lines were drawn in the 1920s and 1930s, not along class lines but along the divides of race and politics.

## The Limits of the Leisure Corridor

Not all of New York was welcome to the waterfront playground of the leisure corridor. Orchard Beach campers used political connections to the Tammany machine to control access to environmental amenities. At the same time, New Yorkers fought the efforts of black real estate investor Solomon Riley to secure a place for African American excursionists. Community membership and legal and political maneuvers constrained the social rights of would-be recreationalists. All three issues dictated the physical spaces recreationalists could occupy or to which they were denied access. As competing interests debated access to the shore, another layer of limitations unfolded in the leisure corridor. Unlike the manipulations of regulatory government and cultural stereotypes of black leisure, the limits imposed by the environment derived not from conscious choices but unfolded as the unintended consequences of regional development. Water pollution brought ecological degradation associated with the urban-industrial core to the city's coastal hinterlands and destabilized recreation patterns at North Beach and across the East Bronx.

In the first decades of the century, Tammany Hall controlled the Bronx Department of Parks and rented Orchard Beach camps to important borough Democrats. Eugene O'Kane, a camper in the 1910s, recollected one acquaintance who "was the brother-in-law of the leader of the political district there. So [he] had one or two tents up there."[115] Once installed at the beach, favored renters treated camps as personal property, not municipal concessions. Enabled by a complicit parks department, tenants circumvented the law forbidding permanent campsites on parkland. Campers received summer permits running from June through mid-September. Once summer ended, however, renters maintained control of their campsites via additional permits running from mid-September to the following June. When these leases expired, the department issued new summer permits, effecting long-term leases.

Orchard Beach became a permanent private settlement for more than 530 families totaling 3,500 residents on public land.[116] Camper Jerry Golino treated his camp as personal property. In 1926 he advertised his summer bungalow for sale for $1,100 and assured potential buyers that his permit could be easily transferred. This was not an isolated case. Bronx news-papers frequently contained advertisements of Orchard Beach camps "for sale": sixty-six appeared during June and July alone in 1926. The practice of reselling permits for exorbitant sums surpassing a thousand dollars—which went to the pockets of renters, not the parks department—meant only a privileged few or the well connected could afford a camp. Families returned to the same camps for four and five consecutive years.[117] Further-more, having secured long-term access to campsites, campers formed so-cial clubs that declared exclusive bathing rights at Orchard Beach. Daily visitors to Pelham Bay Park, as many several hundred on busy weekends, were left scrambling to find bathing space elsewhere.[118]

In 1927 James J. Tobin, a resident of nearby City Island, initiated a tax-payer action against Bronx Park commissioner Joseph P. Hennessy claim-ing that the camp licenses prevented public use of the park. Under its char-ter, the city could issue only temporary licenses for the erection of homes on parks department land if it was considered as of yet undeveloped for park purposes. This status was a subjective benchmark: by the late twenties Orchard Beach featured a department-built comfort station and water taps, the trappings of a developed park. The concept of land "awaiting develop-ment" allowed the department to assuage potential criticism by promising private leases would eventually end while maintaining the status quo. The Bronx Supreme Court sided with Tobin, declaring that this practice con-stituted a "subterfuge of temporary permits to circumvent the law."[119] The court found that permits were unfairly issued and treated as vestments of permanent rights that were in fact illegal. The court ruled "there is no claim whatever that this large area is being used for any public purpose."[120] *Tobin v. Hennessy*'s ruling directed the removal of existing structures and the closure of the camp. But Hennessy successfully appealed on the provi-sion that he would adopt a new, more democratic rental plan. The appeal gave his political associates on the Bronx Municipal Assembly enough time to enact Local Law No. 11 of 1927. The law empowered commissioner Hen-nessy to continue the leases.[121] By signing the law, Tammany mayor Jimmy Walker preserved the political machine's control of the lease process and effectively invalidated the court's ruling. The camps and their biased rental system would remain. Campers and the Tammany machine mobilized po-litical connections, the city charter, and license loopholes to control ac-cess to Orchard Beach. Such restrictions reveal that despite the rhetoric of democratic access, all public spaces have limits.[122]

In the early twentieth century, urban public amusements became demo-cratic arenas that unified recreationalists regardless of gender, class, ethnic identity, or nation of origin. Race, however, continued to matter. When George E. Bevans interviewed more than eight hundred New York City laborers about their leisure practices, he found three-fourths of his subjects were foreign born, hailing from twenty-seven countries. Immigrants from Russia, Poland, Italy, Greece, Hungary, Romania, Ireland, and Germany all participated in mass leisure by the turn of the century.[123] Amusement owners welcomed immigrants and native-born Americans en masse but defined their "publics" in part by the exclusion or segregation of African Americans, elevating the social status of those with access over those denied it. To temper white fears of commercial amusements unraveling social and sexual mores, park owners sanitized the most ribald amusements, gated and policed parks, and prohibited "undesirables," a category that included black patrons.[124]

The restrictions blacks faced in the leisure corridor reveal that discrimination was used to constrain participation in New York's leisure market, often with the implicit support of municipal officials. Greater New York's beaches and amusement parks did not legally impose segregation, but entrepreneurs rationalized de facto segregation as the best method by which to assure white patronage. While a cornerstone of the city's leisure culture, segregation in the state was in theory illegal. The 1913 New York State Civil Rights Act prohibited discrimination in places of public accommodation or amusement. First forwarded in 1907 by Jewish civil rights advocates, the law only partially mitigated discrimination. The state courts' conservative construction of the law and hotelkeepers' and resort owners' manipulation of the legal definition of what constituted a place of public accommodation enabled segregation to continue.[125] At North Beach, for example, even though black church groups attended vaudeville shows at College Inn, other establishments discouraged black patronage. When a black employee of a Chute-the-Chute ride ordered a beer at an Irish-owned dance hall, the bartender stood over him while he drank, smashed the glass on the floor when he finished, and forced him to leave. "No shore" existed, according to an observer, where black vacationers "would possibly be welcome."[126]

In 1925, a New York Evening Post reporter shrewdly observed that the lack of good recreation facilities was part of the problem of segregation-induced overcrowding. Harlem residents, the paper declared, required "air and resorts and recreation like the rest of the Six Million."[127] New York City's African American population grew exponentially in the early twentieth century, more than doubling in the first two decades and then redoubling in the 1920s. By 1930, the city's 340,000 African Americans were 20 percent of its total population. Harlem held 72 percent of Manhattan's and half of the entire city's black population.[128] Westchester also housed a grow-ing black population by 1930, particularly in its Sound and Hudson River satellite cities. The manufacturing centers of New Rochelle, Yonkers, and White Plains had attracted an ethnically and racially diverse working-class in the second half of the nineteenth century. Although blacks continued to account for less than 5 percent of the total county population by 1930, in the three decades earlier, the county's black population had risen by 334 percent to 23,000. The county's suburbs also included communities with industrial pockets and clusters of middle- and working-class housing and industrial centers like New Rochelle with economically and racially diverse populations.[129]

Black real estate investor Solomon Riley aimed to meet (and profit from) African American demand by opening first an amusement park and then a bathing beach on the upper East River within ten miles of both New Rochelle and Harlem.[130] White civic groups, municipal employees, and neighborhood organizations rallied to prevent racial integration on the coast and blocked Riley's ventures through property condemnation, nuisance injunctions, and manipulating city licenses. Riley's plans, and the reaction of residents and municipal government to them, illuminate how racial segregation shaped greater New York's working-class coastal communities.

In 1925 Riley announced he was opening a bungalow community and amusement park expressly for African Americans on Hart Island, in the East Bronx. The resort was to have all the characteristic trappings of the leisure corridor's amusement parks on the four southernmost acres of Hart Island, land he had purchased in 1923. Surrounded by the Sound on three sides, the spot offered panoramic views. The northern end of the island, however, rendered Riley's land marginal, if not inferior, recreational real estate. Since 1868 the city Department of Charities and Correction had run a pauper's cemetery, a women's correctional hospital and workhouse, and a jail on the island. By 1924, 860 prisoners squeezed into the eight-hundred-person capacity dormitories. These institutions were less than de-sirable company: the press deemed the prison "the human dump heap of New York," and the cemetery provided "250,000 skeletons as neighbors."[131] Yet Riley was forced to see Hart Island's environment selectively. Despite its disadvantages, a Bronx real estate investor nevertheless observed that the island was "an isolated spot for colored people, and there are very few places of that character that they can utilize."[132] With few options to choose from, Riley focused on the island's best environmental traits—panoramic vistas and the clean waters of Pelham Bay. What the *New York Evening Post* deemed a "hard-won recreation resort" was to be an important first step toward full black participation in the city's leisure economy and access to its environmental amenities.[133]

Riley faced multiple challenges to his project. Municipal officials attempted to obstruct his ventures via the city's regulatory bureaucracy, and citizens' groups questioned the morality of a Hart Island resort. The Department of Correction refused Riley's request to connect to the existing municipal water, sewerage, and electric systems on the island, forcing him to build urban infrastructure such as a water reserve and an electricity generator from scratch. In 1924 the Prison Association, a reform group, asked the Department of Licenses to refuse Riley a commercial operating permit, citing its unwanted neighbors as sufficient cause. The association's secretary E. R. Cass wrote, "it is easy to imagine [the prison warden's] problem with a crowded amusement park within view of the prisoners as they work, a park visible from the windows of the dormitories."[134] Two sets of values collided on Hart Island. Riley valued the natural world of the island's southern point for its beaches, breezes, and views. But the island clashed with white upper-class society's ideas of scenic environments and appropriately healthful recreation spots. A disdain for the city's human refuse, both prisoners and the buried bodies of paupers, compounded the prevailing stereotype that African Americans recreation spaces were de facto immoral and degraded environments.[135]

As Hart Island's scheduled July 4, 1925, opening neared, the Parks Conservation Association joined the Prison Association to launch a publicity campaign to force the city to condemn the resort. The president of the Parks Conservation Association told the *Times* that because of an innate "susceptibility" to immoral and illegal behavior, African Americans should be kept away from criminal influences. The resort "would not only make trouble between the visitors to the resort and the inmates of the juvenile reformatory . . . but could in the long run react upon the feelings and sensibilities of our colored people who might be tempted to visit the place."[136] The island's location off Long Island's exclusive North Shore made this corruption all the more worrisome to critics. The *New York Evening Post* insinuated concern about the uncomfortable proximity of black persons to the elite white playground. The *Post* explained "Great Neck lies on the other side, the beach pompous with rich estates. In sight . . . are the towers of August Belmont's place. It is a swim to the homes of actors and artists."[137] New York's industrial magnates had established palatial private playgrounds beyond the city expressly to insulate themselves from the teeming urban underclasses. A black amusement park was an unthinkable neighbor for the white enclaves of the North Shore.

The city board of alderman caved in to the campaign and condemned

Riley's property just two weeks before its grand opening. The board declared that civilian use of Hart Island was "a continual source of trouble to the department and a constant interference with the management of prison properties."[138] During the ensuing condemnation proceedings, the park sat vacant; excursionists never visited the island. The board denied race was a factor in its decision. In light of Riley's subsequent struggle to run a black bathing beach on Throgs Neck, however, the decision appears to be the first of a series of municipal rulings in which racism and recreational segregation benefited from implicit government backing.

In July 1929 Riley again attempted to open a black shore resort, purchasing an old beachfront cabaret near Edgewater Park on Throgs Neck. He again met with resistance. Neighbor George C. Crolius repeatedly manipulated municipal regulations and government bureaucracy, in the form of nuisance injunctions and health inspections, to realize racial segregation in the coastal leisure corridor. Crolius owned one of the most expensive homes in the neighborhood directly behind Riley's establishment. Crolius headed a group of white homeowners who disliked the idea of a black resort in their white community. When Riley purchased the property, white residents posted signs warning "Negroes, Beware."[139] Crolius lodged a complaint about the reopening of the dance hall with the Department of Licenses, which subsequently denied Riley a license renewal.[140] Undaunted, Riley applied for a permit to run a bathing beach instead. This second application was also denied, this time on claims that his beach was polluted and unfit for bathing. License commissioner James F. Geraghty decided to withhold Riley's permit pending a full health inspection. Riley appealed on the grounds that the charges masked racial hostilities.[141]

George Crolius manipulated municipal governance as a tool of spatial exclusion, but Riley was a veteran of this type of obstruction based on his experience at Hart Island, and he strategized to sidestep license filibustering.[142] Riley incorporated the Intrafraternity Council in late July 1930 to operate his Weir Creek beach as a private club. In response Crolius initiated a new legal offensive based on his rights as a property owner. He claimed that the depraved conduct of club patrons threatened the morality of his neighborhood and constituted a public nuisance, and he asked the court for an injunction to shutter the venue. Riley's health department license application was again shelved and his beach shuttered for the ensuing trial.[143] The prosecution accused Intrafraternity Council members of everything from harassing local women and children to illegal drinking to revving car engines in the early morning hours. Twenty-four neighbors and witnesses submitted affidavits bolstering the claim. The defense submitted over twenty affidavits denying charges of misconduct. Intrafraternity Council members, largely professional men from Harlem—surgeons, dentists, lawyers, and religious leaders—testified to the club's respectability. The club's caretaker declared, "nothing about the premises would constitute a menace to morals." A. F. Harding of West 135th Street agreed. The injunction, Harding testified, was calculated "to oust colored people from the premises."[144] William T. Andrews, Riley's attorney and a National Association for the Advancement of Colored People (NAACP) activist, declared the prosecution had no legal right to secure an injunction since the club had previously operated as a white-run cabaret without complaint. Crolius's charges, he said, were simply racist. "The fact about the entire matter is that the premises is owned and controlled by Negroes and they are the users thereof while all the persons complaining are white persons."[145]

George Crolius sustained a two-pronged attack against Riley. He manipulated the mechanics of government regulatory licenses and nuisance investigations. He also tapped into public discourse that framed black leisure as inherently immoral and black leisure environments as inherently degraded. But in December 1930, the Bronx Supreme Court avoided addressing morality charges by ruling only on the bathing beach's status as a private club. Crolius produced an admission ticket as proof that the club was in fact a commercial establishment. It appeared, the court ruled, "that the [fraternity] was created as a membership corporation for the purpose of avoiding the necessity of obtaining licenses from the local authorities to operate a public resort."[146] Riley was operating a public venue and required the requisite licenses.[147]

By ignoring immorality allegations and ruling that the beach resort was a commercial enterprise, the Bronx Supreme Court forced the fight over Riley's club back to the licensing department. Nuisance law connected to morality was far more subjective than bacteriological-based bathing beach standards, and Crolius's failure to achieve a nuisance injunction based on a moral argument left him to pursue racial segregation via municipal regulation. Riley was ultimately free to operate a public resort if the Department of Health affirmed that bathing on the northwestern shore of Throgs Neck was safe. And in fact the beach passed all necessary pollution tests in the summer of 1930. But municipal employees again effectively backed segregation. Department of Health employees dragged their feet so much in supplying Riley with the permit he was due that in May 1931 the Bronx Supreme Court stepped in to force license commissioner Geraghty to issue Riley the license that he was owed.[148]

Solomon Riley's white adversaries must have been delighted when an obstinate Geraghty backdated Riley's license to when bathing at the beach first passed its health tests in 1930, and the license expired a month after it was issued. As soon as Riley's long-awaited license expired, Crolius filed yet another injunction claiming the beach had become recently polluted, and Geraghty suspended Riley's license yet again pending yet another health investigation. Crolius continued to drag Riley into court to dispute injunctions through the summers of 1931 and 1932. Riley faced a series of administrative and regulatory challenges deployed to prevent the creation of black recreational space. These challenges reveal a pattern of municipal ruling in which segregation benefited from implicit government backing and how municipal governance and law emerged as a particularly powerful tool of exclusion. Nuisance injunctions and health board inquiries permanently stalled Riley's Weir Creek venture, since every injunction and ensuing investigation temporarily closed the venue. From the human refuse of skeletons and prisoners to accusations of environmental degradation, racial stereotypes became associated with physical space and undermined Riley's ventures.[149]

Westchester became the prototype prosperous suburb in the early twentieth century, known nationwide for its elite suburban enclaves such as Milton Point. But as sociologist George Lundberg pointed out in his 1930s study of the county, suburbia was inherently diverse. By the late twenties the *New York Amsterdam News* ranked New Rochelle, home to the county's largest black community, as one of the nation's earliest integrated neighborhoods and one of the best African American communities in the nation because of its well-off black neighborhoods, homeownership, and strong NAACP chapter.[150] But a 1937 state investigation also conceded that "of course the majority of those who live in New Rochelle are domestics, the same being true of the rest of the county."[151] Westchester's blacks lived in a geography of inequality that historian Andrew Wiese has argued was intrinsic to African American suburbs.[152] State investigators found that many of Westchester's African American residents lived in "inadequate dwellings," faced discrimination in theaters and restaurants, and had trouble getting municipal jobs.[153] Public leisure in the coastal corridor was part of this geography. Segregation to inferior spaces continued to threaten to marginalize African Americans. The fight for equal access to public spaces occurred not only at Riley's East Bronx amusements but at Rye Beach as well, following its redevelopment as a county-run park in the late 1920s.

The coastal environment became a mechanism of social and racial inequality. Riley's experience with the Department of Health was symptomatic of regulatory discrimination that restricted property use to maintain racial and spatial divides on the waterfront. The conflict also pointed to the greater harbor's ecological decline. In the early twentieth century, sewage pollution threatened not just Riley's beach club but the entire upper East River. Intensive urbanization did not cross the Bronx River on the north shore of the leisure corridor. The environmental challenges created by dense urban living in the metropolitan center arrived in the form of sewage and the city's first comprehensive refuse landfill program, which turned marshes into dumps in the twentieth century. The peripheral landscapes of the coast were an evolving series of temporary environments that had different users who had their own visions. The coast was an evolving landscape that drainage patterns, tides, and shifting sands (including landfill operations) altered. Environmental change and the rise of the leisure corridor were intertwined. Changes to coastal land use reshaped the nature of the upper East River.

For three centuries New Yorkers embraced coastal land making. Since buildings erected in marshland were known to settle and sink, marsh was marginal, inexpensive real estate. Private developers supported landfill operations. Barnum characterized the Sound's tidelands as "mosquito-inhabited and malaria-breeding;" and Steinway deemed upper East River marshes "waste lands" that should be filled.[154] From bureaucrats' and planners' perspectives, the wetlands of greater New York made ideal landfill sites. Landfill could improve maritime infrastructure, remove wetlands seen as worthless real estate or nuisances, and make new space for development. Landfills concentrated along the heavily populated shores of Manhattan and northwestern Brooklyn, and until 1895 the city annually dumped up to 80 percent of its solid waste in the Atlantic Ocean. Offshore dumping continued until the mid-1930s, but between 1891 and 1924, improved waste transport methods brought the outer reaches of the Bronx and Brooklyn into the purview of landfill activity.[155] At the end of the century, the city opened a dump on Rikers Island, which sits between the East Bronx and North Beach. It later expanded its upper East River dumping to Clason Point.

In 1896 a reporter for the *New York Times* had claimed Bowery Bay was an "absolutely healthful and pleasant sheet of water."[156] Dumping on the upper East River wrought environmental change that eventually undermined the area's coastal recreation patterns. Scum and trash floated on the tide from Rikers to pollute nearby beaches. In 1903, because of complaints from Queens residents, the street cleaning department stationed men in row boats to gather refuse dropped from sanitation dredges and ran twenty-person shore patrols along North Beach and Astoria twice a day to collect refuse that washed ashore. The department's commissioner promised "everything possible would be done to make the nuisance as light as possible," but admitted that "it could not be entirely abated."[157]

The localized problems of coastal dumps combined with widespread wastewater pollution to severely degrade the waters of the leisure corridor and the harbor more generally. No sewage disposal plan existed for the East Bronx, and sewage entered the upper East River and Harlem without treatment. Despite the limitations of early sanitary science, by the early 1900s, it was clear to the sanitary experts that there was not "a square foot of the waters" surrounding the city that remained unpolluted by sewage.[158] In 1910 the harbor received the excrement of over six million New York City residents, approximately six hundred million gallons of raw and minimally treated sewage daily.[159]

The leisure corridor suffered environmental problems not only from increasing citywide water pollution but also because of localized problems created in part by piecemeal development. In 1916 the New York City Department of Health had noted with approval "that the old style tent is vanishing and being replaced by the more substantial bungalow with its individual . . . sewer or cesspool connected plumbing fixtures."[160] Sewers reshaped the relationship between summer colonists and the coastal environment. While plumbing was a boon for public health, it had unanticipated consequences for the natural world. In the late nineteenth century, Andrew Haswell Green and the city department of parks warned of the pollution challenges piecemeal drainage planning would create.[161] Green fought for comprehensive public works planning to systemize municipal infrastructure and improve management effectiveness. He also worried about environmental degradation of New York Harbor. The unanticipated conditions of the leisure corridor helped keep Green's vision from becoming a reality, producing a regional city with environmental conditions that were very different from what he envisioned. When campers winterized their bungalows, new ad hoc sewerage discharged untreated sewage directly offshore. Into the 1960s thirty-two private sewers dating from the early camp era emptied from Harding Park into the East River. As a matter of economy, public- and private-sector builders alike constructed sewers that discharged at the shore or just past it, at the mean low tide line. Long sewers releasing deep underwater, far from the shore, cost more. Each peninsula on the upper East River had a municipal outfall. Clason Point's was the largest of the East Bronx sewers, with a drainage district of 7,500 acres.[162] Bungalow colonies contributed to harbor pollution by treating its waters as a sink for waste with infinite self-purification properties.

Pollution ultimately trumped social limitations imposed at Orchard Beach and on Riley's resorts by creating environmental conditions that discouraged leisure along the upper East River. Pastoral landscapes, particularly well-cultivated ones, were long understood as particularly healthy environments in America.[163] Urbanites assigned this cultural value to the green peninsulas and islets of the upper East River and Sound. When the leisure corridor first flourished, visitors and the city Department of Health agreed about the social and health benefits of time spent alongshore. Over time, however, the intensive use that such values engendered unintentionally created pollution and strained the ecological limits of the coast. For example, the pool at Clason Point Amusement Park took its water from the East River. It became so dark and dirty that it garnered the nickname "the Inkwell." Vivian Cavilla, who grew up in the Bronx at the turn of the century, recalled that by 1920s the "rivers were filthy."[164] Development transformed the natural world into a place where cultural values of environmental health were no longer to be found. Pollution radically changed ideas about the environmental health of the city's outer wards and ultimately destabilized blue-collar leisure patterns there as people continually sought new and comparatively unspoiled coastlines.

## Conclusion

The temporary and shifting nature of the coastal environs was a major feature of the leisure corridor in the Bronx and along the Sound. Eventually the dynamism that characterized the leisure corridor brought an end to recreation on the upper East River. North Beach was the first park to close. While pollution caused the resort's popularity to wane, it closed because of the dual threats of World War I–era anti-German sentiment and prohibition. George Ehret was in Germany when hostilities broke out and was caught abroad for most of the war. Although he was a naturalized American citizen and his loyalty was never questioned, the federal government confiscated his forty-million-dollar business as alien property during his absence. In addition to such discriminatory actions, Germans also faced fierce wartime prejudice that merged with the temperance crusade. A member of the prominent German beer-making Ruppert family declared prohibition rode "a wave of clamor, hysteria and mistaken patriotism."[165] The 1919 Volstead Act dealt a deathblow to beer-garden culture and North Beach. The son of one concessionaire recalled, "as soon as signs of Prohibition came, 'pfft' the Beach Company walked away."[166] The 1920s also marked the end of Clason Point's amusement district. After a 1922 freak windstorm that wrecked the hundred-foot-tall Ferris wheel and killed half a dozen people, attendance waned. "Between the sewage, people getting automobiles . . . and all the other situations, including prohibition," a Harding Park resident summarized, the resort "died a natural death."[167] By the midthirties, squatters in the crumbling buildings had replaced crowds of revelers at the abandoned parks at Clason Point, and North Beach lay vacant.[168]

The land-use patterns of amusement parks, picnic grounds, and summer colonies dictated how and when urbanism arrived on the urban periphery. Summer colony development brought sewer and water systems, electrification, trolley and ferry lines, and roads to the city's remote corners long before official public works projects reached these areas. Without conscious planning, local leisure patterns, vernacular neighborhood development, and landownership strategies functioned as space-structuring tools, developing a distinctive geography and built environment. Between 1910 and 1930 Manhattan's population fell 19 percent even as the total population of the five-borough city increased by 45 percent. The boroughs beyond Manhattan boomed; the Bronx and Queens experienced the most radical growth. Queens's population grew by 280 percent to more than one million; the Bronx's by 194 percent to 1.2 million.[169] The intermediary threshold between suburban, urban, and rural landscapes settled more firmly into suburban development tied to citywide real estate and public works interests. During this period, however, the bungalow suburbs of the East Bronx ultimately proved more resilient than the leisure corridor that spawned them. Bungalow colonies fostered a ground-lease/homeownership property regime that kept the communities inexpensive and encouraged unique, collectively owned suburbs. The East Bronx and Rye preserved sections of the shore for working-class recreation in the face of widespread privatization by wealthy landowners.

While upper East River resorts declined, Rye Beach underwent a different type of transformation in the twenties. Supporters of a domestic, quiet, and elite Rye waterfront failed to close amusements but embraced the Westchester County Park Commission (WCPC), established in 1922 as a means by which to exert control over the waterfront. Like Milton Point's estate owners, park officials distinguished between "good" and "bad" leisure spaces and activities, valuing officially planned spaces over the vernacular working-class leisure corridor. One of the first areas targeted for redevelopment was Rye Beach. In the early twentieth century, Rye Beach epitomized the vernacular, blue-collar spaces of the Sound. In the 1920s it became the keystone of the county's Progressive Era park system. Orchard Beach in Pelham Bay Park underwent a similar transformation. Pelham Bay Park, created in the 1880s as part of the vanguard park system for mainland New York, would inspire WCPC members and Robert Moses. And Orchard Beach's municipally run tent colony became the focus of Moses's ambitious restructuring of greater New York's environment via modern public beaches.

Private-sector development gave rise to the leisure corridor that first linked the upper East River to the social networks and real estate market of the regional city. In the 1920s New York State, county, and city governments created new realms of political power that brought government oversight to the informal leisure corridor through park planning. Planning officials gained substantial influence over the form of the regional city in the planning commissions of the 1920s and 1930s through comprehensive park and parkway systems. Regional park commissions such as the WCPC employed a different definition of greater New York than that fostered by the leisure corridor. Shorefront parks became keystones of regional park systems, but their planners worked at a larger scale framed not by the coastal corridor but by countywide, citywide, and transcounty perspectives.

Regional park planners who advocated for new public beaches and parks imbued the coastal environment with new cultural values based on their vision of modernized, mass leisure. In 1926, the WCPC condemned Rye Beach's amusement district and a portion of the adjacent bungalow community. The piecemeal informal resort amusements and bungalows were razed and replaced with a professionally designed and planned public amusement park. County officials billed the new "sanitized Playland" and bathing beach as "America's Premier Playground." This proved a harbinger of the modern regional park planning that would transform the public leisure landscape of greater New York.[170]

# 4

# Designing a Coastal Playland around Long Island Sound

In the 1920s, landowners of Westchester and Long Island viewed holiday crowds with trepidation and thought picnickers were a roving threat. Lured by quaint roads, wooded valleys, and rocky beaches, urbanites from the city roamed the nearby countryside at will, ignorant that these spaces were private property. In 1929 the Russell Sage Foundation's Committee on the Regional Plan of New York (CRPNY) completed a recreation survey as part of its groundbreaking plan for the metropolis. Suburbanites reported that recreationalists trespassed, started fires, polluted roadsides, and injured fences, bushes, and trees. The CRPNY discovered both careless urbanites and the suburban response to them. Holiday crowds, the committee concluded, were "such nuisances that landowners [were] in a mood to keep them off with shot guns."[1] Private property needed protection, and the public needed a designated space beyond the city where they were welcome to recreate. New York's city, county, and state governments created powerful park commissions to meet this need and exert public-sector authority over leisure.

Leisure was intimately connected to urbanization and the rise of regional planning in greater New York. The landscape of the working-class leisure corridor grew from unstructured activities forming vernacular street plans and amusement parks. Yet small-scale, owner-built spaces could not meet the needs of the city's great recreating public. In Westchester in the 1910s and 1920s, park planners identified two concurrent forces that spurred a county park program: "the development of the automobile, with resulting demands for good roads and parkways . . . [and] a widespread movement toward outdoor life and a demand for public recreation grounds."[2] In the first decade of the new century, the *New York Times* published daily automobile and roadhouse advertisements, a response to and impetus for motoring tours.[3] Automobility expanded the ranks of recreationalists just as outdoor life became a central part of popular culture. Historian Peter J. Schmitt explains that at the turn of the twentieth century, middle-class urbanites embraced a reinterpretation of nature and wildness through the back-to-nature movement. The popular movement eschewed traditional rural economies and labor to focus on leisure settings such as hunt clubs and an arcadian, romantic vision of nature's benefits.[4]

"Instead of being faced with a small leisured class," Lewis Mumford observed in *The Urban Prospect* (1968), by the 1920s American cities faced the challenge of providing "recreational facilities for a whole leisured population."[5] Yet no governing body existed to regionally structure recreation, parks were unequally distributed and of unequal quality, urbanization increasingly limited the space available for new public parks, and private automobiles crowded existing highways. Columbia University sociologist George A. Lundberg declared that city officials learned the hard way that public recreation opportunities and facilities "*could not be adequately achieved by individual effort or by sporadic informal cooperation among casual groups.*"[6] The reporter for the *New York Times* who declared that residents in search of land-use controls and public parks suffered from "a universal malady, which must be treated as a whole before it can be cured in any part" captured the regional scope of the challenge.[7]

The desire to systematize, professionalize, and modernize parks coalesced with the new demands of greater New York's growing leisure population. Park development and recreation reform emerged alongside additional city-focused Progressive health and welfare programs.[8] Public leisure facilities and parkways became the skeleton for regional development. The result was a planning policy that reconfigured both park design and the shape of the urban periphery. The need for parks ushered in a transformative period of large-scale planning of the urban fringe and experiments in public administration of planning via supervisory bodies and government authorities at multiple jurisdictional levels. Such work, while often associated with the singular influence of master builder Robert Moses, required collaborative and cooperative planning. The reformers, civil engineers, and landscape architects of New York's new park commissions self-consciously assumed responsibility for bringing fringe development under official oversight and identifying the goals of the field.[9] Oversight and coordination of major public works, a pillar of the planning field, became the formal responsibility of government.

When New York City planned its first comprehensive park system for the North Side in the 1880s, boosters were still grappling to understand and shape the emergent regional city. As planner John Nolen later explained, urbanization was erasing the boundaries of the nineteenth-century city. At the turn of the century, geographic expansion wrought a "radical change" in planning by making the county rather than the city "the logical planning unit."[10] In the 1920s and 1930s, park planners turned their attention to the suburbs of Westchester and Long Island. Although on the geographic edge of the city, regional park programs represented a central shift in planning's response to urbanites' mounting recreation needs and the evolving city-suburb relationship.

County and interjurisdictional park commissions created a comprehensive regional planning ideology in greater New York. This planning took place in an identifiable geographic region, the inner ring of the city's hinterland in New York State. The region's park authorities crafted a shared planning approach by incorporating the frameworks and goals of preceding commissions. Planning originated in the North Side park system (established 1884) and evolved through the creation of the Westchester County Park Commission (WCPC, established 1922), the New York State Council of Parks (established 1924), the Long Island State Parks Commission (LISPC, established 1924), and the consolidated city Department of Parks (established 1934). Collectively, powerful twentieth-century commissions translated the growing desire for popular leisure into a regional policy. A cohort of planning professionals held multiple appointments across the region's commissions, further encouraging standardization. Moving among various commissions, these planners articulated a vision of the hinterland territory as a unified park system catering to the region's residents. The germination of planning principles across Westchester, Long Island, and the Bronx in the 1880s–1930s fostered the ideology and practices that enabled Moses to advocate for parks on Long Island and then so profoundly reshape the city via the Department of Parks, a pivotal moment of regional planning that ushered in a new era of suburban development.

In addressing the distribution and quality of parks, planners approached the environment less frequently in terms of the material nature of ecosystems and more often in terms of balancing land uses. Greater New York's park planners circulated new ideas on landscape design, park accessibility, and the "buffer" function of parkland that reshaped the scale of the debate and reflected planners' judgments about New York's relationship with its hinterlands. The circulation of planning ideas defined the urban edge and convinced officials of the need to buffer new suburbs from urban land uses with nature reserves. Large, unpolluted, and evenly distributed parks and beaches could balance dense, private development: the environment functioned as a space-structuring tool.

Social control emerged as a hallmark of park design in the 1850s and remained a central focus of the new professional class of park and playground designers. "There is no problem before the world today," the statesman Elihu Root declared, "more important than the training in the right use of leisure."[11] The unregulated spaces of the working-class leisure corridor grew from localized, vernacular initiatives. Confident in their bureaucratic-technocratic orientation, planners enshrined and pursued a top-down perspective. Leisure became "a public problem" requiring government oversight.

## Pelham Bay Park and the Bronx Park System

New York City annexed lower Westchester in two sections in 1873 and 1895. In the intervening years, city officials and park advocates conceptualized the upper East River as a section of the growing city where open space might be preserved. The predominance of estates in what became the East Bronx appeared particularly conducive to parkland conversion. Timothy Dwight, eighth president of Yale College and resident of coastal Fairfield County in the late eighteenth century, had celebrated the many handsome waterfront villas of the district. While appreciating the territory's beauty, an 1851 visitor observed more critically that the area was thoroughly and "aristocratically fenced up."[12] By the 1880s, however, this concern over privatization was moot: the area's once-elegant mansions moldered in states of decay. Seasonal campers made the most of abandoned estates. The city's parks department also valued the upper East River shore as a site for recreation and platted 450 acres of parkland in the annexed territory and at Pelham Bay in Westchester. If converted to parkland, the former estate district could offer healthful coastal environments, introduce the urban masses to suburban-like landscapes, and foster a regional perspective on recreational reserves irrespective of municipal boundaries. Yet the parks department failed to obtain the majority of the land, to the dismay of John Mullaly, park booster and editor of the *New York Herald Tribune*. In the 1880s, Mullaly organized the New York Park Association to advocate for action to turn these woods, marshes, and fading estates into the city's first comprehensively planned park system.

The decline of the North Side and East Bronx's desirability as estate districts made the purchase of a generous mainland park system economically feasible. The North Side park system that Mullaly and the New York Park Association proposed was signed into law by Governor Grover S. Cleveland in 1884. The system was seen as more than a local amenity. A local-benefit tax plan was proposed that required residents of the Twenty-Third and Twenty-Fourth Wards to pay half the cost of the new parks, but North Side boosters successfully argued that costs should be distributed citywide via general taxation.[13] Despite legal challenges over condemnation costs, the constitutionality of condemning Pelham Bay Park in Westchester, and attempts to block the issuance of bonds by the board of aldermen and the mayor's office, by 1888 the city acquired the full extent of parkland specified in the law. The total bill for the nearly four-thousand-acre system reached nearly ten million dollars.[14]

The North Side park system marked the first time the city uniformly distributed public space in advance of settlement. Until this time parks had been built individually, often by independent commissions. The vanguard system included Bronx, Crotona, Claremont, St. Mary's, Pelham Bay and Van Cortland Parks. Pelham Bay Park was the first urban waterfront park in the city, and in the region, it was second only to P. T. Barnum's Seaside Park in Bridgeport. The Mosholu, Bronx and Pelham, and Crotona Parkways linked the four largest parks. The parkways ranged from 200 to 400 feet wide, which exceeded the widest boulevards south of the Harlem River by fifty feet and were intended to function as both connecting links and extensions of the parks. This innovative design meant visitors could experience North Side parks with few visual or physical breaks. The 1898 edition of the popular *Leslie's History of New York* championed as unusual this dedication of land to public use, since New York was a city "where commerce reign[ed] supreme."[15]

The parks department went to great efforts to advertise the new publicness of the former estate district. While the parks department dismantled buildings with clear titles and sold them as raw material, they also temporarily rented former estates to park workers, helping to defray park upkeep costs. The City Club and Metropolitan Park Association complained to the *New York Tribune* that holdout tenants and park policemen in Van Cortlandt and Pelham Bay parks were effectively "living on their own country estates."[16] Officials assured the public, however, that tenants with monthly

4.1. *Opposite*: North Side Parks. The circles on the map represent miles distant from Grand Central Depot (rebuilt as the Grand Central Terminal in 1913) in midtown. The gray strip bisecting the North Side represents the Bronx River, the city's mainland municipal boundary. From west to east, the three large parks at the top of the map are Van Cortlandt Park, Bronx Park (along the Bronx River), and Pelham Bay Park, more than three times larger than Central Park. *Source: Sketch Map of the City of New York and Vicinity, Showing the Sites and Approaches to the Parks.* New York Public Library Digital Collections.

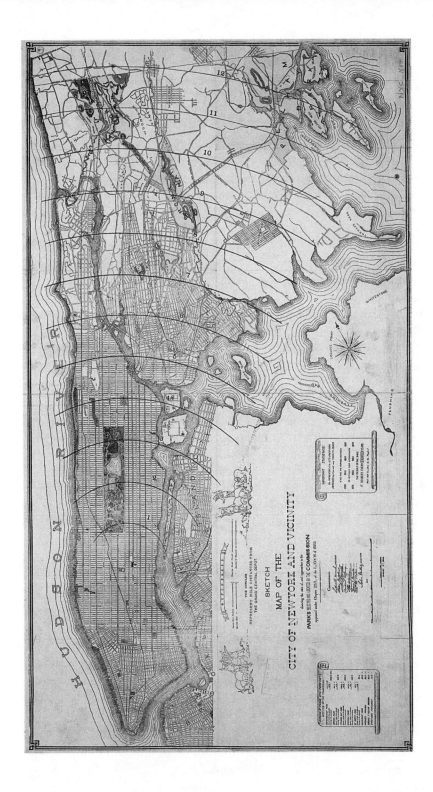

SKETCH
MAP OF THE
CITY OF NEW YORK AND VICINITY

leases lacked any proprietary rights to surrounding territory. While visitors might still find fences and lawns that looked private, "there is absolutely not a foot of private ground in the park and not a fence that one is not at perfect liberty to jump over or crawl under." The *New York Times* proclaimed, "the public has as much right on the lands about as it has on Union or Madison Square . . . if any tenant of the big house tells you to keep off his grass you can gayly sing a song of defiance."[17]

The North Side park system was a regionally scaled effort to ensure public access to a wider range of scenic nature. City residents and officials could not deny, however, that the North Side park system's location had its disadvantages. Until the Interborough Rapid Transit Company extended its subway line to Pelham Bay in the 1920s, the park remained a long trip from the urban core. Visitors rode the Harlem Line of the New York, New Haven, and Hartford Railroad to either the Bartow Station in the park or the nearby Baychester Station. In 1910 fare reductions and an increase in the number of summertime eastbound trains on the Harlem Line improved accessibility; a visitor could subway and then take a train to Bartow for ten cents total. Transit options to Pelham Bay included hack drivers, the City Island horsecar line, a short-lived monorail (1910–1914), and from the 1910s on, the Third Avenue Railroad trolley. The trip, however, remained an all-day affair.[18] High-speed, high-capacity subways were built in the Bronx between 1905 and 1920, but they did not reach Pelham Bay Park until the twenties.[19] Despite transportation trials, Pelham Bay Park boosters argued the park's far-flung location was one of its greatest attributes. In fact, boosters celebrated the park's isolation as providing rural and suburbanesque experiences.

Boosters declared that the evolution of North Side estates into parkland offered workers access to spaces usually denied to them: landscapes filled with country estates and clean environments. In the 1870s, Frederick Law Olmsted and J. James R. Croes presented a plan for the mainland wards to secure the North Side as a district of curving subdivisions, well-landscaped universities and seminaries, and bucolic scenery preserved in parks and botanical gardens.[20] The city rejected the Olmsted-Croes plan, but the preservation of estate grounds as parks evoked the lingering vision of the North Side as a suburban district. Unlike the total environmental reconstruction required to build Central Park, little landscape redesign was needed to turn the woods and lawns of former estates into parks. Van Cortlandt Park incorporated forty-five estates and outbuildings; Pelham Bay Park was an amalgam of fifty-three, judged to be "little more or less than a succession of fine old estates along the shore."[21] North Side booster

Fordham Morris celebrated the new parks as "great suburban parks," and Mullaly enthusiastically agreed, proposing Pelham Bay Park would offer a surrogate suburban or country experience for its urban visitors.[22] He even claimed the estates of Long Island's North Shore, just across the bay, as park features. Since owners carefully guarded the privacy of Long Island estates, the masses would never have access to or enjoy this leisure landscape. But Mullaly pointed to Pelham Bay Park's enjoyable views of "handsome private residences built at unequal distances from the water . . . with smiling skies above and dancing waters below" as a surrogate experience.[23] The use of former estates to build parks turned once private spaces into public amenities for the urban middle and working class, "bringing the real shore-and-country scene into the city proper."[24] Mullaly dubbed Pelham Bay Park the "Newport of the Toilers," associating the park with the social cachet and pristine beaches of the nation's most elite seaside resort. "The classes" could go to Newport, but "the masses" could find quality seashore recreation on half-day or Sunday trips to Pelham Bay.[25]

Park boosters' celebration of the upper East River's environment anticipated the arguments made in favor of the twentieth-century coastal leisure corridor. Advocates of a North Side park system argued that Pelham Bay was not only safely removed from the industry of Manhattan and the Harlem River but also that the bay's tidal patterns were particularly conducive to recreational use of its waters. The East River received tides from Long Island Sound and the bays of New York Harbor. These two sets of tides met at Throgs Neck, the East Bronx's northernmost peninsula. The tidal waters that flowed up the East River from New York Bay ran alongside Manhattan and collected its pollution. The tides that entered from the Sound were much cleaner. Throgs Neck marked a dividing line in water quality, for the strength of Sound tides blocked the East River's polluted waters from traveling any farther. Pelham Bay sat northeast of this dividing line; twice a day tides from the Sound brought clean water to its shores. The North Side park system's planning commission identified this environmental condition as being "of great importance in selecting a site for a park where the waters [border] it, and [form], so to speak, a marine extension of it."[26] Sanitary investigations furthermore revealed that "the intensity of contamination [was] less" in the bay, and bathing remained popular in the park, as well as farther up the Sound in Westchester, for the next fifty years.[27]

Mullaly encouraged urbanites to make the most of Pelham Bay Park's health benefits. Reformers and physicians touted the benefits of the shore's sun and salty breezes, which were thought to contain especially pure oxygen in this era.[28] Like many social reformers, Mullaly blamed Manhattan's dense immigrant wards, in particular their foul air and "anti-commodious habitations," for high death rates, especially among children. He also expressed hope that the North Side system would speed movement to new, less crowded areas, such as the cottage communities near its parks.[29] Even if Manhattanites did not relocate permanently to the North Side, time spent at its parks might improve their health. Sea air and water were considered curative for sick children, particularly those suffering from tuberculosis. Boosters emphasized nature's purported power to heal. Mullaly focused specifically on needy groups he found deserving, inviting settlement houses and philanthropic organizations to make use of the old estates and lawns. He avowed that the Fresh Air Fund, a Protestant child welfare charity, could find no better place to help the poor and infirm than Pelham Bay, where a "great reservoir" of invigorating breezes blew across the "purifying waters of the Sound." A number of charitable organizations visited the park. For a nominal annual fee of one dollar, the Society of Little Mothers, the Working Girls Association, the Guild for Crippled Children, and the Riis Neighborhood Settlement leased estate lawns as camp territories and old estates as boarding houses from the parks department.[30] In 1889, when Mayor Hugh J. Grant proposed to open the park to a wider range of the city's less fortunate, however, Mullaly feared the park's restorative nature was under threat.

Grant endorsed two bills pending in the state legislature that empowered the Department of Parks and Commissioners of the Sinking Fund, a municipal source of capital expense funds, to repurpose the park as a site for the municipal charity hospitals, asylums, and penal institutions that were housed on Blackwell's, Hart, and Wards Islands in the East River.[31] Grant proposed moving these institutions to Pelham Bay as a means of improving patient conditions. His proposal aligned with environmentally determinist beliefs that built and natural environments together determined patient behavior: well-ventilated and curative asylums required healthful and restful pastoral environments. Administrators of nineteenth-century insane asylums argued that madness was a threat to the family unit, and Victorian Americans moved the mentally ill to asylums to preserve the sanctity of the home.[32]

Mullaly protested Mayor Grant's proposal as an attempt to make the park a repository of the city's indigent, insane, and felonious. Mullaly and the New York Park Association perceived the mayor's proposal to transfer asylums to Pelham Bay as a threat not only to the park's aesthetics but to the health and morality of users. Mullaly's protests anticipated by nearly forty years the contest over "appropriate" leisure setting that Solomon

Riley faced when he attempted to open a black resort on nearby Hart Island. Mullaly argued Pelham Bay Park was not intended for the type of people confined to asylums. He claimed to speak on behalf of "workers and toilers, for whom this great pleasure ground was intended" when he condemned Grant's suggestion as a "perversion" of public space.[33] Facing sustained criticism from the park's boosters, Grant abandoned his proposal.

The North Side park system thus proved to be the first official step toward reclaiming the urban periphery as valuable new parkland. Bringing socially undesirable populations to the parks would hinder, not advance, edge development. Drawing on nineteenth-century expectations of appropriate pastoral parkland use established by Olmsted and the doctrine of rural environmental health, boosters initiated debates about the public and its behaviors that would flourish in the coming Progressive Era.

The North Side Park system heralded a shift in urban development in the twentieth century in which the region, rather than the city, became the logical unit of planning. Although a half mile beyond city limits until the annexation of 1895, Pelham Bay was included in the 1880s park system to ensure public green space in a territory destined to undergo urbanization, and this was a step toward the conceptualization of greater New York as a regional city. The map that the Park Association submitted to the state legislature illuminated this regional orientation by presenting Manhattan's growing population as dependent on mainland parks. The association argued that the city owed these parks to its citizenry. Depicting the distance from Forty-Second Street to the North Side parks, the map suggested movement northward. Urbanization had breached the trans-Harlem, and so too could the masses in search of parks. A table on the map proclaimed "Important statistics! The Population of the City of New York increases by 117 per cent every twenty years . . . What shall be done for the People?"[34] The commission to select North Side parks argued that new regional parks could meet the needs of Manhattanites.

## Westchester and the Origins of Regional Park Planning

Westchester County's efforts to manage rapid suburbanization regionally led to the first regional park commission in greater New York in the 1920s. By the turn of the twentieth century, county officials and city planners finally recognized that Westchester was essentially tied to the city: its era as a rural hinterland had passed. The county was a suburbanizing edge of greater New York. Westchester's river valleys and Hudson River shores had lured wealthy New Yorkers to establish estates since the 1840s, but the trickle of estate owners was overwhelmed by a flood tide of new residents moving to the railroad suburbs of Pelham, Eastchester, New Rochelle, and Mount Vernon on the city line. Nineteenth-century suburbanization doubled the county's population every twenty years between 1850 and 1900; growth continued apace in the new century. By 1920 Westchester was home to 340,000 people.[35] Through the 1920s and into the 1930s, however, large estates from ten to five hundred acres still made up 40 percent of the county, while public and semipublic open spaces—Westchester boasted more golf, shore, and county clubs than any other county in the nation—made up 16 percent.[36] Preserving this landscape and the area's reputation as a premier suburb motivated the county to integrate transportation, recreation, and land-use planning on the urban-suburban periphery.

While the North Side set the first standards for comprehensive park systems, Westchester built on this work and set new standards for professional oversight of leisure landscapes. The Bronx River Parkway, built 1907–1925, was the bridge project that intellectually and geographically linked the North Side and Westchester park systems and provided the training that turned park designers into vanguard regional planners. In 1895 the state legislature responded to complaints about the sewage-laden, odiferous waterway that ran through the Bronx Zoological Gardens by authorizing a survey of Bronx River Valley pollution. Over time what began as a river cleanup and sewerage project became a pioneering transportation project, the construction of the nation's first modern parkway built to accommodate cars.[37] The autonomous Bronx Parkway Commission (BPC) executed the project across the Twenty-Third and Twenty-Fourth Wards and Westchester. Like many Progressive Era organizations, the BPC was created by the state and was controlled by influential citizens. Although powerful, the commission lacked the authority to issue bonds. The city and Westchester had to issue bonds to jointly fund the project, although the city was repeatedly recalcitrant in committing its share.[38] The BPC and state decided that the city should pay three-quarters of the cost even though only a quarter of the parkway ran through it, reasoning that New Yorkers would use the road more and that the city would benefit most from cleanup since it contained the river's most polluted sections.[39]

The BPC framed its project as the "improvement" of the environment. The commissioners promised that the roadway's parklike features would provide a healthful natural environment. The commissioners declared that both the environment and the population of the river valley were polluting influences, a framing that historian Randall Mason argues was used to justify the eugenics-influenced removal of the river valley's marginal communities. The BPC considered the residents of the valley to be of degraded racial stock and a blighting influence.[40] For example, superintendent of construction Major Gilmore D. Clarke, a civil engineer and landscape architect, blamed White Plains for dumping trash in the river's floodplain.[41] Clarke grew up around his father's greenhouses in the Kingsbridge section of the North Side and remembered the area when it was fields and stables. The erasure of his childhood landscape and the development of factories and low-income residential communities dismayed him.[42] Like Clarke, the BPC valued only certain types of landscapes and populations. The commission set out to sanitize the river's social and ecological landscapes, an approach that was repeated in subsequent planning in the region.

The BPC worked in the tradition established by Olmsted in the Boston Fens in the 1870s, reimagining the urbanizing waterway as a habitat to enhance human welfare and to reconstruct and manage scenic nature with few visual or physical breaks.[43] Defined by river valley topography, the parkway ran north–south across the Bronx and southern Westchester in recognition of the ecological cohesiveness of the urban edge and the importance of treating the area as an integrated whole.[44] The 1,155-acre linear park averaged six hundred feet wide, a swath of an idealized, bucolic suburban landscape, setting the standard for the design of ribbon parks. Subsequent Westchester parkways built in the 1920s and 1930s followed this standard with rights-of-way that stretched from three hundred to fifteen hundred feet or more. Twentieth-century landscape designers later identified a three-hundred-foot right-of-way with 150 feet on each side of a road as the range of the "polluted zone" caused by automobile emissions. To filter pollutants and protect recreational spaces, planners would recommend that roads be built with three-hundred-foot borders planted with lush trees and shrubs, similar to the Bronx River Parkway right-of-way. The extraordinary scale of Westchester's first parkway would have future ecological benefits.[45] While the BPC never anticipated future automobile pollution concerns, the commissioners were nevertheless aware of the importance of the aesthetic standards they established. Clarke, construction supervisor for the project, recalled that he and his fellow commissioners carefully took "winter 'before' and summertime 'after' photographs of the parkway so as to make it more dramatic than it otherwise might be!"[46] Public works turned nature into a scenic amenity.

Appreciative parkway users quickly realized the utility and aesthetics of such large-scale public works. BPC officials claimed the polluted and refuse-strewn river valley and the "practically valueless," "low class of development" along it retarded "desirable community development and normal increase of taxable valuations."[47] The scenic parkway opened the county to new commuters. Park and parkway development was vital to metropolitan decentralization, a forerunner that paved the way for greater suburbanization. The first section of the road opened to traffic in 1922, and the entire route was completed three years later. The parkway was overwhelmingly popular. As early as May 1924, 17,629 cars passed by Bronxville on the parkway in a thirteen-hour period. The thirty-five thousand cars that traveled the parkway on a single summer day in 1927 created bumper to bumper traffic. This traffic rate approached the heaviest in the state, that of the nearby Boston Post Road in Rye, which fifty thousand vehicles traveled daily by 1925.[48] The answer to this congestion was more parkways; Bronx River Parkway engineer Jay Downer observed "it wasn't necessary to argue with the people of Westchester that planning and parkways are a good thing. They were like the babies and a famous brand of soothing syrup. They cried out for it." Downer likened support of the parkway to the way in which P. T. Barnum secured popular interest in his Bridgeport developments. Downer told the story of how, when initial excitement over Barnum's circus elephant faded, he "took it up to his farm in Connecticut adjoining the New York, New Haven, & Hartford Railroad, and every time a train went by a Hindoo was busy ploughing with the elephant. Agricultural societies had papers read about the possibilities of elephants as farm animals in America." "People learned about the Bronx River Parkway," Downer claimed, in a similar way, via pleasurable personal experience.[49]

The popularity of the Bronx River Parkway inspired bipartisan confidence in county-level planning and support for additional large-scale public works in advance of suburban growth. In 1922 Westchester's board of supervisors, looking to continue the success of the BPC, formed New York State's first county park commission. A unit of rural government, county governments in the state did not traditionally provision large-scale services. In the 1920s, however, Republican county chairman William L. Ward expanded county government via charter reform and created the WCPC with expansive powers. Parkways complemented the suburban growth Ward courted for the county and brought his administration national praise as an exemplar of Progressive government.[50] As the governing body for the county's forty-six municipalities, the board of supervisors could institute development independent of state approval. The board's park commission represented a new form of county-level governance, a body able to plan and act at a regional scale with resources, both funds and technical skill, dedicated to development.[51]

The BPC conceptualized parks as large-scale planning tools, and the WCPC continued this approach. Members of both commissions agreed that while the Bronx River Parkway was not laid out as part of a comprehensive plan, it effectively constituted "a main axis, or backbone, for the development scheme of the important city and suburban territory which it serves."[52] As early as 1912, BPC officials imagined the parkway as part of a future outer park system, hypothesizing that a regional network of parks would promote a "growing sense of unity and intimate relationship between the city and its suburbs, resulting in the increasing subordination of local differences for the sake of metropolitan advancement."[53] Successful public parks paved the way for a county-run public works program. Downer pointed to the parkway and Oakland Beach, the town beach in Rye, as the main influences that inspired Westchester's investment in parks: "in retrospect both of these projects, one a stream valley parkway, the other a waterfront park, can be seen as pointing the way toward the comprehensive park plan for the county as a whole." County planners further explained that the WCPC had "a much broader scope than the primary . . . purpose of providing public parks. It is a larger scale of planning scheme necessitated by the present rapid growth of Westchester County."[54] Downer, employed first by the BPC and then by the WCPC, hoped to use "a connected system of parks and parkways" to structure the region and prevent Westchester from being "caught as New York was, when in 1851 it belatedly realized the importance of parks and set aside Central Park. It is planning now . . . for the sprawling city of tomorrow."[55]

With a specific vision for the county's future, a team of talented professionals, and the power to realize plans, the WCPC successfully built parks across Westchester's 448 square miles.[56] The Bronx River Parkway became the nucleus of this system and its designers the core of the new county park commission. The WCPC took control of the parkway, and county chairman Ward appointed to the commission engineers such as W. W. Young and landscape engineers including Clarke and Downer. Young had worked on the Bronx Valley sewer and later oversaw park-related sewerage. Clarke and Downer organized a talented team of designers and engineers to build Westchester's parks and parkways.[57]

Generous funding made the WCPC's ambitious program possible. No maximum was set to limit appropriations, and additional funds were frequently approved if project costs exceeded estimates.[58] The county authorized the commission to issue certificates of indebtedness, paid for by bond sales, for each project. The cost of acquisition and development was borne by the county via property taxes. In 1925, parks cost county taxpayers nearly $470,000, but these costs fell to $290,000 in 1927 and to $60,000 in 1928 because of income generated by concession fees.[59] In the following two years, annual receipts in fact exceeded expenditures, and the profit was used to pay down bonded debt charges.[60] Generous funding and the willingness to take on debt allowed the WCPC to increase park acreage in the county significantly. In 1921, parks covered less than 1 percent of Westchester, leaving its 361,000 residents with a meager ratio of four acres of parks per thousand residents. By 1927, the county's population reached 448,000, but the forty million dollars appropriated for park purchase and development between 1922 and 1927 had added more than sixteen hundred acres of parks to the county. Parkland created by the WCPC constituted 13 percent of county territory and created a stunning ratio of 85.4 acres of parks per thousand residents, or one acre for every twenty-eight persons.[61] Manhattan could offer only one acre of parkland to every 1,304 persons. Nassau County on Long Island, home to suburban and estate districts similar to Westchester's, had just one acre for every 1,179 people.[62] The CRPNY celebrated Westchester's system and set its standards for the aesthetics of open space, recreational amenities, and population to park space ratio for the region by the county.[63]

Officials celebrated Westchester's parks as not just improving access to recreation but increasing the county's wealth as well. Greater New York's park commissions espoused a shared economic development theory that public systems were assets of progressively increasing value.[64] In the early 1900s BPC officials charged that pollution and what they called practically valueless development along the Bronx River had once retarded increases in taxable valuation. The parkway, however, had increased neighboring property values. In 1932 the WCPC reported that since 1910, assessed values of land near the Bronx River Parkway had increased 1,178 percent as opposed to the 395 percent increase outside the parkway's zone of influence. Due to dramatically higher assessed values in the area under the parkway's influence, Westchester collected almost twenty-three million dollars more in tax receipts than from unaffected property. Tax revenues repaid the entire cost of the parkway.[65] The WCPC expanded the economic influence of the Bronx River Parkway across the entire county. In 1923 the total assessed valuation of taxable property in Westchester was nearly 750 million dollars. Four years after the creation of the county's park and parkway system that value had doubled.[66]

In the first three decades of the twentieth century, the BPC, the WCPC, and the Committee on the Regional Plan of New York and Its Environs presented park development as an engine of economic growth.[67] The

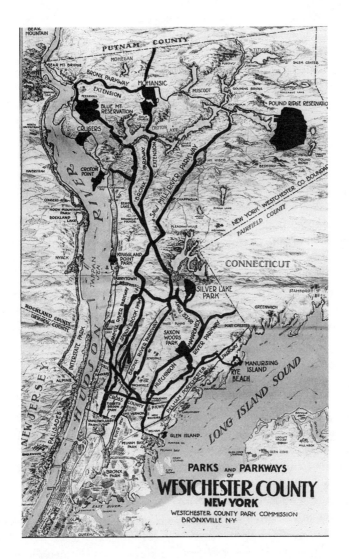

**4.2.** Westchester's parks and parkways by 1935. *Source*: Edward Van Altena, *Parks and Parkways of Westchester County*, photograph of lantern slide, 1935, Parks Commission Lantern Slide Collection, Parks Commission Photograph Collection. Courtesy of the Westchester County Archives.

argument had been made in New York since the 1830s, when park build-ers first promoted parks as raising nearby real estate values and lifting the city's tax base. In the 1860s Olmsted rationalized his park proposals by forecasting the quadrupling of nearby property values. In the 1880s Mul-laly anticipated increased tax income would more than pay back park bonds and interest, declaring that parks would be "entirely free of cost."[68] In the late 1920s, the *New York Times*, reporting on Westchester's recent suburbanization, noted "it is significant to note that, among the residents and officials of the county, the term investment is being used rather than expenditure."[69] Parkways obviously increased land values, but planners admitted that quantifying their exact influence among other factors such as rail line and population growth was impossible.[70] Nevertheless, offi-cials pointed to parks to boast that public works paid. As WCPC president V. Everit Macy avowed, "a considerable proportion of this unprecedented increase was due to the park program."[71]

Westchester's park system featured wooded reserves, three golf courses, six swimming centers with beaches and pools, and Playland, an amuse-ment park at Rye Beach. From its inception, the WCPC prioritized a county beach at Rye. In June 1923, Governor Alfred E. Smith signed a new law for a county park at Rye Beach to replace the amusement district.[72] While Rye Beach's defenders had successfully blocked local political attacks and at-tempts to shutter the beach, county officials triumphed where local elites failed, successfully closing and dismantling the Rye Beach amusement dis-trict. The commission considered the beach to be the only commercial or public beach of consequence along the Sound in the county where it could experiment in new standards of coastal recreation design. It invested more than seven million dollars there to realize a new vision of "appropriate" public space for a New York suburb.[73]

The WCPC justified such work by claiming that Rye Beach was un-planned, unattractive, and a source of pollution, echoing the BPC. Progres-sive Era politics favored regional park planners who claimed the authority and know-how to successfully rebuild the coastal environment. Applying ideas about nature to their evolving urban and ecological settings, park planners demanded the authority to rebuild landscapes they deemed un-attractive, unhealthy, and outmoded. The WCPC's papers and publications in professional design journals deemed patrons and the built environment to be a "double plague" festering on beautiful Long Island Sound.[74] WCPC president Macy declared that Rye Beach's patrons were typical of the "kinds of human blight that can infect public places."[75] Lumping nearby properties into this description of the amusement district, whether or not they were in poor condition, enabled the commission to clear additional space to re-design the shore completely. The WCPC worked with the county's Sanitary Sewer Commission to remove the city-built sewage disposal plant on the east side of the amusement district and divert sewage to a new inland treat-ment plant. The pumping station that sent sewage to the new Blind Brook plant was concealed within Playland's bathhouse.[76] The new system mod-ernized and camouflaged urban infrastructure and gave the commission greater control over the coast's nature. The next step was to reconfigure the built environment and social makeup of Rye Beach. The commission demolished the resorts to build what was called a "sanitized Playland," sig-naling a new era of social and environmental control.[77]

Playland opened in 1928 as the nation's first municipally run and com-prehensively designed amusement park. Clarke toured parks across the country to draw inspiration for his plan for Playland. Clarke later boasted of the originality of this work: "at that time I didn't know anything about amusement parks . . . [but] wanted to create . . . a kind of amusement park that probably had never before been accomplished."[78] Three parks in one, Playland featured a swimming park with a boardwalk, bathhouse, beach, and freshwater pool; a naturalistic lake and walking trails; and an amusement midway with rides such as the Tumble Bug, Derby Racer, 1,001 Troubles, and a carousel.[79] The midway's thousand-foot-long colonnade and grassy mall and 120-foot-tall music tower provided a central axis for the park.

V. Everit Macy declared that at Playland not just environmental but "so-cial reclamation was also carried out by converting a haphazard seaside resort . . . into a unified, publicly operated amusement park of . . . whole-some moral standards."[80] Architect Leon Gillette and landscape architect Clarke rejected the vernacular architecture of the working-class leisure corridor. They declared that most amusement park architecture was visu-ally chaotic—"sometimes tawdry, but always cheap, uninspiring and even depressing—having grown haphazardly over time without unifying archi-tectural elements."[81] Experts involved made exaggerated claims about the powers of bright, modern, and stylish beaches. Like Mullaly and his fellow nineteenth-century park advocates, the WCPC drew a causal link between physical environment and patron behavior. Playland's designers argued that the "certain classes of patrons that [had] been lost to amusement parks in recent years" would return if there were "modern methods of en-tertainment [and] modern and artistic decorations."[82] The award-winning firm Walker & Gillette designed the park's distinctive art deco architecture to be wholesome, moral, and characterized by "finer aesthetics."[83] Walker

& Gillette created a unified and precisely controlled leisure environment for the amusement section. In contrast to the beaches' former ad hoc layout, Walker & Gilette employed an orderly geometric layout. In contrast to the garish colors of the area's former amusements, the pastel palette of Playland's lights and decorative friezes were specifically chosen to create a calming, unified aesthetic, and the nursery rhymes depicted in the friezes were meant to suggest charming innocence.[84] Design and construction supervisor Frank W. Darling, a nationally renowned expert in amusement park administration, declared that the park's expensive manicured lawns and sophisticated gardens paid dividends. Playland was to be an elite landscape that encouraged new behaviors and attracted new crowds.[85]

Through Playland's landscape and marketing, the WCPC prescribed new behavior standards for the crowds once enthralled by Rye Beach's eclectic landscape and ballyhoo barkers. A 1928 volume of *Amusement Park Management* about Playland explicitly connected park design to middle-class propriety: the "'class' was the same as at any public place; a cross section of the great republic." Playland's visitors "were only 'classy' in their orderliness, because they were stimulated to harmonize with their surroundings."[86] In a 1933 report on leisure in Westchester, sociologist George A. Lundberg characterized Playland as an idealized vision of wholesome leisure culture as part of the area's emerging suburban lifestyle.[87]

Despite the WCPC's rhetoric of a democratized public, racial and class biases informed the commission's idea of "the public." Like the East Bronx residents who fought to segregate shore resorts, Darling, who became Playland's general manager, used race to define who was and was not welcome at the park. He instructed ticket takers to pass out instructional pamphlets to black visitors that admonished the black patron "that he conduct himself at Playland as he would in the parlor of his own home."[88] Playland planners and managers conflated orderliness and safety with racial segregation. Darling saw minorities as a liability to his amusement park's carefully constructed leisure environment and denied their right to leisure there. African Americans accused Darling of overcharging them for concessions, segregating Playland's pool and restaurants, denying outing permits to their community groups, and threatening hack companies and ferry operators who transported black patrons, prompting one operator to briefly hang signs "requesting Negroes not patronize the craft."[89]

**4.3.** *Opposite:* Walker and Gillette's rendering of Rye Playland in 1927, the year before it opened. The park proved instantly popular. *Source:* "Architect's Rendering of Playland Park." Courtesy of the Westchester County Archives.

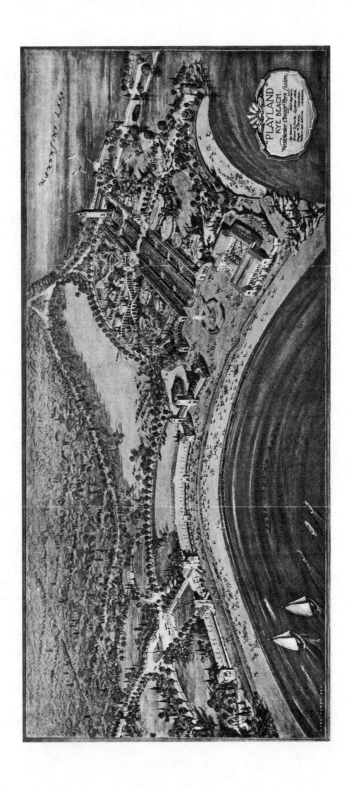

Public leisure spaces became battlegrounds for desegregation in blacks' fight for equality in the twentieth century. Despite the rhetoric of planners who frequently conflated regional infrastructure with democratic planning, the spaces of the regional city were not universally accessible. Playland was no exception. While the WCPC ignored issues of racial discrimination at the park in its reports and did not keep records of the racial makeup of Playland's crowds, the *New York Amsterdam News* tracked charges of discrimination made against Playland's general manager. In 1933 National Association for the Advancement of Colored People (NAACP) activists successfully demanded a special county commission to investigate alleged racist management practices. Benjamin Levister of Mount Vernon, a young black lawyer and investigator for the New Rochelle NAACP, testified that Darling told him that Playland represented "too large an investment to be jeopardized by depreciation through the admission of Negroes."[90] The visual and physical intimacy of swimming made pools and beaches in particular sites of intense contestation. Like Playland, the local YMCA barred black children from swimming in its pool in the company of whites in the midthirties.[91]

In 1935 the Westchester County Committee against Racial Discrimination, working with the NAACP, won civil action suits against Playland for denying black patrons access to the pool. A Playland beach and pool manager and a ticket seller were both found guilty of discrimination.[92] The racism that shaped these actions had similarly constrained Solomon Riley's ventures in the East Bronx. And at Hudson Park in New Rochelle, one of two swimming spots, the older beach functioned as an unofficial "colored" section through the 1940s.[93] But segregation was not universally fixed, and integrationists achieved some inroads. The NAACP reported improved race relations at Playland. In 1936 a group of young women affiliated with the NAACP visited Playland to test whether the park had reinstituted restrictions, and they reported courteous treatment at the bathhouse and pool.[94] In fact, the municipal resort allowed a more racially diverse public than the vernacular working-class leisure corridor of which the former Rye Beach

4.4. *Opposite*: Sanitized Playland. In this image of the mall in 1930, a worker changes one of the colonnade's decorative light bulbs, while another, to the left and wearing a tie, sweeps up litter. Both employees are in uniform. Promotional material never failed to comment on the park's order and cleanliness. Behavior was even regulated in the parking lot. A sign warns that those caught undressing in cars would pay a five-dollar fine. A trip to the bathhouse to change cost fifty cents. *Source*: "Playland Employees" (*top*); "Parking Lot Entrance" (*bottom*), 1930. Courtesy of the Westchester County Archives.

was a part. Residents nevertheless continued to advocate for racial and residency restrictions to limit access to Westchester's new public spaces.

County planners saw Westchester's parks as regional attractions. Not all county residents, however, embraced this perspective. Some board of supervisors members indignantly complained that Westchester was being forced "to play the role of a gracious if embarrassed host" and called for residency restrictions. Westchesterites constituted only half of Playland weekend visitors; a third of the park's patrons resided in Brooklyn, Manhattan, or the Bronx in 1930.[95] "Westchester for Westchesterites!" proclaimed the *Mamaroneck Times*. "If the New Yorker is to enjoy all of our privileges, let him come here and live. Let him pay his taxes and help foot the bill for the upkeep of all these things."[96] A Rye resident went even further, claiming that even his fellow Westchesterites had no right to the beach. In Frederick W. Sherman's view, residents of Rye Beach were "victims of public officials . . . forced to pay tribute to those strangers who are exploiting her for the county at large."[97] Residents of White Plains denounced Rye residents like Sherman who would "build a wall of spikes around . . . Rye and Oakland beaches."[98] The WCPC rejected localism at both the neighborhood and county level. The commission had built parks and forecast profits with a regional public in mind. When thunderstorms threatened Playland on a crowded July Fourth, general manager Darling contacted licensed hack companies to send extra buses, but Rye village police arrested the drivers for lacking necessary permits. Outraged, Darling wrote to the local paper announcing that he ran Playland "solely out of consideration for those thousands who come here for wholesome recreation, whether or not their pleasure runs counter to the demands and complaints of Rye citizens."[99]

The WCPC's vision of the ideal public, the regional nature of its park system and open access to it, and the celebrated influence of the planning establishment became the model for park planning in greater New York. The WCPC earned national renown, and its approach exerted direct influence on subsequent park planning on Long Island and in New York City, where the regional planning of "buffer" parks evolved as a powerful tool with the potential to segregate and control both land use and public behavior on a regional scale.

*"Buffer" Parks for the Queens-Nassau Border*

In the 1920s, state officials, eager to replicate Westchester County's successful park commission, encouraged the germination of park planning ideology across the Sound to Long Island. Park officials envisioned a border-crossing system to safeguard barrier beaches and establish large inland reserves. The framework would preserve the rapidly developing island's remaining open space for recreation. The island became a testing ground that both further solidified the recreation and planning ideals developed by Westchester's planners and led to collaborative and interjurisdictional planning on the Queens-Nassau border. Park planning on Long Island also exposed the limitations of both state park planning and New York City's decentralized parks department. Park planners continued to systematize and professionalize their field, but the challenges that arose on Long Island revealed the ways in which local communities and purely advisory planning roles could retard comprehensive planning.

State officials and park experts such as architect and planner Charles Downing Lay envisioned "more or less all of Long Island . . . as a playground" for greater New York. The island was particularly well suited to residency and recreation since, unlike Westchester, the island was not subject to metropolitan traffic or significant urbanization beyond western Queens.[100] Yet into the 1920s, the region lacked both sufficient public recreation amenities and a unified public park plan. Between 1900 and 1920, the borough's population increased from 152,999 to 469,042, and speculators eyed the woods of northeastern Queens and western Nassau County. The rate of suburbanization inspired a journalist to drolly advocate the lawnmower as the official symbol of Queens.[101] But park development did not keep pace with residential growth. The borough did not even create its own park commission until 1911, leaving Brooklyn to oversee park development. By the twenties, parks made up only 1.9 percent of Queens. The borough's expansive cemeteries provided a type of open space, but park advocates worried that burial grounds might eventually be sold to developers under the pressures of suburbanization.[102] Across the city line in Nassau, planners lamented that parks were entirely "left to chance with the result that there [was] not even the nucleus of an existing system."[103] By the 1920s Nassau's North Shore was divided into nearly six hundred estates overlooking the Sound. Unlike the South shore, where the island's broad, glacier-made outwash plain and barrier islands formed wide, flat beaches, a series of narrow bays characterized northern Nassau. Town and village parks on these narrow beaches covered only seventeen acres, far too small a territory to serve the regional public.[104] In 1923 parks made up but 0.07 percent of Nassau, or 123 of the county's 145,000 acres.

In 1922 the first comprehensive park plan for the island appeared. The New York State Association, a citizens group dedicated to progressive civic reform, published *A State Park Plan for New York*. Written by its secretary Robert Moses, the report drew heavily on the WCPC's model of regional park planning. It was widely read by park planners across the nation and hailed by the nation's leading planners as a seminal document in the field.[105] Two years later, in 1924, with the support of Governor Alfred E. Smith, Moses translated this report into state policy when he wrote the legislation establishing the State Council of Parks. When the bill passed, Smith wrote Moses jovial congratulations: "Dear Bob: It looks to me as though the entire State is to be made into one grand park. Please do not include the premises in the rear of my house as I have some excellent rose bushes there which I would like to preserve," closing his letter, "Yours for trees and streams and running brooks."[106] Moses and Smith looked to the creation of the council as a turning point in park planning. This legislation made New York the first state in the nation to develop a centralized park planning agency and action plan. Named president of the council, Moses became the intermediary between state and local park authorities.[107] Long Island was one of the council's primary concerns, and in the very year it formed, the council lobbied the state for the creation of a state park commission for Long Island. Smith appointed Moses head of this commission as well. Moses's dual appointments ideally positioned him to plan a state park system for Long Island while considering regional suburbanization and park needs.

The State Council of Parks encouraged a WCPC-like approach to the Queens-Nassau border with the goal of bringing the untended urban edge under government oversight. In 1924 Smith authorized the Metropolitan Conference of City and State Park Authorities, providing the State Council of Parks with a forum in which to discuss balancing suburbanization and open land preservation across the city and adjacent counties. As head of the State Council of Parks, Moses chaired the conference, which brought together members of the New York State Association, borough park commissioners and their staff engineers and landscape architects, and members of the Long Island, Westchester, Taconic, and Palisades Interstate Park commissions to devise an interjurisdictional park plan. Conference attendees said that there was little hope of solving greater New York's park problem on "a local or borough basis or by any method of local assessment. It is a city-wide problem which must be passed on and financed on a city-wide basis."[108] Like the CRPNY, the conference was advisory in nature and lacked power to implement plans. While this fact might appear to hamper planning, the CRPNY countered that advisory commissions were more likely to employ a regional scope. Greater New York crossed state, county, and borough lines. Hundreds of public authorities, utilities, and private corporations such as railroads also operated in the region. No official group was likely to achieve absolute planning powers across such a jurisdictionally complex region. Based on two years of research, the conference outlined two fundamental—and interrelated—park concerns in Queens. First, urban recreationalists might spill into Nassau, wreaking havoc on private land and overwhelming small communities unprepared and unwilling to open local parks to nonresidents. Second, suburbanization in Queens between Hillside Avenue and Long Island Sound could potentially gobble up all remaining open space in the borough within the coming year.[109]

The solution lay in the formation of a "buffer system" of park preserves for Long Island. Moses pointed to John Mullaly's North Side parks, which covered 16 percent of the borough, as the type of forward-thinking, comprehensive planning needed on Long Island.[110] A buffer "bearing the same relation to Nassau County that the park system of the Bronx bears to Westchester," the Long Island State Parks Commission (LISPC) explained, would address concerns associated both with recreation and suburbanization.[111] Conference members proposed a system of peripheral buffer parks that included twenty thousand acres around Alley Pond and Creedmoor State Hospital in northeastern Queens near the elite Nassau community of Wheatley Hills.[112] Not only did Queens lag significantly behind the Bronx in total park area, the buffer parks of Van Courtland and Pelham Bay lined nearly half of the borough's border with Westchester. Alley Pond was on the official city map as a potential park, but the land was never purchased. By 1925 more than half of the proposed park was bought for subdivisions at public auction, as was the land adjoining Creedmoor State Hospital for the Insane. Echoing the concerns expressed in the 1880s by Mulally and more recently by the BPC and the WCPC, conference members worried about the unwelcome proximity of the asylum to parks and homes: "it requires little imagination to visualize the problem which will result if houses are built all over this area up to the Nassau border even surrounding and hemming in an institution for the insane . . . with inadequate grounds for the inmates."[113] Political interests anxious to control urbanization informed the way nature was set aside in parks. Alley Pond Park represented planning ideology but also a localist perspective that disparaged certain attributes of urban life. Buffer parks satisfied both planning interests and the propensity for political and social isolationism in local Nassau communities. They could ensure distances between disparate land uses and delineate the edge between borough and suburb.

The LISPC and the CRPNY supported the proposed buffer system to unite and balance land use on the borough-county border. The plan fit the shared goal of balancing growth across the region, particularly on the metropolis's edges, where opportunity existed to institute a long-term park plan ahead of intensive suburbanization. No district, an editor at the *New York Times* declared, "needs a buffer more sorely than that where the Borough of Queens empties its tides of traffic into the region of private estates in Wheatley Hills"; "City dwellers complain because they have so far to go in reaching the countryside. Residents of the surrounding country complain that excursionists jam the roads, litter the wayside and invade the privacy of their estates. A buffer park takes up the shock."[114] A successful park system would buffer disparate land uses, like estate colonies and asylums, from each other and funnel the public into "appropriate" recreation spaces.

The boosters, planners, and municipal figures who established the regional park systems of greater New York evidenced a shared and long-standing interest in the relationship between urban expansion, population growth, and new parks. Smith and Moses characterized traffic on Long Island, the situation at Alley Pond, and the privatization of open space as problems of overcrowding that could be solved in part through parks. Suburbanization and region population growth never led to a formal park mandate, but boosters and planners regularly employed population metrics to campaign for parks, beginning with Mullaly in the 1880s. In 1928 Lee Hanmer, a recreation expert for the CRPNY, calculated park space per capita of population to project future population growth to argue the necessity of new large parks on Long Island.

Both the LISPC and the CRPNY assumed that since metropolitan growth spanned multiple jurisdictions, so, too, should land-use planning. The CRPNY counseled village and town governments to think "of the totality rather than the parts" and "recognize that the great whole is a living thing, with . . . something like a functional physiology."[115] Moses thought local governments that rejected this perspective and discriminated against nonresidents compounded park problems.[116] Regionally planned new parks, not exclusion, would remedy overcrowding. The LISPC assumed its public works would keep excursionists from dispersing across the island and minimize the practice of carving out unofficial recreation spaces on random roadsides and help locals protect private property. In 1925 the commission proposed the Northern State and Southern State parkways to alleviate the traffic jams of summer weekends when urbanites sought relief at suburban beaches and resorts, and the CRPNY endorsed the plan.[117]

Financial independence and broad land acquisition powers made the

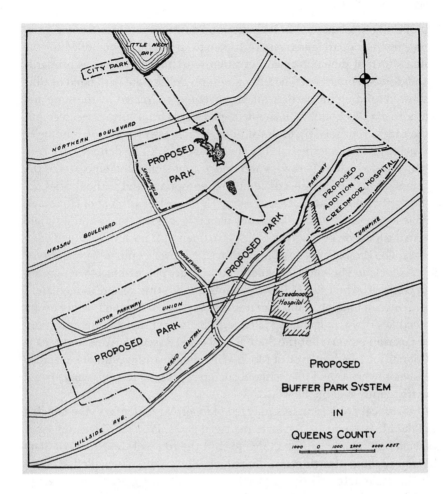

**4.5.** Buffer parks from the LISPC's perspective. This detail of the proposed buffer park at Alley Pond and the land surrounding Creedmoor Hospital appeared in the LISPC's first report. The LISPC had no authority in Queens, but the commission considered parks in that borough as an essential part of the system that they were building directly to the east. *Source: First Annual Report of the Long Island State Park Commission to the Governor and the Legislature of the State of New York, May 1925* (Albany, NY: Long Island State Park Commission, 1925), 18.

LISPC likely to succeed in its buffer and parkway program. Before 1924, state bonds for parklands were restricted to land acquisition, reflecting the state's turn-of-the-century conservationist approach to parks as primarily undeveloped preserves. The fifteen-million-dollar state park bond of 1924 allowed funding for permanent improvements. Created by the state and funded via this bond, with supplemental funds from concession receipts, the LISPC did not rely on local budgets and assessment for funding.[118] When Moses wrote the bill that created the LISPC, he defined park construction to include parkways and bridge approaches and empowered the commission to purchase, condemn, or appropriate land without prior approval of county boards of supervisors. The State Department of Highways, which stringently guarded highway planning and law, said a county could veto the location of a highway; the department had no authority over parkways. The Bronx Parkway Commission (BPC) had acquired the majority of parkway land via lengthy condemnation court proceedings to set prices. In contrast, when the LISPC condemned land, title vested immediately, and the commission could take possession at once. Payment would be decided later by a judge. Long Islanders challenged the commission's condemnation powers, but the State Finance and Assembly Ways and Means Committee ruled in favor of Moses's bill after hearing testimony from Jay Downer, who argued that without such powers private landowners could permanently stall public works.[119]

Skeptical Long Islanders rejected planners' arguments on the benefits of the LISPC's powers and state-managed parks. Protests by wealthy landowners eventually led the LISPC to shift the route of the Northern State Parkway south of their estate district, and the commission failed to establish a single state park on the North Shore until the 1970s. Westchesterites celebrated their county-level park commission as an example of progressive home rule. By contrast, the exportation of the Westchester model to Long Island severed the critical link between local support and regional park planning. With power over two counties, the LISPC was the first truly regional planning body in metropolitan New York, but its state-granted power proved to be a liability when locals rejected the commission as a foreign, invading agency.

Nassau's resistance to a state-imposed park plan reveals the influence of local communities on regional development. Such opposition, however, never derailed either Moses or the larger germination and maturation of regional planning ideology around greater New York. The Conference on Metropolitan Parks, the creation of the State Council of Parks and the LISPC, and the work of the CRPNY reflected a shared awareness that urbanization profoundly affected regional land use. Moses became a powerful figure in this work, expanding government park programs. The 1920s-era collaborative work that resulted in Long Island's parks gave the region's planners, and Moses in particular, what he deemed to be "sufficient confidence [and] sufficient public support . . . to move into the city and begin building."[120] In 1934, Mayor Fiorello H. LaGuardia appointed Moses commissioner of the newly consolidated city parks department. By the end of the decade, he completed the buffer park system for the Nassau-Queens border through the construction of Cunningham, Alley Pond, Kissena, and Flushing Meadow Parks.[121] Moses's appointment to oversees all city parks marked the final chapter in the germination of planning through the region. Moses set his sights on Pelham Bay, the keystone of the region's first large-scale park system, to commit the city to an ambitious modern park program.

*Robert Moses and City Parks*

Before Mayor LaGuardia consolidated the borough parks departments in 1934, a centralized park board had theoretically coordinated citywide development. In reality, park work occurred independently in each borough. Decentralization had hampered park modernization. By 1932, a member of the WCPC surveyed New York's municipal parks and declared "what we cannot understand, is why the city of New York cannot provide parks for its own people. They have the parks down there, but evidently they don't know how to develop them so that people will like to use them."[122] Moses first began planning city parks in 1924 as chairman of the Metropolitan Conference on Parks, although the advisory commission had no power to execute his plans. Mayor LaGuardia appointed Moses parks commissioner of all five boroughs and Governor Smith secured an amendment to state law that allowed Moses to hold leadership positions in the city parks department, the LISPC, and the State Council of Parks simultaneously. In 1938, a city charter revision removed land acquisition and improvements from borough president control in favor of the parks department. As a result, Moses wielded immense power over public land use and public landscape design in greater New York, continuing the tradition of nineteenth-century figures such as Green and Mullaly.

Moses saw himself as guardian of large open public spaces in the face of municipal neglect, indifference, and inappropriate subdivision practices. Moses demanded the city reclaim the waterfront from private control and hold it in trust for the public, demonstrating a dedication to beach and marshland preservation.[123] The CRPNY had recently concluded that public recreation opportunities were deficient along the city's 191 miles of water front bulkhead line. Public parks made up twenty-five miles of this line, with less than five miles in beaches. The city's meager beachfront was inconveniently located far from the city center on the south shores of Brooklyn and Queens.[124] Long Island had Jones Beach and Westchester had Playland, but New York City lacked modern beach facilities. Commercial amusements and bathing venues privatized the city's beachfront.

In the Bronx, Moses looked to reconfigure antiquated waterfront recreation spaces to equal Playland and Jones Beach. Moses had little appreciation for the camps of the working-class leisure corridor, which included the municipally run Orchard Beach camp. A characteristically vernacular landscape, the upper East River leisure corridor included both shaped space, such as fenced-in amusement districts and bungalows, and unshaped space, such as open fields.[125] Like the WCPC, Moses considered shore camps to be nuisances similar to beachside amusement parks and detrimental to the shore's scenic value. When Moses investigated the former Bronx parks department's files, he found the Orchard Beach permit records unorganized and incomplete. "It's a fine mess. We have tried to find out on what basis the leases were originally awarded," Moses told the city newspapers, "and there is nothing in the Parks Dept. files to give us this information." Given the 1925 court findings on the subversion of the law regarding camp licenses and the legal loophole regarding permits on lands "awaiting improvement," the lack of records was unsurprising. "Politics entered into the awarding of the leases, but from now on," Moses declared, "the camps and cottages are out."[126]

Canceling Orchard Beach camp leases would, in theory, minimize conflict over access to and use of Pelham Bay Park's desirable shoreline.[127] Despite the rhetoric of planners who claimed public parks were a manifestation of democratic principles, campers found they had little say in the planning process. Campers petitioned the mayor, demonstrated at city hall, and launched a lawsuit claiming the relationship of landlord and tenant, which afforded tenants substantial rights, rather than that of licensor and licensee, but Moses prevailed. The Bronx Supreme Court justice who handed down the decision noted that campers' attempts to remain until the "beach actually is developed . . . put a 'strained construction'" on the law.[128] When some campers refused to leave, city police officers evicted them. The era in which campers shaped the beach to their own ends had ended. One approving reporter enthused that Moses had removed parks

**4.6.** The Eviction of a bungalow occupant, Orchard Beach inspection tour, July 1934. *Source:* Courtesy of the New York City Parks Photo Archive.

from past political patronage "which is a notorious park squatter, and restored them to their rightful owners": the masses.[129]

Camps were but one of many obstacles that Moses believed hindered use of Pelham Bay Park: he also thought the park suffered from mismanagement and neglect. The removal of a naval camp had scattered leftover slabs of concrete and carved "great gashes" into the ground, while Depression-era cutbacks in park spending left roads in need of repair.[130] Moses also disparaged renovations at Orchard Beach by the Civil Works Administration (CWA) that president Franklin Delano Roosevelt had authorized under the National Industrial Recovery Act of 1933. Moses declared the CWA project "a monstrosity, atrociously and inadequately planned."[131] The bathhouse lacked proper ventilation, and a reporter recalled the beach itself was but a "dark strip of gravel-filled sand."[132] Of additional concern, the CWA had apparently failed to investigate local tidal patterns or the effects of its new beach on coastal morphology. Its poorly designed seawall, rather than protecting the beach, submerged most of it at high tide. "We will see to it," the parks commissioner promised, "that Orchard Beach is a real beach," and he ordered a complete demolition and reconstruction.[133]

The new Pelham Bay Park featured a beach aesthetic based on ocean-front rather than estuarine environments. Architect Aymar Embury II, whom Moses also hired to design the Henry Hudson and Triborough bridges, formalized Moses's preference for surf beaches as aesthetically and functionally superior to the area's estuarine and more steeply pitched coastal landscape. While the edge of the city was still seen as an area in which recreationalists were more likely to find a clean environment, Moses thought man-made nature offered a more beautiful beach. Embury redesigned the shore to create a mile-long, crescent-shaped beach connecting the landforms of Rodman's Neck, Hunters Island, and the Twin Islands, a beach made not of marsh grass or rock but a wide expanse of "pure white sands floated . . . from Sandy Hook."[134]

While central to the city's coastal infrastructure development policy between 1924 and the midfifties, land making wrought new pollution challenges due to fill materials.[135] Clean ash made the best fill, since it minimized risks of uneven settling and avoided the noxious odors of organic waste and problems of solid waste fill. The parks department contracted with the Department of Sanitation for clean ashes, but when the ash arrived mixed with garbage, it was nevertheless used. In March 1935, the boom installed to corral fill either broke or was left open—Moses offered both explanations. Whatever the cause, the outgoing tide deposited orange peels and cabbage leaves on Westchester's shores. Moses generally argued that the benefits of improvements outweighed the annoyances of refuse, soot, and the stenches of park landfill. He was usually quick to condemn landfill protestors, as he had when Clason Point residents complained that Soundview Park landfill "caused much unpleasantness" because of odors and rats.[136] The trash released at Orchard Beach caused such a blatant problem that even Moses admitted "it is true that at times our citizens had something to complain about" when it came to coastal dumps. The Department of Sanitation spent fifty thousand dollars to employ street cleaners to scour Westchester's beaches and to dredge clean mud to cover the problematic fill.[137]

City newspapers declared Orchard Beach was rebuilt "in accordance

**4.7.** Orchard Beach in Pelham Bay Park. Aymar Embury II's imposing bathhouse and pavil-ion in 1937, a year before its completion. The design earned the beach the name "The Bronx Rivera." *Source:* Gottscho-Schleisner, Inc. / Museum of the City of New York, 88.1.1.4585.

with the Jones Beach ideal."[138] In reality, however, both facilities owed their aesthetics and environmental redesign to Playland. Like the beaches that preceded it, the rebuilt Orchard Beach was "conceived as a complete experience," from the grand bathing pavilion to the massive parking lot lined with flower beds.[139] Clarke of the WCPC, appointed by Moses as con-sulting landscape architect to the Department of Parks, designed the dra-matic entrance parkway, replicating Playland's entrance mall. The oak and poplar-lined axial approach road stretched along a grassy 250-foot-wide, fourteen-hundred-foot-long mall. The design signaled the entrance to a new suburbanesque order. Embury's ninety-thousand-square-foot colon-naded pavilion, which evoked the grand Trocadero in Paris, indicated a new era of municipal grandeur. The complex included changing rooms, baseball fields and tennis courts, play areas for children, picnic groves, and concessions. By hiring many of the same architects and landscape archi-tects who had designed Westchester's and Long Island's new parks, Moses further ensured a unified aesthetic for metropolitan New York's public lei-sure landscape.

Moses replaced Orchard Beach's privatized summer colony with a land-scape dedicated to certain types of play, physical betterment, and public access. Through the thirties, Bronx Park commissioner Thomas Dolan and Borough president James Lyon campaigned for an amusement park to compete with Playland, but Moses distinguished between recreation and amusement and barred the latter from the park.[140] Moses believed that the garishness of amusement parks tricked the urban populace into accepting morally degenerate recreation, degraded the waterfront, and overrode the basic purpose of beach going: healthful rejuvenation in a clean environment. "Orchard Beach was planned for the families of the Bronx," he told Mayor LaGuardia, "and I will see that they enjoy the full use of it without interferences from rowdies."[141] Moses enforced sanctions for behavior of which he disapproved: flower picking, littering, walking on lawns, faking drowning, and undressing in the parking lot. Violators faced a judge immediately at the seasonal court opened in a nearby police station, even on weekends.[142]

Park planners subscribed to Progressive Era park ideology that environment could be used to control behavior. Landscaping, architectural designs, and crowd psychology explicitly prescribed family-oriented, middle-class behavior in greater New York. When sociologist Edward Ross coined the term social control in 1901, he emphasized this increasingly institutional-ized aspect of park planning. In 1931 Moses recounted a story in the *Saturday Evening Post* to prove the LISPC's strict behavior expectations. A state trooper had visited a New York City woman to return a bag of chicken bones, tomato skins, and half-eaten hard-boiled eggs that she had thrown from her car on a parkway and issued her a fine. The policeman tracked the litterer from a letter amid the trash, a publicity stunt designed to teach the "unappreciative, messy or rowdy" recreationalist better behavior.[143]

Pelham Bay Park encouraged refined behavior standards while emulating the upper-class leisure landscape of private golf and shore clubs. When the park first opened in 1888, its former estate landscape had offered urbanites access to the idealized genteel landscape of the country manor. In the 1930s, the golf course in the affluent suburb was the apogee of class, and while the sport was generally regarded as a pastime of the wealthy, by 1934 Lundberg declared it "one of the last strongholds to crash before the plebeian encroachments on the aristocratic ways of life. The war-prosperity and conditions following in its wake resulted in a tremendous stampede to golf."[144] Moses offered urbanites a public version of that landscape by restoring the 1901 Pell Golf Course and opening Split Rock Golf Course. The Pelham Bay and Split Rock clubhouse featured the details of a private club and made the exclusive suburban experience of the private golf course, particularly its low-density recreation landscape, available to the masses.[145] Marble pilasters faced a circular drive, and an even grander facade of a patio framed by white marble columns fronted the golf course. The interior, with its leather club chairs, was similarly ornate. The use of marble, even as an accent, was unusual on a Depression-era public building.[146] Based on the success of user fees in Westchester and Long Island public parks, Moses departed from previous city policy and introduced golf fees at Pelham Bay Park. A "less ambitious, cheaper plan, [and] poor design and flimsy construction," he said, would have reduced the need for charges but would have also sacrificed excellence.[147] While Moses's assumption of a universal set of recreation values was presumptive, he nevertheless steadfastly championed the public's right to high-caliber recreation facilities. Even architectural critic Lewis Mumford, a frequent opponent of Moses, praised his aesthetic focus. "The very spot that his architects and planners touched bears the mark of highly rational purpose, intelligible design, and aesthetic form. No spot is too mean, no function too humble to exist without the benefit of art."[148]

## The Limits of Regional Park Planning

Although on the geographic edge of the city, regional park programs in Westchester and on Long Island represented a central shift in recreational policy and the scale at which it occurred. The North Side park system inspired the regional approach of professionally trained planners who asserted themselves as specialists in landscape design, and policy makers increasingly deferred to their expertise. New, powerful permanent park commissions structured the urban fringe with modern, public spaces. These successes in regional park planning led to regional decentralization and the growth of suburbs in Nassau and Westchester. In 1930, planner John Nolen reflected that the automobile had transformed the American city into a metropolis, a "radical change" that was forcing city planning programs to look to the county as the logical planning unit.[149] This planning work made the metropolitan region a more concrete concept and a new focus of public policy at multiple levels of government. Regional parks allowed power brokers like Moses to gain influence. Park planners experimented with advisory, county, state, and regional jurisdictions to find an effective level of government to bring oversight to large-scale land-use change. The resulting regional authority was in itself a product, although beyond the city, of centralized infrastructural development. The landscape features often associated with New York's emergence as a modern city in this era—the comprehensive highway network and quality large-scale public parks—originated in the hinterland projects of regional commissions.

In building parkways and buffer parks, regional park planners did not so much merge city and suburb as create self-contained recreational outlets in New York's environs through limited-access parks. Thousands of acres insulated Orchard Beach from nearby communities; in Rye, Playland Parkway connected the park with the Boston Post Road, alleviating traffic on residential approach streets and buffering noise. The town maintained Oakland Beach adjacent to Playland for residents only, effectively creating a parallel restricted waterfront.[150] Moses proposed a similar system for Long Island. "Which would you rather have," he claimed to have asked a wealthy resident, "another hot-dog alley . . . or a chute which will carry people swiftly through without your even having to see them?"[151]

While Moses frequently took credit for conceptualizing the regional works he oversaw, behind the scenes Moses, the region's park commissions, and the CRPNY worked in sync if not toward explicitly collaborative ends. Clarke pointed to the WCPC as the "impetus that led to the formation of the New York State Council of Parks," which in turn advocated for regional parks and empowered Moses on Long Island.[152] As city parks commissioner, Moses linked city parks to LISPC projects, an opportunity that an admirer declared to be "one of the most exciting and pleasing items in years to a student of administration and of regional planning."[153] Planners were hired consecutively by the various regional park organizations or held multiple positions in the region at once. This overlap helped keep localism at bay. For example, the WCPC worried that banning city residents from county parks might spur Moses, as head of the State Council of Parks, to retaliate and exclude Westchester from future state funding.[154] Clarke also recalled that the CRPNY followed the work of the WCPC and the LISPC: when the WCPC chose a new park site and announced plans to the CRPNY, "it would invariably show up in the . . . maps. From that moment on it became theirs and we got little credit for our work. I always found it amusing."[155] Half of the CRPNY's eighty-nine proposals for major open space reservations were realized in their entirety or in part.[156] The influence went both ways. Moses implemented many of the *Regional Plan of New York and Its Environs'* projects in the 1930s. By incorporating existing plans into their proposals, Moses and the CRPNY guaranteed at least partial success for their construction programs and imbued others with a sense of prophetic inevitability.[157] This merging of public- and private-sector interests included a shared emphasis on parks that served populations beyond local boundaries.

The vanguard regionalism of the Bronx's 1884 home-rule park system initiated this era of comprehensive park planning in New York's suburban hinterlands. In the Progressive Era, planners refined this ideology in plans for Westchester and Long Island in the 1920s and in the city following the 1934 reorganization of its Department of Parks. Planners came to value parks as a form of environmental management and boasted that their work reserved large swaths of scenic open space from urbanization and suburbanization. This era of regional park development reflected planners' belief in the capacity of large reserves to segregate and control both land use and public behavior. In the process, park builders in New York's environs rethought the city-suburb relationship, conceptualized a recreating public, and facilitated the germination of planning theory through regional park plans.

# 5

## "They Shall Not Pass"

### Opposition to Public Leisure and State Park Planning

In the summer of 1916, wealthy artist and designer Louis Comfort Tiffany found little rest at his Long Island retreat. He was angry. A new public bathhouse tarnished the panoramic view of Cold Spring Harbor from the hilltop patio of Laurelton Hall, his fantastic Moroccan palace set among sixty acres of luxurious gardens on the island's North Shore. Laurelton Hall was meant to be a place of privacy and artistic inspiration. The bathhouses, built by the Town of Oyster Bay, were a dark, plebian smear at the base of Tiffany's work of art. Since June, as many as one hundred people had frequented the bathhouses daily. Outraged that the town would build its only public beach at the foot of his property, Tiffany ordered his employees to dynamite the underwater groin that supported the beach. In response, Oyster Bay instigated legal proceedings.[1]

Tiffany's legal battle reveals the essential concerns of greater New York's estate communities in the Progressive Era contest between private rights and public power. In this fight, estate interests manipulated property law and local government to reject the public authority and idea of the general public at the heart of regional planning. In 1902, Tiffany had purchased 550 acres and a beach two-thirds of a mile long on Cold Spring Harbor. The area had until recently boasted a fashionable summer hotel and casino, and clammers and picnickers had traditionally enjoyed open access to the beach. Tiffany successfully petitioned the State Land Office for a "beneficial enjoyment grant" to extend his development rights across land under water four hundred feet into the bay. His plan to build a private beach required this grant; the foreshore, land exposed during low water and submerged at high tide, was considered public unless granted to an individual

**5.1.** Louis C. Tiffany's Laurelton Hall. The fountain terrace overlooks Cold Spring Harbor. *Source*: David Aronow (twentieth century). Laurelton Hall, Oyster Bay, New York, home of Louis Comfort Tiffany; Exterior, Terrace, looking toward Oyster Bay, from a series of forty-two interior and exterior views of Laurelton Hall. The Metropolitan Museum of Art, New York, NY, U.S.A. © RMN-Grand Palais / Art Resource, NY.

for development.[2] Oyster Bay challenged the state's authority to issue such a grant by claiming ownership under a 1677 patent. The town informed Tiffany that his construction of a privacy wall constituted a trespass on public land and illegally blocked baymen's shore access. Ignoring this accusation, Tiffany built a groin to support landfill to widen the beach and then took the town to court when it tried to dismantle it.

The state supreme court originally supported Tiffany's grant; in 1913, however, the court reversed its decision, recognized the town's title, and ruled Tiffany's claims to the shore void.[3] In 1905 the foreshore was too narrow for a bathhouse, but thanks to the designer's land making, the town gained the expanded public beach and built the bathhouse that so angered him in 1916. Incensed, Tiffany returned to court with a new argument, protesting that the beach (which he had built) was in fact a nuisance that compromised navigation.[4] After five injunctions and appeals between 1910 and 1922, the court decided in Tiffany's favor and restrained Oyster Bay from further building on the beach.

Tiffany's fight with Oyster Bay introduces the tools with which the rich transformed preferences for exclusivity into regional development policy in greater New York. In 1926 Tiffany set out to disenfranchise the town by redefining the terms of local government. Using State Village Law, he oversaw the uncontested incorporation of the Village of Laurel Hollow from a group of contiguous estates surrounding Laurelton Hall. The new municipality immediately restricted beach access to town residents only, excluding the people of Oyster Bay. Twenty-one years after he purchased his estate, Tiffany finally barred the public from the shore.

Beneficial enjoyment grants epitomized the proprietary hegemony over leisure landscapes that inspired the creation of Nassau County's elite North Shore estate district. But they were only the first step in the manipulation of property to constrict public access and challenge regional park plans. Elites like Tiffany extended the privacy of their property first across public beaches with beneficial enjoyment grants and then across contiguous estates through village incorporation. The incorporated village epitomized the privileged, exclusionary localism of the North Shore. The legal mechanisms of land grants and incorporation combined with home rule and weak county government fostered an extraordinary period of hinterland development. Local proprietary interests effectively barred the public from the millionaire colony of Long Island's North Shore and allowed estate owners to resist regional community building.

Antipark contests erupted in the suburban hinterlands on both sides of Long Island Sound in the first three decades of the twentieth century. The industrial barons of New York City's banking and investment industries built estates in Greens Farms, Connecticut, and on the North Shore. In both communities powerful business magnates challenged the expansive visions of public rights forwarded by the Connecticut State Park Commission (established 1914) and the Long Island State Park Commission (established 1924). From 1910 to 1932, North Shore barons made their enclave inaccessible to outsiders in search of recreation. From 1914 to 1937, Greens Farms elites mobilized "with jaws set, teeth clenched, and one slogan, 'They Shall Not Pass,'" blocking public use of Sherwood Island State Beach.[5] These battles provided a counternarrative to successful government intervention in land use as the New York metropolis grew. Greens Farms and North Shore estate owners mobilized a defensive localism and property-rights rhetoric, and they shared a belief that access to scenic environments was a private not a public right in order to challenge external interventions.

*Long Island's Gold Coast*

The concentration of Gilded Age wealth on Long Island's North Shore gained the region the nickname the "Gold Coast" by the early twentieth century. The Gold Coast was home to the largest geographic concentration of wealth and power in the United States. The nearly 110-square-mile district was discernible because of the distinctive serrated hills and fjord-like harbors that stretched twenty miles east from Queens and south to the Hempstead Plains of central Long Island. The shoreline was not conducive to commercial ports and remained largely devoted to localized agriculture, fishing, and estates through the 1900s. The Long Island Railroad stretched the length of the island by 1895, but it did not significantly affect local economies. None other than the nation's sixty richest men built estates on the North Shore, including J. P. Morgan, Vincent Astor, Henry Clay Frick, Jay Gould, Henry Ford, Pierre DuPont, William Whitney, Charles Pratt, and William K. Vanderbilt.[6] Nick Carraway's cheeky summary, "I had a view of the water, a partial view of my neighbor's lawn, and the consoling proximity of millionaires" in F. Scott Fitzgerald's *The Great Gatsby* captured the region's defining characteristics.[7]

The Gold Coast encompassed more than six hundred estates virtually undisturbed by industry, public parks, schools, or subdivisions. The districts' estate owners belonged, in the words of Lewis Mumford, to "a small leisured class" that sought palatial country retreats to seclude itself from the "whole leisured population."[8] The private leisure spaces of homes and golf and country clubs outnumbered public parks on the Gold Coast and across Nassau more generally. By June 1923, private clubs covered more than four thousand acres, or 2.3 percent of Nassau; the following summer a count of public recreation spaces revealed that parkland made up a meager 0.07 percent of the county. Nassau ranked first in the region for private club acreage: there was nearly thirty-three times as much land in golf and country clubs than in public parks. The Committee on the Regional Plan of New York (CRPNY) calculated that across New York and its environs, each one thousand residents had access to between 0.5 and fourteen acres of parkland, while each one thousand golf club members enjoyed access to two hundred to twelve hundred acres of club property. The Gold Coast

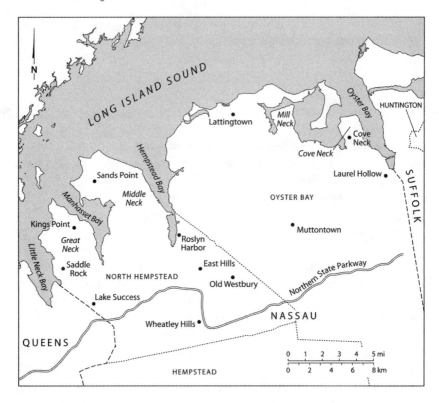

**5.2.** The North Shore, Long Island's Gold Coast. A region of estate grounds, golf courses, and bridle paths, the Gold Coast was in fact a landscape of private greenery. The nation's captains of industry valued and worked to preserve the region's exclusivity. North Shore barons established village governments known as *estate villages* to control development. The communities identified on this map, in addition to numerous others, manipulated estate villages to block commercial activities and municipal investments. *Source*: Map by Bill Nelson.

was the epicenter of this privileged recreation landscape.[9] Expansive private estates further compounded the effect. In the twentieth century, elites across America moved to suburbs enveloped in pastoral greenery and worked to preserve this desirable environment as a private amenity. In Nassau, utilities magnate John E. Aldred and William D. Guthrie, lawyer to the Rockefellers, bought and demolished sixty homes on four hundred waterfront acres north of Glen Cove to create space enough for two large estates. As Aldred later recalled, "Mr. Guthrie and I destroyed the village of Lattingtown to get the view we wanted."[10] Standard Oil cofounder Charles

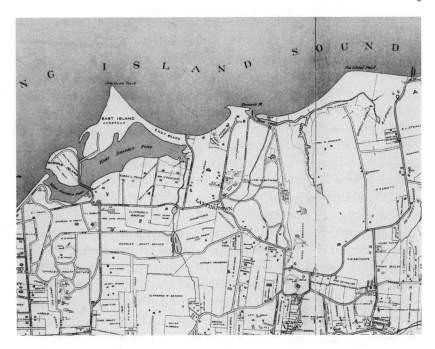

**5.3.** Gold Coast Estates. Palatial estates and estate villages blocked intensive suburbanization. John E. Aldred and William D. Guthrie bought and demolished the former shore village of Lattingtown to build their private, exclusive waterfront retreats. *Source: Glen Cove—
Locust Valley—Bayville—Oyster Bay, etc.* Lionel Pincus and Princess Firyal Map Division,
New York Public Library Digital Collections.

Pratt alone owned 1,100 acres that included three-quarters of a mile of waterfront and forty adjacent acres of land under water. Estate owners were
willing to go to great lengths to preserve this landscape of private leisure.

The obstructionism of Gold Coast millionaires dominates the historical narrative of the North Shore in the 1920s. Popular attention to Robert
Moses's incendiary battle with millionaires over parks and the Northern
State Parkway distracts from a more complicated understanding of Moses's
conflict with New York's industrial aristocracy over Long Island's role in the
region.[11] Before the fight, Moses enjoyed support for his work on the State
Council of Parks and the Long Island State Park Commission (LISPC). But
advocacy proved easier than execution of an island-wide park plan. Neither
wholesale criticism nor praise of Moses adequately addresses the battle
between regionalism and localism of which he was a part. Regionalism demanded intraregional projects in consideration of shared public interests.

Localism favored a multiplicity of small, autonomous publics and a narrow vision of shared responsibilities. Reckoning with the localism mobilized by private property interests refocuses the narrative of Long Island development on the power of elitist home rule to shape public land-use patterns and reveals the limits of regionalism—and Moses's power—on the island.

Long before Robert Moses took on the Gold Coast elite and fought to make Long Island a public playground for New York City, landowners had begun insulating their beaches from public use. Beneficial enjoyment grants, which empowered riparian landowners to privatize and build on tidelands, were the first step to Gold Coast isolationism.

In 1850 New York's land board, which managed state public lands, created beneficial enjoyment grants as part of its management of the Public Trust Doctrine. Because of this doctrine, a legal trust established at the nation's founding, the government is required to preserve public use of the shore. Although best known for its post-1960s application by environmental activists, the Public Trust Doctrine is not an inherently environmentalist doctrine. The trust derives from government protection of maritime commerce, navigation, and fishing. In the nineteenth century, it was generally interpreted as a way to support economic development, which included conveyances to riparian landowners.[12] The trust imbues tidelands with a jus publicum and a jus privatum, unlike regular property, which is generally endowed with a singular title. In theory, the trust allows conveyance of the jus privatum to a proprietary interest as in a beneficial enjoyment grant, but this title remains subservient to the jus publicum: the state's first duty is always to preserve the public interest. In practice, each state has the authority to define the doctrine, and in New York the limits of jus privatum conveyances were ill defined.[13] Between 1880 and 1920, the state land board effectively managed state-owned foreshore as property liable to divestment. Land board grants to land under water conveyed the development rights of vast stretches of the North Shore to private citizens: grants encompassed the entire western shore of Hewlett's Point; much of the eastern shore of Hempstead Harbor; and nearly all of the western shores of Oyster Bay and Cold Spring Harbor.[14] Tiffany's transition from riparian land grant to village incorporation was not unusual. In the early twentieth century, beneficial enjoyment grants became stepping stones to even greater privatization as estate owners looked to control not just riparian beaches but the nearly forty miles of shoreline through home-rule governance.

In 1910, Louise and Roswell Eldridge pioneered estate incorporation in New York State on the North Shore. Speculators had developed subdivisions in the southern section of Great Neck peninsula near the Eldridge's estate Udalls, which sat on a hill overlooking the Sound, in the preceding decade. These new residents, largely of modest means, called for the incorporation of villages and special districts drawn to include estates, such as Udalls, to impose property taxes to cover the costs of new municipal services.[15] To shelter their estate from inclusion in any new special tax districts, the Eldridges preemptively incorporated the territory around Udalls as the Village of Saddle Rock. The 126-acre estate made up 90 percent of the new village. Before 1910, state village law required a minimum population of two hundred persons in a square mile or less, constraining incorporation to territories with moderate or high population densities. In 1910, however, the legislature amended the law to allow the incorporation of districts less than one square mile inhabited by at least fifty persons. Roswell Eldridge's influence in state politics probably enabled the passage of the amendment, which made possible the transformation of Udalls into a municipal entity.[16]

The creation of Saddle Rock initiated an era of Gold Coast incorporation in which estate owners drew village boundaries in the service of personal interest—usually the aesthetic preferences and privacy expectations of individual estate owners. The intensive incorporation of individual estates as villages created a millionaires' district across the North Shore between 1911 and 1932, the period during which it was possible to incorporate small areas with populations as small as fifty persons. These "estate villages" were composed of contiguous large estates without traditional village centers. In the two decades after 1910, more than two dozen estate villages were incorporated, including Lake Success, Laurel Hollow, Old Westbury, Saddle Rock, and Sands Point.[17] They formed, in the words of a local newspaper, a multitude of "independent 'little kingdoms'" that blocked development by town government.[18]

Gold Coasters wielded incorporation as a powerful regional development tool, albeit in the service of exclusivity rather than the progressive reform traditionally associated with regionalism. Good government advocates deemed the early twentieth-century proliferation of local jurisdictions on Long Island a grotesque form of home rule. Moses predictably joined the critique, claiming "the greatest curse of suburban communities is the establishment of special districts which are too small, or which attempt to do work which should be done by the town, or by the county as a whole."[19] In 1920 Nassau included 171 local governments and nineteen villages. The count increased to a staggering sixty-three villages and 173 special districts by 1933. Yet this variety of localism led to more than a multitude of autonomous, small publics unconcerned with any large, shared project.[20] An alternative regionalism emerged from estate village special interests that preserved low-density residential development.

Wealthy individuals shaped development first through claiming rights as property owners and second by aggregating this privatism to construct an estate village. The tactic enabled isolationist estate owners to rally behind a shared vision of the right to private property as the right to exclude. Tiffany's efforts to block public bathing in front of his Gold Coast estate captured both impulses. In 1926 Tiffany and his neighbor, the prominent lawyer Henry W. DeForest, sponsored the incorporation of Laurel Hollow. The two families made up seven of the fifteen people who voted on the incorporation of the community of less than a square mile. Village government developed under Tiffany's thumb: DeForest's son-in-law became mayor, DeForest and Tiffany's son-in-law received two of the three village trustee positions, and Tiffany's architect became road commissioner.[21]

Estate villages achieved homogeneity not only by linking continuous estates but also by avoiding suburbanizing districts. Service communities that supplied estates with labor and provisions and the commuter railroad stops at Glen Head, Locust Valley, Oyster Bay, and East Norwich remained unincorporated. In nineteenth-century America, incorporation was generally employed to supply suburbanizing districts with urban municipal infrastructure of streets and sewerage. In contrast, estate owners used incorporation to avoid the public works taxes and assessments that accompanied suburbanization.[22] Since public works traditionally consumed most of a municipalities' budget and governance responsibilities, elite estate communities minimized these expenses. After incorporating in 1931, the "millionaire village" of Roslyn Harbor, which included Brook Corners, the country home of former vice president of the New York Edison company Arthur Williams, reduced taxes to a minimum: Roslyn Harbor was taxless.[23] Whereas infrastructure such as streets and utility services tied homes to a community, Gold Coast villages rejected both these services and a sense of community obligation.

Land transfers between estate-holding families were a common practice in the process of village incorporation. Estate owners sliced parcels from their property and sold them within their cohort to create the minimal population required by law for incorporation without including subdivisions in their new villages. Leading up to the vote for the consolidation of the villages of Barkers Point and Motts Point into the village of Sands Point, a local reporter observed, "from the real estate transfers recorded in the County Clerk's office . . . one would think a boom had struck the Point section. But it was only to create a few more freeholders . . . as will be readily understood by a careful reading [of the records]."[24] To enable consolidation, the Guggenheims, Kingsburys, and Laidlaws sold family members land. Of the twenty real estate transactions, all carried nominal prices, and all but four unfolded between family members. Representatives of the excluded subdivisions speculated, with probable accuracy, that school tax avoidance drove the land transfers for incorporation. The three villages successfully consolidated in 1912.

Incorporation made a charade of the democratic process. Charles E. Ransom, the town clerk of Oyster Bay, conducted ten incorporation elections in the twenties. "In most comparatively few home owners were eligible to vote," he recalled. "On several occasions the entire vote was cast in the first hour . . . in almost every instance the election was held in luxurious surroundings and the hosts did everything possible to make the hours pass pleasantly."[25] Ransom oversaw the vote, and Winslow S. Coates usually acted as the attorney for the petitioners. Contestation was rare. Local businessmen who depended on estate business tended to vote with Gold Coast barons, as did the large portions of would-be villagers who worked on the estates. Estate owners were simultaneously voters' employers, campaigning politicians, and election hosts. Voting against such figures was at the least uncomfortable. In Saddle Rock, for example, Roswell Eldridge was mayor from incorporation in 1911 to his death in 1927, when his wife Louise succeeded him—in an election on Eldridge property. She subsequently held the office through the 1930s. In the words of a *New York Times* headline, incorporation could be easily summarized: "Millionaire Village Born as Iced Drinks Clink; 13 Voters Create Muttontown, L.I., Unanimously."[26]

Incorporation withdrew land from town oversight. Exclusionary laws ensured privacy through spatial and social distance, the inherent purpose of estate village governance. Home rule limited state lawmaking in favor of municipally generated laws that narrowly defined community access.[27] Control of public space was a motive for suburban autonomy. Keeping picnickers off private roadsides was only the start. The Gold Coast villages of Lake Success, Kings Point, and Sands Point all passed ordinances restricting parking near parks and swimming spots to restrict users to residents within walking distance.[28] Following incorporation, Tiffany's Laurel Hollow immediately restricted beach use to residents; Lake Success prohibited "meeting on sidewalks."[29] The result was a legal localism defined by local government rules that aligned with cultural ideas of localism. One Sands Point resident who supported closing local roads to prevent pleasure parties from parking and walking to the beach explained, "we're trying to stop having outsiders come in."[30] John E. Aldred similarly valued the inaccessibility of the Village of Lattington and bluntly declared, "We, Mr. Guthrie and I, the Pratts and the Morgans wanted to keep it so."[31]

Estate owners additionally used municipal status to defend against outsiders and bar commercial land use through zoning. Legal incorporation and zoning made cultural ideas of localism, self-determination, and a conservative preservation of the status quo municipal policy.[32] Zoning and planning powers, sanctioned through state enabling acts in the 1920s, further empowered estate villages to define local material and class values through land regulation.[33] Land-use zoning allowed the village to directly formalize existing land-use patterns while also indirectly protecting existing social and class differences. Cove Neck banned the erection of "amusement concessions and 'hot-dog' stands" on the peninsula in its first official ordinance.[34] When incorporation was proposed for Lake Success in 1927, one resident complained the proposed village zoning was "so rigid . . . as to deprive the property owners therein all the free use of their property." A majority of residents, however, welcomed restrictions barring city recreationalists. They approved incorporation.[35]

Beneficial enjoyment grants and restrictive village ordinances enclosed traditionally public beaches without substantial challenges until the 1920s. In 1911 the New York attorney general expressed concern that beneficial enjoyment grants facilitated alarming privatization, since recipients treated them as fee-simple permits that bestowed full and exclusive ownership rights. Yet his declaration that the proliferation of the grants had engendered a "radical departure" in the preservation of public beaches was never critically examined.[36] Gold Coast privatization was compounded by estate villages' legal closure of old rights-of-way as well as unlawful private encroachment. After its initial survey of the shore in 1924, the LISPC concluded that lax government oversight and "pre-emption by private owners and the closing up of old rights of way" that provided beach access between estates made the shore practically inaccessible.[37] North Shore town and village parks made up less than seventeen acres of public land. The only medium-sized land preserve was the seventeen-acre Roosevelt Memorial Park, but it was owned by a trust, not the public.[38] In a 1925 speech to the legislature, Democratic governor Alfred E. Smith lamented that the state's practice of granting lands under water to private citizens had occurred with "apparently no thought of the future on the part of [the land board] directed towards retaining in the public possession for recreation, health and numerous other public purposes."[39] Having abdicated its sovereign trust of the foreshore, the state was in danger of squandering an essential public amenity.

In the first two weeks of June, 1925, environmental conditions and the propark movement collided in greater New York. During the first week of the month, "vicious" temperatures soared into the nineties, setting record highs and driving thousands of New Yorkers from their crowded, over-heated apartments. The heat wave drove urbanites to camp in the city's parks as they searched desperately for cooling breezes. Some even slept on the steps leading to subway stations, hoping for breeze produced by passing trains. Others fled to Long Island: the *New York Times* reported that "out in the country and along the beaches people camped in their cars or spread bedding on the sand and slept under the sky . . . every beach within driving distance of New York was lined with cars."[40] By Saturday, June 6, the fifth day of blistering heat, Long Islanders did not welcome the influx of breeze seekers. Local police barred nonresidents from the village of Huntington's beach. Forty people were arrested.[41]

The heat wave proved politically expedient in the fight for Long Island parks. The editorial page of the *Times* pointed to the weekend crowds as it proclaimed its support for the LISPC's Northern State Parkway. The fol-lowing week was pivotal in the fight for parks. On the tenth of the month, Governor Smith announced a special summer session of the state legis-lature to demand action. Smith capitalized on the still-fresh resentment of the general public that had suffered the previous week's heat without access to adequate outdoor facilities. On the same day the LISPC officially published its first annual report. The commission declared Long Island's parks priceless resources, advocated for the recreation-seeking public, and condemned private interests encroaching on Long Island shores.[42] The physical parameters of New York's environment, in this case the disad-vantages of its muggy summer climate, had brought into stark relief the importance of the park fight.

Governor Smith followed his June 10 announcement with an evening statewide radio broadcast in which he condemned Gold Coast barons for monopolizing the Long Island waterfront. "After you leave the city line . . . you can ride in an automobile about fifty miles and you cannot get near the water." The governor went on to condemn local government for parochial isolationism of restricted park and beach access. People who "stood back, with all the arrogance that comes with great wealth, and said: 'I don't care whether it takes any of my land or not. I don't want to even see it from the porch of my house,'" Smith declared, were not the people the state was going to serve.[43] Hailing from the poor Irish neighborhoods of Man-hattan's Lower East Side, Smith resented estate owners who complained parks would invite "the rabble" to the North Shore. "I am the rabble!" he declared.[44] For nearly a decade Smith, a well-known urban machine

Democrat, was the principal figure of New York State's powerful Progressive Party. Smith implemented widespread civil service and social reform in his four terms as state governor between 1918 and 1926. His staunch support of urban working-class rights included the right to public recreation. New York State, Smith declared, would not bend to wealthy residents who deemed the general public "undeserving of the superior views" of the North Shore. He concluded "private rights must yield to the public demand." Smith thus declared war on Gold Coast localism.[45]

Estate villages and beneficial enjoyment grants predated state park plans, yet in 1924 North Shore privatization became the LISPC's main target in a fight that revealed this phenomenon to the region. The 1924 creation of the LISPC was part of Smith's sponsorship of public recreation and regional planning. Smith declared that "the cure for the evils of democracy is more democracy," to which Robert Moses added, "when rich and poor can play side by side at a state-controlled resort, that theorem is demonstrated."[46] Governor Smith and Moses, the first president of the LISPC, claimed the island's expansive waterfront was a natural playground for New Yorkers. The commission immediately increased Nassau's parkland. By 1925 there were nearly 230 acres of small parks and playgrounds and 2,600 acres in parks over one hundred acres, but all five large parks were located on the South Shore. The North Shore had but seventeen acres of small parks.[47] The LISPC recommended constructing parks, beaches, and parkways along both sides of the island. The commission argued that the Northern State Parkway through Wheatley Hills would do little damage to local aesthetics or property values, since the acres needed for the parkway right-of-way represented but a fraction of the average estate. Governor Smith rationalized, "the same boulevard which carries the millionaire from his office to the threshold of his golf club or estate should carry the City man in his small car out to parks and the shorefront in the open country."[48]

Gold Coasters, however, valued the North Shore's inaccessibility. The proposed Northern State Parkway, although platted along the southernmost section of the millionaire colony, was thus seen as a threat to the entire region. This was not the first time New York State authorities clashed with local communities over how land and resources should be managed. Friction had emerged between park officials, elites, and rural people in Adirondack state park in the nineteenth century.[49] While Progressive Era park opponents enjoyed far greater social and political influence, the fight was similar. In both instances locals challenged highly trained technocrats who insisted on the need for experts to manage state parkland.

Gold Coast barons wielded their influence in both local and county government to contest the LISPC's North Shore plans. The Nassau Republican Party, the party of estate owners, rendered the potentially powerful county government of New York toothless. The towns of North Hempstead, Hempstead, and Oyster Bay had voted against consolidation in 1898 and formed Nassau County in 1899 in rejection of city oversight. Resistance to annexation was a national phenomenon. Municipal organization to stop annexation or the encroachment of unwanted people and land use was employed nationwide. At the turn of the twentieth century, Chicago's North Shore suburbs and then Long Island's district of the same name incorporated.[50] From Boston to New York to Chicago, suburban incorporation on the urban periphery was frequently born from a desire to remain apart. Estate villages formed a powerful, homogeneous, and autonomous political block in Nassau politics.[51] In New York and neighboring New England, county government traditionally served as an intermediary between state and local governments, but on Long Island counties lacked the power to mediate. In the early twentieth century, Boss Wilbur Doughty's Republican machine took control of Nassau and fostered a decentralized, one-party system that let county powers lie fallow while incorporated villages dictated regional policies. Estate villages benefited from this disempowered county government.

In 1924 and 1925, Nassau's Republican representatives in the state legislature moved to disempower another unit of government and further increase the power of local municipalities. The representatives attempted to subordinate all state parkland acquisitions, and thus all LISPC plans, to the state land board.[52] The LISPC could purchase, condemn, or appropriate land without approval from county boards of supervisors and could use the power of eminent domain without immediate payment. Nassau Republicans hoped to make the land board the first bureau or office with the ability to exert checks and balances on the park commission. The board seemed a likely forum in which Nassau Republicans could place individuals whom they could influence to block LISPC plans. Of the board's two appointed appraisers, one was brand new, the former owner of a paint shop in Buffalo, and neither had experience in park planning. Governor Smith, however, condemned this attempt to subject park planning to local "influence and manipulation." He vetoed the bill.[53] When Gold Coasters failed to block parks at the state level, estate owners mobilized the Nassau County Committee (NCC) to co-opt regional planning to support North Shore isolation.

The North Shore's elites had recently organized the committee to fight Moses's park and parkway proposals. The committee declared that its 264

members, owners of an aggregate eighteen thousand acres, spoke for regional residents who resented "interference of the state in local affairs" and
wished to be "freed from the LISPC," which made park and parkway plans
"without regard . . . to local needs."[54] In 1925, the NCC hired respected
landscape architect Charles Downing Lay to survey Long Island's beach
and parkway needs independently. Lay's *A Park System for Long Island* re-
iterated many LISPC suggestions, calling for parks in western Long Island
to relieve the congestion plaguing existing facilities and thoroughfares. He
disagreed with the LISPC, however, over its proposed North Shore projects and for making plans that lacked local support. *A Park System for Long
Island* recommended that "the whole territory of the northerly part of Nassau County be omitted from any plans for parks or parkways" until the district was "ripe" for development—an unspecified and distant future date.[55]

Charles Downing Lay made two economic arguments about parks and
real estate values to challenge the LISPC's plan for northern Nassau. First,
despite the fact that park commissions touted increased real estate values near parks to justify their work, Lay countered that in special circum-
stances the opposite was true: North Shore parks, he said, would "mean a
great loss for owners and for the County" since the estate district gained
value from its exclusivity. The Gold Coast was thus an exception to the rule
that the LISPC needed to recognize and work around.[56] Lay argued that
parks should always be built on cheap, unimproved, and rough property
not easily subdivided, since such real estate would experience the largest
returns.[57] The state had a responsibility, he said, to build parks in a way that
would produce value via increased assessed values; the extraordinary value
of the Gold Coast made it a unique place in need of special treatment.

In the fight over North Shore parks, estate owners gained crucial support from the CRPNY.[58] Employing a top-down, bird's-eye approach to park
distribution, the LISPC rarely deferred to local preferences. In contrast,
the CRPNY attempted to mediate between public and private interests,
and in so doing it came to support localist antipark sentiment and home-
rule planning. An impressive range of park planners and landscape architects echoed Lay's call and offered alternatives to the LISPC plan. Edward

5.4. *Opposite:* Parkways for the North Shore. Charles Downing Lay proposed an alternative
Middle Parkway on behalf of the Nassau County Committee. Lay calculated that the Northern State Parkway was an inappropriate use of state funds since the North Shore was home
to less than half the population of the South Shore. Lay recommended the state build parks
to the south near the center of population. *Source:* Charles Downing Lay, *A Park System for
Long Island: A Report to the Nassau County Committee* (privately printed, 1925), 11.

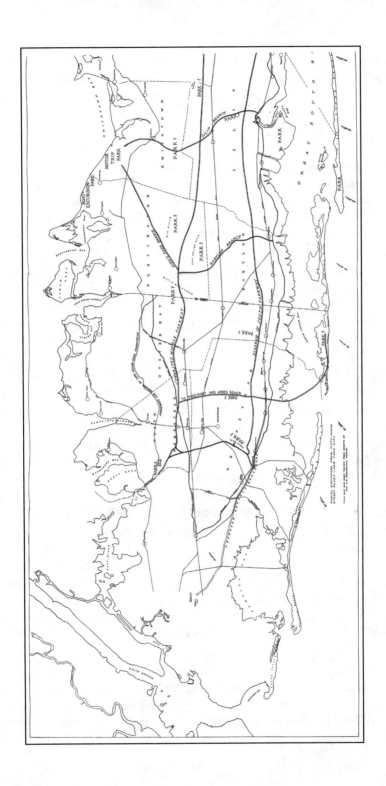

Basset, legal counsel for the CRPNY and member of the New York State Association that first proposed a state park commission, and Nelson Lewis, a vanguard of the city planning and zoning movement and a leading municipal engineer in New York City, both supported Lay's call to preserve the Gold Coast. In East Hills and Old Westbury, residents who together owned twenty-six square miles collectively hired H. V. Hubbard, a former partner of the Olmsted brothers' landscape architecture firm and head of Harvard's School of City Planning, to propose a different route for the Northern State Parkway. The landmark 1929 *Regional Plan of New York and its Environs* by the CRPNY called for a regional park system on the periphery but pointedly omitted large parks on the North Shore. Rather than advocating for equal park acreage on both of Long Island's shores, the committee instead proposed more and larger parks for the southern half of the island, urged preservation of the Gold Coast's existing standards of private open space, and opposed Moses's proposed parkway through Wheatley Hills.

Gold Coasters did not reject regional planning outright. Private interest and public enterprise joined in an unexpected but powerful coalition on the North Shore. Regional planning and localism were not mutually exclusive. Lay's *A Park System for Long Island* reveals that plans could forward antigrowth goals.[59] But Gold Coasters did, however, insist that planning occur in service of their privatism and localism and on terms that flattered their sense of importance. Demarcating and policing the boundary between city and suburb was central to this interpretation of regionalism. The NCC used regional planning theory to validate this exclusive claim to the region's best environmental amenities and neutralize the authority of planners who advocated overriding village priorities through a regional public recreation program.[60] One frustrated resident of Old Westbury admitted, however, that it was hard to sustain protests against parks because the LISPC treated such arguments as an "almost irresistible opportunity" to accuse opponents of snobbery. Moses's tendency to label anyone who disagreed with him as undemocratic obscured the fact that not all professional planners agreed with his Northern State Parkway plan.[61]

The NCC and celebrated Scottish planner Thomas Adams, director of the CRPNY, declared estate villages a valuable way to preserve low-density landscapes and a form of environmental protection. Adams celebrated the "open" development of the estate grounds as offering crucial balance to the dense "closed" development of the city's boroughs.[62] In 1925 New York City's aggregate density was thirty persons per acre, although the densest districts of Manhattan and Brooklyn reached 536.[63] In contrast, in 1930 Nassau's population density per acre was but twelve persons.[64] The CRPNY commended "open" development as a deciding factor in establishing the county's desirable low population density. The committee defined "open" development as public parks and reservations, semipublic spaces such as cemeteries and country clubs, and private estates. Replacing the green spaces of the North Shore with homes, roads, and "closed" development, Adams said, would constitute "a public misfortune."[65] Adams accepted estate villages' restrictive land-use patterns as a means to preserving open space in greater New York. The CRPNY accommodated and ultimately sanctioned the elite privacy created by the mutually reinforcing decisions of estate villages.

The CRPNY agreed with Lay that North Shore parks were unnecessary since the gracious landscaping of estates, golf links, and bridle paths constituted veritable parkland. Adams argued that "the hilly land in the north of Long Island is peculiarly adaptable for large residential properties forming what are practically private parks."[66] Parklike private estates preserved more open land than the state could ever expect to purchase for parkland and helped balance land use on the island.[67] "Wealthy citizens inclined to use their money in developing and preserving the natural landscape," a CRPNY report explained, "are creating for the Region . . . something that may be as valuable from a cultural point of view as any collection of works of art."[68] Private estates preserved beautiful landscapes at no cost to the public and indirectly contributed, the CRPNY's recreation specialist said, "to the health and enjoyment" of those living in more populous districts of Long Island.[69] Adams acknowledged that the estates were generally closed to the public and proclaimed a public value derived from the private beauty. He did not, however, acknowledge that this aesthetic was ideological — a manifestation of the Gold Coast's politics of exclusion.

Despite the battle over the Gold Coast, the CRPNY generally supported Moses's park vision and incorporated many of the LISPC's proposed parkways into its regional plan. In fact, the committee approved of the park commission's Jones Beach State Park, which it opened during the stalemate over the Northern State Parkway. Situated on a barrier island twenty miles from the city, the massive complex became the most celebrated public works project of the era.[70] The committee's fight with the LISPC over Gold Coast public spaces proved the exception, not the rule, to Moses's relationship with the committee.[71]

Although Adams pledged the support of the Russell Sage Foundation to both Moses and Governor Smith and offered to collaborate on new park plans for northern Nassau, Moses approached the CRPNY as an adversary. Adams failed to convince Moses that he offered a nonpartisan and expert opinion and that estates were a legitimate land preservation technique. Neither Moses, in his role as park planner, nor Governor Smith, with his reputation as a champion of the urban masses, forgave North Shore residents for walling off the shore and leaving the public with only "dust and dirt."[72] To Moses and Smith, estate village exclusion and antidevelopment politics were inexcusable. Across the Sound, Connecticut's state park planners waged a similar battle against defensive localism and privatism. While dismissed by planners as balkanized and parochial, private landowners in fact organized successful, large-scale challenges to Progressive Era park planning. For more than two decades park planners fought estate owners to open Sherwood Island, one of Connecticut's few state-owned beaches, to the recreating public.

*Sherwood Island and the Connecticut Shore*

In 1914 Albert M. Turner identified 230-acre Sherwood Island as the best site for a state beach in Fairfield County, setting the stage for a showdown between the Connecticut State Park Commission (CSPC) and nearby estate owners who recoiled at the idea of thronging crowds in their midst. Turner knew the Connecticut shore intimately. He knew the rocky beaches of the narrow southwestern Sound, the modest sand dunes of its Rhode Island border, and the omnipresent pungent mud of its salt marshes at low tide. For three months in 1914, he hiked the coast from New York to Rhode Island. Hired by the newly minted CSPC, Turner surveyed the state's 245-mile coastline for a large, scenic beach well removed from the polluted industrial ports of New Haven and Bridgeport. Turner's report, one of the first state park surveys in America, became a foundational document of American state park ideology.[73] Of the miles Turner walked, approximately forty-five were inside city or borough limits, including roughly six miles of city parks. Turner found seventy miles of the shore tightly packed with private beach cottages and an additional forty miles of large, costly residences. Only ninety miles remained available for state beaches.[74]

Connecticut's park commissioners prioritized a state beach program because the majority of the state's population lived in the state's somewhat narrow and rolling coastal plain.[75] In 1914, the state's average density was 231 persons per square mile, but along the Sound this ratio reached 529 persons per square mile. Fairfield County's coastal commuter corridor along the New York, New Haven, and Hartford Railroad (NYNH&H) already housed a quarter of the state's more than one million residents.

"From the date of the first meeting of the Commission it has been plainly evident that the field most urgently demanding attention," the CSPC said, was the "shore of Long Island Sound. Its popularity for purposes of recreation is almost universal, there can never be any more of it, and the rapid development of the last two decades has left very little of it accessible to the public."[76] The CSPC wanted to open five evenly spaced 2.5-mile-long beaches. Turner identified 230-acre Sherwood Island in Westport, a former farmers' collective and tide mill site, as the only potential state beach in Fairfield County, the state's western corner and the eastern end of the metropolitan corridor developing along the Sound.

Turner faced significant obstacles in his search for a state beach site. First, state law privileged private use of the beach over common public use. Connecticut allowed owners of upland property to use the foreshore for docks and other purposes without specific grants.[77] Establishing the high-water line as the public-private boundary provided for the public status of the beach. Yet Connecticut's law of land under water, found entirely in state courts' decisions, favored riparian owners over public claims to access. Connecticut courts furthermore ignored recreation as part of the public's right to use tidelands and limited these public rights to only unobstructed access for navigation.[78]

A second problem was skyrocketing property values. As late as 1898, Connecticut's beaches had often been included free in the sale of adjacent property or priced by acre. By the 1910s beachfront was priced more expensively by foot; land that earlier sold for four hundred to one thousand dollars an acre now sold for fifteen to forty dollars a foot, or three thousand to ten thousand dollars an acre.[79] Real estate values threatened preservation efforts. "Natural scenic beauty and the unrestricted private ownership of land are things apart, and quite incompatible," Turner concluded. "The small landowner fairly clogs the landscape with his wooden dreams, and the big one walls it up."[80] He acknowledged that "to the fortunate few who may have a country house or a shore cottage with an automobile or so," a public beach was unnecessary. Without state beaches, however, Turner worried the majority of the public would soon face only "the dusty highway and the No Trespass sign."[81]

Turner declared that Connecticut had a responsibility to protect the shoreline from privatization. To convince the CSPC, Turner evoked the memory of P. T. Barnum. He referred to the statue of Barnum in Seaside Park, the popular park Barnum personally built in the heart of Bridgeport, to garner support for a public beach. "There in his armchair, watching the rising and falling tides and the passing of the generations, sits the man

**5.5.** Connecticut's privatized beaches. This photograph, which the field secretary included in his first report to the CSPC, captures the privatization that so worried Turner. The sign says "$10 Reward for the correction of any person caught trespassing upon these primises [*sic*] day or night with or without dog or gun." Turner captioned the image "One good reason for State Parks." *Source*: Turner, "Report of Field Secretary," in *State of Connecticut Public Document No. 60, Report of the State Park Commission to the General Assembly for the Fiscal Year Ended September 30, 1914* (Hartford: Hartford Printing, 1914), 55.

who can best answer the question 'What is it for?'"[82] For Turner, Seaside encapsulated Barnum's prescient park planning and his insistence that Bridgeporters deserved an accessible shore in developing Fairfield County. Furthermore, Turner's vision of the public was not limited to Connecticut. The state was lucky, he said, to have a waterfront: only seventeen did. Of these, only four had a shoreline proportionally equal to Connecticut's coast. "We are, in a sense," Turner told the state, "trustees for those less fortunate States."[83]

Turner advocated for a collective public and included recreation in his definition of the public good in 1914, nearly a decade before these ideas gained nationwide credence or animated regional park planning in Westchester and on Long Island. At the end of World War I, fifty-five years after the establishment of the nation's first state park at Yosemite, two-thirds of the states lacked state parks. Not one had successfully developed a comprehensive state park system.[84] Turner envisioned Connecticut as a leader in park planning. Yet his philosophy that the state had a responsibility to protect public beach use from privatization proved unpopular in Greens Farms. Estate owners defined access to the shore as a local amenity and took a stand against use by the larger public. In 1914, Turner reported that most of Connecticut's tidelands had already passed into private hands and was "more jealously guarded each year as its value increases."[85] His fear proved a reality in Greens Farms, where park opponents successfully stalled state beach development for nearly a quarter century.

In the late 1800s, Westport, home to four thousand people, developed as a summer resort for New Yorkers fifty miles from the city on the NYNH&H. Business magnates settled in the Greens Farms section of town around Sherwood Island. Tycoons such as Edward T. Bedford, an associate of John D. Rockefeller, transformed former colonial farms into palatial summer estates—Bedford became Westport's largest taxpayer—and forwarded a powerful vision of privileged localism.[86] The CSPC hoped to acquire all of Sherwood Island. The keystone parcel of its plan was the high ground between the mill pond on the island's western edge and the NYNH&H tracks. In 1921, however, George W. Gair, an executive of one of New York City's most important and wealthiest box and paper goods firms, purchased the site. Gair moved to Greens Farms knowing that the state intended to create a park, but his rising local influence empowered him to challenge the plan.[87] Gair acquired large sway in local politics because of the half-million dollars he paid in yearly taxes and his appointment as chairman of Westport's board of finance in the 1920s. Determined to preserve the seclusion of his new estate, he organized an antipark constituency of powerful Republican estate owners, including his neighbor Edward T. Bedford.[88] Bedford lived just north of the railroad and owned multiple properties in Greens Farms. His Italianate villa and garden overlooking the Sound was a popular town landmark. Park opponents lived in close proximity to each other and Sherwood Island; they viewed the members of the public who lived outside of Greens Farms' socioeconomic, spatial, and jurisdictional boundaries as invaders.

Based on Turner's suggestion, the CSPC first purchased land on Sherwood Island in 1914. By the close of 1917, however, the commission owned just 30 piecemeal acres separated by marsh. Only visitors willing to ford New Creek at low tide could enjoy state-owned Alvord Beach on the southeastern side of the island, a narrow strip of sand backed by marsh. The lack of signage, accessible roads, or clearly marked parking made locating public holdings akin, an early visitor complained to a local newspaper, to

searching for "a light hid under a bushel."[89] The narrowness of Sherwood
Island Lane, the sole approach road, made it impossible for a large car to
turn around or park except by using private driveways. Out-of-towners reported that when they asked for directions they received stony stares and
silence, or else locals feigned ignorance and sent interested parties to town
hall for further inquiries.[90] Because of these limitations, the CSPC considered the park useless until further development.

Greens Farms residents saw a public park as incompatible with their
understanding of property rights and their patrician community. "Home
sanity," Gair claimed, was at risk because of out-of-towners, "with the
usual 'don't-care-a-damn' spirit for the locality," who, he predicted, "would
change Greens Farms, with all its unique charm and quiet home life, into a
Coney Island and kill many places like mine."[91] Green Farms elites formu-
lated an antipark protest based on their rights as property owners. Property
law and a substantial body of American jurisprudence required the state
to police the boundaries of private property and to guard against potential
transgressions.[92] The prestigious Sturges family complained that visitors
violated this boundary by trespassing and building fires on their beach and
lawn on Pine Creek.[93]

George Gair issued a call to arms to preserve the sanctity of the private,
domestic waterfront of Greens Farms. The elitist localism of Greens Farms
prefigured the defensive localism that typified the conservative white
homeowners' movement mobilized around racial, ethnic, and class homogeneity after World War II.[94] Decades earlier, Gair and his cohort conceptualized property rights as including a right to exclude and fight external
interventions. Greens Farms residents demanded neighborhood control of
the shore. Property owners' right to exclude ignored the Public Trust Doctrine and overshadowed the traditional range of rural actors who used the
shore. Greens Farms elites ignored traditional local interests who shared
the beach, such as oystermen, tide mill operators, onion shippers, and a
farmers' cooperative. Estate owners espoused a community identity and
a concept of privacy that diverged from inclusive, regional perspectives on
town resources and ignored traditional coastal land use.[95]

Following Sherwood Island State Park's establishment in 1914, Westport's town clerk predicted that nearby landowners "would try to put the
State in an embarrassing position" by holding up the purchase or trans-
ferring or developing the property.[96] He was right. Arthur Sherwood restricted his thirty-nine acres to residential use through covenants, while
other antipark allies engaged in a flurry of real estate transactions to subdivide nearly two thousand feet of shore, including the area informally used

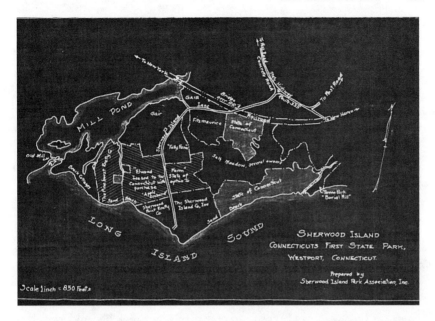

**5.6.** Sherwood Island Park Association properties map of Sherwood Island. While this map was commissioned in 1932, these private parcels were created in the 1920s scramble to block the state park. *Source*: Sherwood Island Park Association, Properties Map of Sherwood Island, Folder 3, Box VII, William H. Burr Jr. Papers pertaining to Sherwood Island State Park. Courtesy of the Fairfield Museum and History Center, Fairfield, Connecticut.

by recreationalists as parking.[97] Between August and December 1924, three real estate companies incorporated, bought Sherwood's land, and platted the center of the island for restricted residential use. These subdivisions, combined with the nearly fifty acres Bedford sold to the Gair family, secured almost one hundred acres from state reach.[98]

Sherwood Island residents doggedly policed property boundaries to prevent the few excursionists who found the public beach from spilling onto private property. They installed concrete walls and high wooden fences. Frustrated visitors found "most of the beach from high to low water mark fenced off in the interest of private owners, with a high barrier of railroad ties and guarded by a big yellow dog, unmuzzled."[99] One excursionist angrily concluded that the typical Sherwood Island resident "has the advantage, due to greater wealth, of being able to own land bordering the beach and thereby thinking he is owner of the beach and the water in front of his property."[100] Estate owners effectively, in the words of the *Bridgeport Sunday Post*, "claim[ed] the foreshores for their own."[101]

Greens Farms did not stop at intimidation and physical barriers to isolate state-owned Alvord Beach. Elite localism functioned as a defensive political strategy. Gair and his like-minded neighbors mobilized home-rule autonomy and vowed to protect local prerogatives on the assumption that state government could not know what was best for a locality. At an October 27, 1924, town meeting, Gair fathered and ushered through an aggressive antipark resolution: "Resolved: That the town of Westport does not desire the state of Connecticut to acquire additional land at Sherwood Island for park purposes. Resolved further that the town of Westport does not desire a state park at Sherwood Island. Resolved, that the representatives from this town to the next General Assembly do their best to prevent an appropriation for any such purpose."[102]

Greens Farms localism became municipal policy through this resolution and in the town's 1929 dredging of New Creek. Gair and cottage tenants on the island's western tip had complained that the mud and stagnant water of New Creek at low tide bred swarms of insects. In the summer of 1929, Westport's selectmen approved and contracted the dredging of the creek on the grounds of mosquito control. New Creek became a twelve-foot channel extending 450 feet.[103] Since fording the creek was the only way to reach Alvord Beach, the dredging effectively disrupted access to the state-owned waterfront. Outraged park supporters castigated Westport and "Gair's ditch" as a blatant attempt to thwart public access to the beach. Demonstrators from inland Redding, Ridgefield, and New Canaan Hill forded the creek to protest the physical and political barriers erected at Sherwood Island.[104] Nothing came of the protest, and it became even more difficult to gain access to the beach.

In February 1931, nearly one thousand people packed one of the largest and most contentious town meetings in Westport's history to discuss renewing the village's antipark platform. The state parks commission had owned Sherwood Island in the Greens Farms section of town since 1914. The "royal families of Greens Farms" stood accused of barring the general public from the Long Island Sound beach.[105] The essence of the disagreement, according to the local press, distilled to "whether the beauties of nature belong to the public or to millionaires." Cries of "shut up," "throw him out," and "we want to vote by ballot" punctuated the two-hour debate. Eventually the moderator declared the tumultuous crowd too unruly for a vote and adjourned the meeting.[106] Park supporters had to wait until the 1932 elections to renew their demands that private preferences yield to public interest.

Greens Farms elites not only aligned local politics with personal interests but also established their community's privileged localism in the state legislature to wield private power over public property. Gair's Greens Farms constituency enjoyed substantial influence in the state Republican Party. Comptroller Frederick M. Salmon, the chief fiscal administrator of state accounting from 1923 to 1932, and fellow Republicans in control of the appropriations committee of the general assembly allied with park opponents.[107] By the early twentieth century the Republican Party, drawing votes from white, Protestant suburbanites and rural residents, controlled the state legislature.[108] Party representatives in the state assembly honored Westport's antipark resolution and consistently denied the CSPC the funds necessary to finish purchasing land at Sherwood Island. In 1916 the CSPC had requested a four-to-five-million-dollar bond, supplemented by appropriations to cover interest and sinking fund charges.[109] The state declined to issue a bond in that year — and for the next two decades.[110] In the midtwenties the commission again lobbied for a bond similar to the one approved by New Yorkers, but its argument that a long-term, forward-thinking bond would save Connecticut taxpayers from the growing costs of annual delays in land acquisition fell on deaf ears.[111] By 1924 the CSPC had held 139 meetings and submitted plans to five successive sessions of the General Assembly without successfully securing additional financial support to finish Sherwood Island State Park.[112] The commission requested funds for the same projects over and over without success. The substantial progress at Hammonasset Beach State Park in New Haven County underscored the state's persistent evasion of funding for Sherwood Island.

By 1924, three years after its opening, the state had spent $130,960 for 565 acres and the construction of first-aid and lifeguard stations and a fourteen-hundred-locker bathhouse at Hammonasset and only $12,959 for the acquisition of forty-eight noncontiguous acres at Sherwood Island. The CSPC was forced to admit that compared with Hammonasset, its Westport beach was a failure.[113] In 1923, park commissioner George A. Parker had in fact resigned to protest the legislature's inaction. William H. Burr, one of the only Greens Farms residents who supported the park, commented on the obstructionism and subsequent resignation to Turner: "Sorry Mr. Parker resigned, but now we know what we are up against."[114]

Having successfully blocked the completion of Sherwood Island State Park for a decade and a half, in 1931 Republicans in Hartford went as far as attempting to totally disempower the state park commission. In that year, the Republican-controlled general assembly created a special subcommittee to relieve the park commission (renamed the State Park and Forest Commission [SPFC] in 1921) of its control of parks in Fairfield County.[115]

The subcommittee toured Sherwood Island under the care of Gair and Westport's first selectman King W. Mansfield, a proponent of beach relocation, and subsequently recommended abandoning the territory in favor of a new park at Roton Point, Norwalk. Protest erupted in Fairfield's cities and from its planning organizations. The editor of the *Bridgeport Post* condemned the state for considering the seven-hundred-thousand-dollar project and abandoning Sherwood Island property that it had owned for seventeen years. The president of the Fairfield County Planning Association accused the subcommittee of neither visiting Roton Point nor appraising the land before issuing a "flabby" and "amorphous" report that served only Greens Farms' antipark agenda. Because of the uproar, the bill was recalled.[116] Nevertheless, that the general assembly created such a subcommittee underscores the power of Republican Greens Farms elites to block state-sponsored regionalism.[117]

In the 1920s, the state legislature acknowledged Westport's obstructionism and its own "policy of inaction."[118] This inaction, secured by Republican interests, was exacerbated by the fact that Connecticut counties lacked the power to mediate between state and local politics. The state's New England town tradition situated governmental authority in local jurisdictions. When Connecticut was founded in 1636, a system of town government was established for the colony, but county government was not created until the 1660s. Town government remained the source of political power and decision-making authority in the state. As a secondary form of government, the county lacked the power to level direct taxes, the authority to create its own budget, and a chief executive who could forward a regional development program.[119] Law professor Albert Levitt accused Westport town officials of violating the constitution by denying the public use of parkland that was paid for with tax dollars for the benefit of the entire Connecticut population.[120] Under the federal Public Trust Doctrine, Connecticut held the beach below the high-water line on behalf of the public but had failed to uphold the trust. Inland residents were left with no effective county authority to demand recourse; they could do little more than angrily write letters to the editor condemning the state's pandering to the "guard of New York commuters" to keep the "common herd from the back towns" off the beach.[121]

In rejecting the idea of a regional public, Greens Farms localism exacerbated the park crisis in Fairfield County that had inspired the SPFC's plan for Sherwood Island. In the 1920s Fairfield failed to meet the ideal of an acre of park for every one hundred persons, a ratio identified by the planners of the *Regional Plan of New York and Its Environs*, and lagged behind its neighbors. Southwestern Fairfield County provided one acre of parks for every 332 residents, while the Bronx provided one acre per every 209 people and Westchester provided a stunning one acre for every twenty-eight.[122] In Fairfield County's cities the situation was even worse. Bridgeport, for example, possessed only one park acre for every 401 people, while Stamford had but one acre of park for every 508.[123]

As an idealized category, a region is defined by a shared geography and structural political and economic processes within which residents find commonalities in political, economic, and social trends. Greens Farms localism, however, revealed the extent to which residents drew internal boundaries that segmented the region. To residents, the regionalism that drove state planning was synonymous with public access and outsider control. Both were unwanted. Greens Farms fought these dual aspects of regionalism—the influence of outsider state planners and a broadly defined recreational public. For example, Frederick M. Salmon, a Westport state representative, did not believe his hometown had any responsibility to provide regional parks. He dismissed park support exactly because it emanated from Norwalk and Bridgeport. "Let those cities clean their own polluted harbors," Salmon said, "and they won't have to depend on Westport for clean bathing waters."[124] Isolationism led residents to deny any responsibility to address the recreation needs of nearby cities.

For more than two decades the state failed to guarantee the rights of the public in the face of privileged localism at Greens Farms. In the words of field secretary Turner, "the constantly increasing burden" of beach acquisition and development had "been persistently evaded."[125] Planners in greater New York, led by the SPFC's Turner, forwarded the idea that Connecticut had a responsibility to provide beaches to the region and called for a regional perspective that locals rejected. Gair's cohort both protested this regionalism and articulated an alternative vision. Their privileged strain of localism bolstered this fantasy by rejecting a regional conception that linked their enclave with industrializing centers like Bridgeport and the right of a broad public to recreate alongshore even though the state held Sherwood Island in trust for the public. Planners predicted Fairfield would become a county of large cities woven together by intensive suburban development. "What Westchester County is today Fairfield County will be tomorrow," the Fairfield County Planning Association urged. "What Bridgeport is today, the other cities of the County will be tomorrow."[126] Prosperous Bridgeport had the state's largest park system, overburdened as it was; suburbanizing Westchester boasted the nation's most celebrated comprehensive county park system. Each was an example of growth and successful park planning. Fairfield's planning association subscribed to Turner's vision, but Greens Farms rejected beach access as a general public right as well as regional planners' vision for the coastal corridor.[127]

## The Limits of Regionalism

The SPFC and the LISPC failed to convince property owners in greater New York's elite suburbs that attempts to isolate themselves were best served by large regional parks to contain urban and suburban recreationalists. Instead, elite estate owners in Greens Farms and on the North Shore fought state efforts to develop regional public recreation amenities. Because of the decades of antipark fights, neither the eventual creation of Sherwood Island State Park in Connecticut nor the LISPC park and parkway plan were entirely successful endeavors for their park commissions.

In 1929, after four years of rancorous negotiations with Gold Coast barons and the CRPNY, the LISPC acquiesced to a five-mile detour to reroute the Northern State Parkway around Wheatley Hills. Robert W. DeForest and Otto Kahn both influenced the final location of the parkway. Kahn shifted the parkway route and preserved his private golf estate by donating $10,000 to the LISPC. Since this rerouting put the parkway on DeForest's land, DeForest subsequently dedicated fifty acres to shift the parkway a second time so it ran on the southern edge of his holdings.[128] Moses found DeForest's role in the negotiations galling, since he was president of the Russell Sage Foundation and a member of the CRPNY. While Moses accused him of the "unconscionable use of his position" to block the parkway, historian David Johnson argues that the foundation participated at most indirectly in the fight for special privileges in the era's regional planning.[129] DeForest's personal interests notwithstanding, the Northern State Parkway still proved the limits of the LISPC and Moses's influence. The LISPC never established a state beach in the region until the 1970s.[130]

State park commissioners in Connecticut similarly faced setbacks. While the SPFC was aware of the limits of its power, the Great Depression brought additional challenges.[131] In 1936 and 1938 the commission sought, without success, a bond issue for the acquisition and development of state parks. In fact, the legislature refused to appropriate the nearly half million dollars necessary to complete land acquisitions at Sherwood until 1937, following the Democrats' capture of the state government during the Depression. The delay forced the state to spend large sums to purchase properties that had been subdivided expressly to block the park.[132]

Both park advocates and park opponents expressed dissatisfaction over Connecticut and Long Island state park planning. Most looked with envy to Westchester's county-level planning, but for different reasons. In a 1925 pamphlet, the NCC railed "Westchester County has Home rule in Park Matters. Why should Nassau be exploited by the State?"[133] In contrast, the committee considered the LISPC to be an invasive foreign agency uninterested in local needs. The LISPC also set Westchester as a model but only celebrated the county's powerful park commission and not home rule. The LISPC focused on the county commission's success overriding localism, such as Rye's unsuccessful efforts to restrict parking on local roads. In Connecticut, William H. Burr, one of the only local supporters of Sherwood Island State Park, also praised the Westchester County Park Commission's (WCPC) progressive work as part of his advocacy for the park. Progress, Burr said, should not stop at the state border; Connecticut needed to both finish Sherwood Island State Park and coordinate with New York. Such work would build on the WCPC's success to realize a truly regional park system.[134]

The potent alignment of property-rights principles, localism, and environmental values rooted in private space empowered opposition to external intervention. Elites in both Greens Farms and on the North Shore created an alternative vision of governance that served private values. State, county, and local government often competed to exert control over public land use. Traditions of decentralized government empowered home-rule challenges to regionalism. The resulting battles led to three important and mutually reinforcing lessons. First, park obstruction underscored the exclusionism inherent in these communities. In Greens Farms, localism meant private consumption of the shore; on the Gold Coast, it meant collective consumption by a narrowly defined community. Such a vision of the public was particularly narrow given the extent to which high property values limited entry to the Gold Coast. Second, the failure of state government to ensure public access to the shore in both places reinforced localism. Finally, home rulers rejected state park commissions as foreign invaders and endeavored to disable states' regional planning powers. Local government was an essential player in the rapidly evolving and uneasy relationship between the urban center and its periphery.

# 6

# "From Dumps to Glory"

## Flushing Meadows and the
## New York World's Fair of 1939–1940

For decades the Brooklyn Ash Removal Company dumped refuse across Flushing Meadows in north-central Queens. In the early 1900s, the company turned the marsh adjoining Flushing Bay into a mile-wide, three-mile-long eyesore. Smoldering refuse burned brightly at night, and sooty clouds frequently shrouded the area. The company's trash barges and rail cars lined the bay. Additional railroad yards and automobile junk shops littered the shore of Flushing Creek. In 1925, F. Scott Fitzgerald introduced the dump to the nation, describing the meadows as a "desolate area of land . . . where ashes grow like wheat into ridges and hills and grotesque gardens."[1] New Yorkers already knew the dump because they suffered from its stench. During the dump's operation, a critic claimed, locals traveled "through that part of Flushing in a car or by train [by] holding [their] nose and praying that the ordeal would soon be over. It was a dreary gray expanse of ashes, junk, and smoldering refuse."[2] Fitzgerald's "valley of ashes" was the city's most conspicuous wasteland, a noxious chasm cleaving western Queens from the celebrated beauty of Nassau County's Gold Coast.[3]

On a late June day in 1936, city officials and reporters braved boggy ground to gather in Brooklyn Ash's dump for the launch of a redevelopment program. With Mayor Fiorello H. LaGuardia at the throttle, a steam shovel clawed into a ten-foot-tall butte of ash. The attendees and cameramen who crowded the abutting hills endured odors emanating from three decades worth of ripening excavated trash. The highest mound, nicknamed Mount Corona, towered nearly one hundred feet above the crowd, but not for long. The year 1939 promised the "Dawn of a New Day" at the world's fair slated for the site.[4]

**6.1.** Robert Moses included this image from the parks department files, which looks north across Flushing Meadows, with the caption "Valley of the Ashes" in *The Saga of Flushing Meadows* "*Fairgrounds—Pre-Construction—Aerial view of site*." *Source*: Manuscripts and Archives Division, New York Public Library Digital Collections.

The Queens's world's fair was a touchstone that united the regionalization of the city, local ambitions of the outer boroughs, and suburban decentralization with contemporary urban design theory. Fair officials frequently made this point and declared that construction of "the Fair itself contribute[d] to the building of the World of Tomorrow." The New York World's Fair corporation took on the work of not only building a fair but building wholesale an urban infrastructure system on an area equivalent to more than two hundred city blocks. Fair construction joined the environmental infrastructure of parkland with the technological infrastructure of highways and utility systems. The corporation boasted that "in its scope, its implications and its results," the fair was a great public works project that would secure the city's position at the forefront of modern metropolitan planning crystallizing in greater New York.[5] Viewed comparatively, fair construction and the World of Tomorrow's most significant exhibits on the futurist utopian city emerge as complementary narratives of peripheral land use and professionally orchestrated regional development to shape the modern metropolis.

In the 1930s, Depression doldrums trapped the nation in economic uncertainty. New York City business magnates and civic leaders, however, looked to reinvigorate national morale with a world's fair. In contrast to the dreary present, a fair could offer a bright, hopeful vision of America's future. It would also crucially bring investment and jobs back to a city where, by 1935, one of every three people gainfully employed in 1930 was out of work.[6] George McAneny approached parks commissioner Robert Moses to discuss a location for the fair. As former Manhattan borough president and head of the Regional Plan Association (RPA), the successor to the Committee on the Regional Plan of New York (CRPNY) and an advocacy organization for its *Regional Plan of New York and Its Environs*, McAneny had long advocated for professional planning and ambitious public works. Moses agreed with this approach to city building. He seized on the public-private venture as the means to fund Flushing Meadows development, a site he first identified as a potential park in the 1930 Metropolitan Conference on Parks.[7] Having dedicated the city parks department to modernizing the waterfront, the fair would further empower Moses to unite shoreline development with his infrastructure and public-space programs.

A cohort of influential New Yorkers with corporate roles and government appointments organized the powerful nonprofit New York World's Fair 1939 Inc. in October 1935 to build the fair.[8] Members included leaders from Chase Manhattan Bank and Consolidated Edison; influential politicians such as John J. Dunnigan, majority leader of the state senate; and New Deal administrators such as Harold M. Lewis, former CRPNY member and coordinator of the state's Public Works Administration.[9] The redevelopment of Flushing Meadows was a unique moment of collaboration among scientists, designers, planners, city businessmen, and officials from nearly every level of government, from the Queens Bureau of Sewers to New Deal agencies. Moses became the city's point man for all work connected with the fair, a powerful role made possible by his overlapping appointments. Since 1935, he concurrently helmed the Triborough Bridge Authority, the consolidated city parks department, the Long Island State Park Commission, and the State Council of Parks.[10] The combination of fair investments and the federal relief funded the comprehensive planning agenda circulating in greater New York since the start of the century.

The 1939–1940 New York World's Fair was the largest reclamation project ever attempted to that time in the eastern United States. The US Geological Survey's Bureau of Reclamation defined *reclamation* as "regulating the water supply for a given area of land, which, under natural conditions, has an excess or deficiency of moisture" that diminished land values. The term and the bureau's work were generally associated with irrigation of arid lands, but in greater New York the term was used in its second meaning concerning drainage and keeping waters off overflow lands or swamps. Coastal land making had a long history in New York City; since the colonial era urban growth had meant shoreline development. The popularization of the idea of reclamation potently coincided with the city's need to expand and modernize its garbage landfill operations in the 1930s. When advocating both park creation and tideland fill, Moses frequently used the word reclamation, often in conjunction with conservation, and at times interchanging it with restoration.[11] Despite Moses's shifting vocabulary, coastal reclamation did not attempt to restore an untouched or "natural" state. The Long Island State Park Commission (LISPC), the city parks department, and the fair corporation approached reclamation as the work of reshaping coastal nature and effecting control over it. Dumps and piecemeal sewerage infrastructure, natural drainage patterns and topography, open mudflats and salt hay fields would all be reshaped. Greater New York's planners continued to imbue nature with values concerning the social makeup, behavior, and health of the city's recreating public. Their vision of what a "reclaimed" coastal environment should look like focused on clean, modern, public space.

The fair site captures the key components of the large-scale reshaping of New York City in the 1930s. When planners became municipal figures in the twentieth century, they elevated their values and methodology, securing sweeping powers and funding. The environment, public works infrastructure of bridges, highways and parks, and increasing suburbanization tied the urban core and periphery together and gave shape to the region. The fair brought automobility and technological rationalism as well as a new, man-made nature to Queens. In 1932, preeminent city planner John Nolen summarized his profession's vision of metropolitan growth when he explained, "the future city will be spread out, it will be regional," the "natural product" of modern roads and the automobile.[12] The fair spurred a new regional approach to coastal reclamation planned through public-private cooperation in advent of the "City of Tomorrow."

## The Future at Flushing Meadows

The New York World's Fair opened on April 30, 1939, the 150th anniversary of George Washington's presidential inauguration on Wall Street. A statue of the president stood on the site's central mall, but beyond the anniversary opening and this towering plaster figure, the fair paid little attention to the past. Instead the fair looked toward the future. Visitors arriving by the newly extended subway disembarked onto a wide wooden boardwalk lined with flags that led to entrance gates. "If under the flags, close to the railing, you stop and look down," one observant visitor explained, "you will see a remnant of the original marsh, thick slime and green springing grass, from which all else has been at one time redeemed. 'From Dumps to Glory' is the phrase of the guidebook. Before me lies the World of Tomorrow."[13]

The New York World's Fair 1939 Inc. funded the project, but industrial designers, not businessmen, were the true planners. As the last and largest of America's six Depression-era world's fairs, the World of Tomorrow stood out for its synthesis of regional planning, futurist industrial design, and new consumer goods. Stephen A. Voorhees, president of the American Institute of Architects, helmed the board of design for the World of Tomorrow, which included a second industrial designer, Walter Dorwin Teague, as well as architect Robert D. Kohn, a member of the reform-focused Regional Planning Association of America (RPAA). Gilmore D. Clarke, Jay Downer, and architect William A. Delano rounded out the board. Because of this leadership, the exposition functioned as a brainstorming platform for the nation's leading industrial designers, who looked to expand their profession into the realm of social engineering.[14]

The fair was an experiment in modernist design and planning. Seven thematic sectors made up the fairgrounds: amusement, communications, community interests, food, government, production and distribution, and transportation. The fair corporation claimed that the sector zoning, which integrated architecture, color, and landscaping, saved visitors from experiencing a "hodge podge of unrelated and confusing impressions," presenting instead "a simple story, clearly told."[15] Using the language of zoning, the fair's designers evoked contemporary planning theories on scientifically regulated land use. The fair's industrial designers' great faith in total urban and environmental planning was similarly captured in the exposition's most popular exhibits, Democracity and Futurama.

The fair revolved thematically and spatially around the Trylon, a six hundred-foot-tall obelisk, and Perisphere, a giant globe. Both were the official symbols of the World of Tomorrow. The Perisphere housed the Theme Center, distinctive from most fair exhibits in that it did not house a visiting nation or corporation; the exhibit represented the official vision of the board of design. The Theme Center presented the RPAA's garden city ideology in Democracity, a massive diorama, and the film *The City*. A comedic segment of the film depicted an exodus of weekend travelers clogging highways for excursions that were neither restful nor scenic. Motorists find only billboards, weeds, trash, and gaudy roadside attractions. Caught in unending traffic, frustrated drivers honk their horns. The urbanite in search of nature was forced to picnic roadside, hemmed in by walls; a small girl cries as she looks at a "no trespassing" sign reminiscent of the Gold Coast. The film, its narration written by Lewis Mumford, a prominent theorist on cities and urban architecture, promised to save urbanites from this fate via the healthy environment, sociability, and sense of community of planned suburbs.[16] The Perisphere offered a second dramatization of the benefits of regionally planned garden cities. From revolving balconies suspended inside the globe, visitors looked down on Democracity, the metropolis of 2039. The diorama's urban core was devoted to business and culture. The model city's 250,000 workers commuted daily from beyond the greenbelt that bounded the city. Democracity's designer Henry Dreyfuss, a celebrity industrial designer of the era, explained the exhibit as a city "built in greenery, with a perfect traffic system" and strictly controlled population distribution in mill towns and residential suburbs. Farms ringed the periphery of these development nodes.[17] The Theme Center emphasized that without planning suburbs, whether for the daily commuter or weekend excursionist, the city would succumb to chaotic, unchecked urbanization.

At General Motor's Highways and Horizons pavilion, Norman Bel Geddes's Futurama presented a remarkably similar if corporate-sponsored vision of the future. The large-scale diorama simulated the experience of traveling by airplane through the world of 1960. Bel Geddes pointed to the work of the Bronx River Parkway's engineer Jay Downer, the person responsible for eliminating grade crossings and preventing unattractive roadside encroachments, as an antecedent to the exhibit.[18] Employing the argument first popularized by Westchesterites to celebrate the parkway, Bel Geddes claimed expressways guaranteed increased commerce and land values and thus benefited communities.[19]

Futurama and Democracity presented a planner's vision of the futurist city in which the urban core and its surrounding area formed an integrated whole. While based on differing chronological scales—Futurama predicted America in thirty years, while Democracity presented a far-off future—the exhibits shared defining characteristics of bounded urbanism, large, regulated highways, and suburban decentralization. The General Motors film *To New Horizons* accompanied Futurama to explain that because of comprehensive regional planning, fresh air and green parkways characterized both the center city and the surrounding residential districts. Democracity similarly boasted that "air-light-movement" could be successfully obtained "in a planned city."[20]

Futurama and Democracity rejected the unplanned and patchwork development that characterized the urban peripheries of American cities. In both, explosive urban growth had become a thing of the past. Bel Geddes focused on the rebuilt metropolitan periphery and hinterland by 1960. Planless "fringeland"—a jumbled landscape of shabby realty and marginal farms with streams polluted by "outlying factories, auto graveyards, dumps, and the roadside shanties"—had once marked the city edges. In the future, however, the edge and city center would be "entirely replanned together— the first built up, and the second built down."[21] The planned suburban community could offer a new hybrid landscape on the metropolitan periphery. While the fair is most often characterized in terms of its futurist theme, the redevelopment of the Flushing Meadows fair site was grounded in the very real, contemporary planning forces of Progressive land-use planning, urban infrastructure systems, and federal investment in civil engineering experiments. In fact, the comprehensive redevelopment of the urban borderland that Democracity and Futurama championed mirrored the work done on Queens's waterfront and across the regional city in preparation for the fair.


*Corona Dump*


The World of Tomorrow was the finale in a series of twentieth-century transformations of Flushing Meadows. Until the first decades of the century, the immense meadow, one and a half times the size of Central Park, remained largely in its natural state. At Flushing Bay, a bar of hard material deflected tidal ingress. Silt accumulated in the former inner harbor to form the thousand-acre, U-shaped marsh. As the natural drainage basin for the surrounding highlands, the marsh remained wet, soft, and nearly impossible to build on. For more than one hundred years, farmers harvested salt hay and shellfish, traversing the boggy land by scow and light draft boat.

Strong's Causeway, a rickety series of embankment roads and bridges, was the territory's only major route. The inhospitality of the territory was captured in the 1930s in a poem as "limp miles of swamp, dump-flower, shack, / At dead end of distance and gutter: or mud-trough / Sucking at soft mouth of sound and bulrush cut."[22] The instability of the meadowlands and locals' dependence on ferries for access to Manhattan retarded urban growth in greater Flushing through the nineteenth century. The Army

Corps of Engineers dredged a channel in the bay between Rikers Island and Flushing Creek in 1881, but the channel's relatively shallow depth and the lack of docks discouraged industry.[23]

At the start of the twentieth century, a number of investors envisioned a great industrial port on Flushing Bay inspired by the successful development of Queens's East River waterfront. In 1909, prominent Long Island City developer Michael J. Degnon and the president of his Flushing Bay Improvement Company wrote the House of Representatives in Washington, DC, to solicit federal coastal improvements. The men proposed a canal between the East River and Jamaica Bay on Long Island's southern shore via Flushing Creek. While well publicized, nothing came of the proposal.[24] Through the first three decades of the century, additional potential investors proposed a range of plans—from a railroad passenger terminal to an aviation field and dirt track speedway—but like the Flushing River Canal, they, too, remained unrealized.

The most successful development in Flushing Meadows was the ash dump associated with Degnon's industrial port plan. Degnon enjoyed renown for his work on the Williamsburg Bridge, the Interborough Subway, and his Long Island City industrial park. He looked to replicate this success on Flushing Bay, but to do so he needed to stabilize the adjacent marsh with fill. He contracted with the city sanitation department and the Brooklyn Rapid Transit Company to collect Brooklyn's ashes for landfill.[25] In 1907, when the transit company's subsidiary, the Brooklyn Ash Removal Company (BARC), began transporting ash to Degnon's land, the *New York Times* complimented the work as mitigating the sanitation crisis caused by the proliferation of coal-burning furnaces in downtown Brooklyn in the late 1800s. The dump was declared a positive good. Echoing the praise attributed to landfill completed by William Steinway and anticipating the arguments Moses would use in the 1930s, the *Times* claimed Degnon would reclaim "acres of land long looked upon as worthless."[26] The expectation that humans could and should improve the environment to build real estate values was a constant during greater New York's century of coastal city building. By 1916, Degnon appeared to fulfill this promise: the construction of two miles of bulkhead and substantial landfill was underway. With the nation's entry into World War I in 1917, however, the project lost momentum, and Degnon eventually lost the property. The BARC, however, continued dumping.[27]

For twenty-six years, the BARC, under the direction of Tammany man John "Fishhooks" McCarthy, deposited an estimated fifty million cubic yards of ash across three hundred acres at the northern edge of the marsh.[28]

The territory, covered with shifting hills of ash as well as piles of rusting boilers, baby carriages, and car bodies, took on a surreal moonscape quality. As the dump grew, public opinion shifted away from the early praise bestowed by the *Times*. Dumping came to function, in the words of one observer, as "a man-made glacier . . . spreading death and destruction, depositing a moraine of ashes, the waste of civilization. While not recorded by geologists, these deposits came to rival in majesty the efforts of the Wisconsin ice [glacier] elsewhere on the island." Ash landfill destroyed the healthfulness of northern Flushing Meadows. Twenty years after the BARC opened Corona Dump, Fitzgerald declared the area "a dismal scene."[29] Excepting a municipal asphalt plant at the tidewater and a handful of industrial buildings teetering on stilts, the marsh remained largely empty beyond the dump. The only adjacent development was the Long Island Railroad's yards and Gasoline Alley under the "ash plague" that the dump cast to the west, a strip of junk yards and automobile service stations.[30]

Despite three-hundred-acre Corona Dump, locals still used and found pleasure in Flushing Meadows. For some, the dump and marsh offered profits. Truck farmers grew crops in the ash, and trappers pursued muskrats; the BARC paid ragpickers to collect paper, cloth, glass, and metals from the waste.[31] For others, the nature of the tidelands appealed. Helen Keller, a resident of Forest Hills, wrote of the affection she and Anne Sullivan, her instructor and companion, had for the meadows, fondly pointing to "its delicate beauty in the spring, its long bright summer grasses, the wild birds that flew over it on their autumn migrations." Due to the proliferation of Audubon societies, sportsmen activism, and the 1920s identification of the nation's four major waterfowl flyways and associated wetlands, the call to protect waterfowl habitat was also gaining steam.[32] Yet in the debates over Corona Dump, scenic and habitat benefits remained largely unexamined by planners, bureaucrats, or the fair's corporate interests. Each interpretation of the marsh was revealing of the community from which it came—such as well-off nature enthusiasts—and reflected different visions of what a restructured environment should look like—particularly the vision of the region's powerful planners. For example, Charles Downing Lay had written of the beauty of New England tidal marshes and acknowledged that a marsh was "a delicately balanced organism," but even he ultimately suggested marshes could be made "more usable" by reclamation via drainage.[33]

Even as some locals found value in the marsh, most nearby communities railed against the environmental degradation caused by the Corona and the Rikers Island dumps. Early dump disposal techniques were crude. While ash and street sweepings were the primary fill materials, trash sent to the Rikers Island dump, which the Department of Sanitation opened in 1895, was unsorted and contained organic waste. As head of the Commission on Street Cleaning and Waste Disposal, George Waring decried the inclusion of garbage in landfills as "unsanitary and objectionable." In 1907 his commission reported that rubbish made landfills unsightly and liable to combustion: "a fire, once started, in a fill containing rubbish burns into the mass and smolders for lack of oxygen, emitting a very disagreeable, pungent smelling smoke. These fires are difficult to extinguish."[34]

Powerful stenches from decomposing matter swept across North Beach on westerly breezes, while Bronxites suffered from the smells of rotting garbage when the wind shifted.[35] Concerned citizens of the Bronx gathered to discuss the smells emanating from the Rikers dump as early as 1894. Coastal land uses affected more than just local shorefront landowners. Judge Ernest Hall, who lived two miles inland from Rikers in the Twenty-Third Ward, testified that the stenches from the island had driven his wife from their home. A neighbor confirmed the problem, testifying that breezes from its dump woke people from slumber with "stench that filled [their] nostrils beyond endurance. . . . The silver on the door knobs, and other metallic substances in and about the houses in that section were turned and spotted. . . . If metal can be so affected, what must the effect be upon a healthy lung?"[36] Ashy smoke, the stench of rot, and hydrogen sulfide, which smells of rotten eggs and blackens doorknobs, appeared to threaten the quality of life in the city's outlying districts. Recognizing that garbage would continue to be dumped on Rikers, residents affected by the by-products of the decaying organic matter demanded that it at least be deodorized, although chemists, Bronx politicians, and Department of Sanitation officials debated the effectiveness of chemical deodorizing.[37]

Residents of the coastal periphery claimed dumps compromised local comfort and health. Breezes from Rikers upset Bronxites. The practice of using tidelands as a sink for waste at Rikers and in Flushing Meadows bothered the Queens communities of College Point to the northeast of the meadows, Corona and Corona Heights to the west, and Forest Hills to the southwest, depending on the wind's direction. Even communities up to five miles west, such as Jackson Heights and Woodside, suffered from the smoke and stink of these dumps on the bay.[38] But the distances across marsh and bay and ward boundaries separated protesting groups such as the Corona Community Council and the Flushing-United Association. As a result, objections against dumps remained localized, and coalition building did not arise to turn complaints into collective action. Lawsuits leveled against the BARC failed to block the renewal of its lease.[39]

When brought to court for alleged sanitation violations, the BARC's lawyer argued that ash fill rid the neighborhood of stagnant pools of tidewater and mosquitoes and served a common good.[40] In the twenties and thirties, the BARC embarked on a publicity drive to characterize the dump as innocuous rather than polluted. The dump's superintendent acquired rat-catching dogs and claimed that a fourteen-man firefighting force patrolled the dump seven days a week. The BARC also pointed to workers' application of large quantities of coal oil by-product, which gave off a pine smell, as offsetting odors. To further quell local opposition, in November 1930, the BARC's president announced the company would build three golf courses. The first of these, the nine-hole Corona Park Golf and Country Club, opened the following September.[41]

The BARC continued dumping operations in Flushing Meadows with but minor changes until the midthirties, when dump protests and demand for the reform of machine politics—the city paid exorbitant fees to the Tammany-run BARC—finally spurred action. On January 1, 1934, Mayor LaGuardia's fusion government ended Tammany's twenty-year monopoly on Brooklyn refuse removal and placed waste disposal under municipal operation.[42]

New York City and the New York World's Fair 1939 Inc. agreed to jointly develop Flushing Meadows. The city arranged to buy and condemn more than one thousand acres, including Corona Dump, and lease the territory to the corporation for the exposition. James J. Halleran, public works commissioner of Queens, complained that the court set unfairly low prices: he said his condemned land was worth five dollars a square foot, fifty times the eleven cents a square foot assigned. This economic assessment of coastal land captured the common Progressive Era valuation of the environment as a profitable resource, in this case in terms of urban real estate profits. Halleran hoped to turn a substantial profit on real estate made from tideland fill, an echo of frequent ambitions made in the environs of the East River–Sound corridor for nearly seventy-five years. The fair corporation's lawyer countered that the property had "never been useful for any purpose. It is swamp land . . . and, unless filled in as planned by the Park Commissioner, the only things that will grow there are cat-tails and mosquitoes."[43]

The lack of infrastructure and the pollution of the meadows led one journalist in 1924 to declare the area a formidable barrier to the further growth of Queens. Surveying the acres acquired from the BARC, city officials lamented the company dumped ash "without any planning or grading, without the slightest idea of what would happen." In the 1920s and 1930s, professional planners empowered with government positions pointed to landscapes like Flushing Meadows to insist on the importance of their work. The unregulated coastline of the city's outlying districts was a place to prove the potential of professional authority over city building. Redevelopment boosters declared that Mount Corona towered over the polluted meadowland as a "monument of indifference and careless city management" and demanded control of the unplanned periphery.[44]

Located far from New York's urban core, where land remained relatively cheap and where regulatory oversight was minimal, Flushing's marshland was more valuable as a dumping site than as developable real estate or a clean environment. Yet the marsh was the basis of both the territory's current disfiguration and its prospects for improvement. The marsh, despite Corona Dump, was also "the last great open space left anywhere near the geographical center of New York City."[45] This second perspective pointed to the ideological reframing of the meadows underway in the 1930s. This reassessment was both practical and based in idealistic planning theory. In terms of planning, this perspective redefined the wasteland as an opportunity to experiment in city building and environmental engineering. Pragmatically, coastal landfill would create new property in an island-bound city long desirous of more land. Joint intervention by the public and private sector could chart a new course for the urban periphery in the name of the World's Fair.

## A Man-Made Littoral

The powerful World's Fair coalition effectively pioneered the first comprehensive environmental cleanup in New York City. "The lease between the fair corporation and the city," park commissioner Robert Moses explained, "reflects the predominant idea of reclamation."[46] For Moses and his cohort, reclamation represented New Deal natural resource management and the drive for increased efficiency in land use, a goal that is most often associated with federal agencies such as the Tennessee Valley Authority and the US Army Corps of Engineers. "Reclaiming" the shore meant reshaping Flushing Meadows to meet the city's environmental need for clean park space. Scholarship on Moses's career in the 1930s focuses on his great arterial highway projects, but Moses and his cohort of public officials understood these projects as being as much about reclaiming the city's peripheral waterfronts as they were about the public infrastructure of parks and roads. Moses blamed "corporate and individual selfishness" and "planless and feeble government" for the city's polluted beaches and obsolete waterfront infrastructure.[47] But Moses was uniquely empowered to reestablish New York's shores as places of public leisure and automobility. The regionally scaled work done in preparation for the fair facilitated large-scale environmental reclamation.

When the plan to reclaim the marsh first gained publicity, reporters declared it a foolish project and nicknamed the site the "Mad Meadows."[48] No park-building precedent existed in the city. Pieced together from old country estates, the creation of the city's largest park, the nearly two-thousand-acre Pelham Bay Park, did not require any landscaping. Central Park's 843 acres were totally reconstructed, but the territory was solid land. Moses himself recognized the challenge of transforming Flushing Meadows and Corona Dump into a modern park. He previously attempted to reshape the meadowland and Flushing Bay coastline through a road project. In the mid-1930s, he used Triborough Bridge Authority (TBA) funds to extend the Grand Central Parkway west as a bridge approach road. Patterned after the Port of New York Authority, the TBA was a stand-alone government agency with an operating budget independent from state or local government, a way to sidestep city politicians' oversight as well as municipal credit limits. Empowered to issue bonds secured by tolls, the agency built, operated, and maintained facilities through toll receipts. While the road ran directly through the Corona Park Golf and Country Club grounds, Moses declared that this brief visual and olfactory oasis offered only "the pathetic beauty and fragility of a single rose in a dung heap."[49] Nonetheless, the Depression-strapped city refused to fulfill his request for additional funds to reconstruct the rest of the marsh abutting the parkway. The publicity and the funding afforded by the fair would make possible, in Moses's words, "the long awaited dream of a clean, unpolluted waterfront in the vicinity of Flushing Bay."[50]

The scale and complexity of the project also encouraged the collaboration that came to characterize the entire organizational partnership. The division of labor and overlapping jurisdictional appointments of the fair builders facilitated cooperation. Parks commissioner Moses wrote the four-year lease between the city and fair corporation. The lease gave the parks department, and thus Moses, the power to vet fair construction.[51] General superintendent of the parks department W. Earle Andrews's appointment as general manager of the fair further insured cooperation. In December 1936, the fair corporation created a construction department. Colonel John P. Hogan of the Army Corps of Engineers was named director, overseeing site design, soil engineering, and foundation load testing. Hogan supervised thousands of federal, state, and city engineering and

construction workers, from the corporation employees to private consul-
tants from the city's leading research institutions and engineering firms.
The city organized a parallel supervisory committee that brought together
engineers from the state-run Department of Public Works, the LISPC, and
the local Queens Topographical Bureau. Because of this cooperation, the
project's scale, and innovative land making, the fair corporation declared
the construction of the World of Tomorrow to be nothing less than "a romantic saga of modern engineering."[52]

The successful cleanup of Flushing Bay hinged on the ability of sanitary
experts and civil engineers to reclaim both polluted waterways and land. In
1933, sanitary expert Dr. George A. Soper, a former member of the Metropolitan Sewerage Commission, declared water supply (both drinking water
and waste water disposal), refuse disposal, and smoke abatement to be the
three "great sanitation questions" confronting the city.[53] All three concerns
plagued northern Queens and the upper East River. From East Bronx bungalow districts to northern Queens, pollution due to piecemeal sewerage
revealed that localism could bring benefits as well as challenges to the city's
borderland communities. The combination of sewage pollution and refuse
dumping in Flushing Meadows epitomized the failings of the city's increasingly obsolete industrial waterfront, land-use patterns that elevated industrial uses to the detriment of ecological health and recreation. Creating
a new coastal environment required a threefold cleanup program.[54] Such a
project first involved the related projects of making dumps less objectionable and regulating future disposal. This work included ameliorating air
pollution caused by dump fires and water pollution caused by sewage and
aesthetic redevelopment of the shoreline. The environmental problems
caused by dumps and sewage treatment ignored jurisdictional boundaries.
Only large-scale, comprehensive public intervention in private-sector land
use could reclaim the coast.

Sewage pollution epitomized the unintended environmental changes
created by the lack of a comprehensive public works plan in Queens:
each district had built infrastructure by considering only immediate, local
needs. In the words of the fair's chief sanitary engineer, the needed sewerage work amounted to "planning sewers, drains and water distribution" of
"an unknown number of pipes to serve an unknown population with an unknown amount of water supply and sewage disposal, at an unknown cost."[55]
Dramatic and sustained population growth prevented accurate estimates
of utility needs. The electrification of the Long Island Railroad from 1905
to 1908 and the extension of the subway to Queens via the Steinway tunnel in 1915 transformed swaths of Queens into suburban districts. During

1910–1920, Queens's population grew by 65 percent; the following decade it gained 130 percent.[56] As historian Richard Harris notes, the development of bridges, the expansion of trolleys and subways, and higher wages and shorter hours for workers all contributed to the extensive suburbanization of the boroughs beyond Manhattan before World War II.[57]

Suburbanization expanded housing opportunities for the working class; it also inspired concern in city officials. The tension between localism and regionalism that appeared in Queens did not hinge on a rejection of government as it had on the Gold Coast. Instead, the rise of regionalism in the borough hinged on the recognition of the environmental challenges wrought by city building under home rule and private-sector interests. In 1928 Mayor Jimmy Walker lamented that New Yorkers "in their anxiety to live in Queens have rushed ahead of civilization itself. In some districts they are living on dirt streets, and in large sections they have no sewer connections. There is a constant threat of epidemic."[58] Five years later planner Clarence Perry dolefully observed, "practically the whole western half of Queens Borough is a real estate hodgepodge" due to premature, unplanned development.[59] Market forces drove fringe infrastructure development: token public-sector regulation had nominal effect.

Mayor Walker questioned whether New York could be considered the greatest city in the world as long as the infrastructure, like sewers, of outlying sections remained woefully inadequate. His fears were well founded, but the idea of public works built in advance reflected a shift in opinion as to the timing of urban infrastructure development. In the late 1800s, Andrew Haswell Green was conflicted over the need for comprehensive public works and the costs associated with them. He had long advocated for infrastructure investment and a park system for the Bronx. Yet in the 1870s Green ultimately concluded that because of the city's fiscal straits and the need for city-center reinvestment, infrastructure in outlying areas should only be provided when needed, and local property owners should be assessed for the costs. He said that the general public should not pay for what he claimed were benefits for individual property owners. Similar thinking shaped Queens's early twentieth-century public works program. Sewers in the borough were until the 1930s designed to serve only for a limited period and were built to provide just half, or even just a fourth, of the ultimate required capacity in order to avoid heavy tax assessments on nearby landowners.[60] The lack of infrastructure in Queens hampered growth.

Because of piecemeal sewerage construction, by the 1930s, thirty-seven million gallons of minimally screened and raw sewage entered Flushing Bay daily.[61] Proposals for new treatment plants made by the Metropolitan

Sewerage Commission (MSC) went unheeded. Dissolved oxygen levels fell from 1909 through the early 1930s, indicating declining water quality. In 1929 the city finally formed a department of sanitation to orchestrate a sewerage program, but by 1936 sanitary researchers reported that only 20 percent of the total sewage discharged per day received any form of treatment at all, and of this less than 1 percent received adequate treatment. Sewer line technology limited treatment effectiveness. The city built combined sewers that collected both storm runoff and wastewater despite the MSC's recommendation for separate systems. During dry weather, treatment plants received sewage, but heavy rain caused overflow, which mixed human waste with storm water that ran not to plants but directly into surrounding waters.[62] Treatment technology was also limited. The city built its first treatment plant at Coney Island in 1886; the basic technology of this and subsequent plants provided simple screening and sedimentation that removed only large solids.[63] The North Beach plant, the sole installation near Flushing Bay, featured the same outmoded technology and additionally lacked the pumping equipment necessary to discharge sewage at high tide. Officials proposed new treatment centers for the area before the fair was proposed, but the exposition prompted plans for thirty-eight new plants to modernize the city waste treatment system.[64]

The fair spurred the sanitation department to unify, restructure, and modernize north-central Queens's sewerage and wastewater treatment infrastructure. The department built new treatment plants at Rikers Island and North Beach and at Tallman's Island on the eastern side of Flushing Bay. The new North Beach plant completed the area's transition from amusement district to modern, urbanized shoreline; workers built the plant on the footprint of Steinway's old trolley park. All three new plants featured chlorination and an activated sludge procedure, a bacteriological process that removed most suspended and colloidal matter as well as a portion of dissolved organic matter. Raw sewage would no longer collect on tidelands, and the decomposition of existing sludge on the bay bottom would progress so that the "generation of gas" would "be at least greatly diminished, if not eliminated, and the stench of sewage exposed by low tide would no longer greet travelers" on the Grand Central Parkway Extension.[65] Fair builders predicted that surrounding waters would noticeably improve within a year.

City builders on the coastal margin of northwestern Queens struggled to accommodate themselves to the physical parameters of the environment well into the 1930s. The low-lying meadows made sewage difficult to collect and dispose of effectively. Natural drainage patterns dispersed sewage across the marsh; tides, which traveled five miles across the shallow bay and through its marshy headlands, did not fully flush sewage away.[66] The Queens Sewer Bureau had planned to use Flushing Creek as an outlet for storm water, but the decision to close the river for the fair site necessitated a new drainage system, a hallmark approach of New Deal–era environmental management. The borough sewer bureau flanked the site with two of the largest water drains in the world, each equal in diameter to the Holland Tunnel. The drainage system kept sewage and storm water from inundating the fair site and, in drying out the land, provided additional stability for construction. Working decades before the establishment of federal regulations concerning wetland construction, the fair's builders did not question the potentially destructive side of landfill. Later studies proved that altering intertidal marshes undermined natural hydrologic processes. Such practices curtailed the capacity of the coastal environment to mitigate pollution in adjacent water bodies because it filled marshes that in the past absorbed hundreds of pounds of heavy metals and thousands of pounds of hydrocarbons and nitrates yearly.[67] The fair's chief sanitary engineer appraised this work, however, in terms of pollution and drainage control and triumphantly boasted, "modern sanitation protects New York World's Fair."[68] Technological modernization was vital to the human management of the Flushing environment on which the fair depended.

Fair builders addressed the sanitation issues of both refuse and sewage disposal to improve the water quality of Flushing Bay and the upper East River. Sanitary infrastructure could fix water pollution, but something also had to be done about the city's use of marshes as dumps. Flushing Meadows's dump was not unusual. Since marshes absorbed more fill than solid land (because refuse sinks), they were attractive for dumping. Marshland pollution and landfill practices were two sides of the same problem. The city parks and sanitation departments collaborated with the fair's construction department to orchestrate a new program of refuse disposal at Corona Dump and the nearby Rikers Island dump, which, since it sat at the mouth of Flushing Bay, would be seen by all fairgoers arriving by water.

In the early twentieth century, New York City produced a prodigious amount of solid waste, making dump space a mounting concern. In the first three decades of the century, population increases, the rise of mass consumption, an ascendant 'throwaway culture" that depended on the continuous disposal of old things and replaced the previous system dedicated to the reuse of trash, and the popularization of new packaging designed to be thrown away increased per capita waste generation.[69] In 1932, the quantities amounted to 10.8 million cubic yards of garbage and

8.4 million cubic yards of ash. The city burned about six million cubic yards of trash, buried an equal amount in landfills, and dumped half a million cubic yards at sea annually. But in 1933 the US Supreme Court ordered the city's ocean dumping discontinued by July 1, 1934.[70] The court ruled in favor of New Jersey that the city created a public nuisance when it dumped refuse at sea because the trash eventually polluted regional beaches. The ruling forced the city to change its garbage disposal program. Even with its existing and under construction garbage incinerator plants, without off-shore dumping, the city needed to expand landfill practices by opening the Soundview dump on Clason Point and intensifying dumping at Rikers Island.[71] By the mid-1930s an estimated six million cubic yards of waste arrived yearly at Rikers, making it the city's largest garbage landfill. Landfill raised the island's elevation of five feet to 120 and expanded its original sixty acres to over 330 acres.[72]

While Corona and Rikers dumps had long bothered locals, by the mid-1930s city officials realized that the dumps were a liability to the fair. Rikers cleanup began in November 1935, a month after the New York World's Fair 1939 Inc. organized and seven months before the city and corporation finalized the fair site lease. That November, Moses urged Mayor LaGuardia to reconsider the island from the point of view of the coming exposition. The work he recommended was twofold: to ameliorate air pollution em-anating from the dump fires of the "notorious East River Vesuvius" and improve the island's aesthetic appearance from the water.[73] Moses seized on the suspension of dumping at Rikers and orchestrated the rerouting of formerly Rikers-bound refuse to Soundview Park; the plan exemplified how Moses frequently manipulated various city departments in pursuit of his own goals. Department of Sanitation scows transported the fill, and the Department of Docks became responsible for construction of a bulkhead for the fill, providing Moses with the infrastructure needed to further his Clason Point waterfront plans.[74]

The Department of Sanitation, aided by the Department of Correction, which had opened a prison on the northwestern corner of the island in the mid-1920s, began a plan provided by the Department of Parks to build an "outer landscape rim" of "green knolls and meadows."[75] Depression relief forces under the direction of Sanitation employees reshaped the shoreline to hide the dump, while inmate squads took on the responsibility of seed-ing and planting the island's new shores. This work entailed considerable difficulty because the landfill reeked and was often on fire. The removal of four million cubic yards of fill to construct a 220-acre municipal air-field at North Beach reduced the smoldering mound to a uniform ten-foot elevation. Landscapers planted rapid-growing and drought-resistant black locust, gray birch, and poplar to "soften [the island's] bleak appearance" and screen the dump.[76] The Department of Sanitation's landscape architect oversaw similar designs at Tallman's Island. This landscaping marked the sanitation department's first beautification work.[77] Work reconfiguring Rikers Island combined with new sewage treatment plants across the upper East River built a new coastline intended for public viewing during the fair.

The redevelopment of Rikers and Tallman Islands, while important to the East River cleanup, paled in comparison to the work required to reconfigure Flushing Meadows. The fair's builders described shoreline and meadow development as nothing short of a "war against stubborn nature, to wrest from the meadowlands, the gem" of a modern park.[78] The language of war with and conquest over nature was typical for the era. Planners who called for reclamation turned to scientific expertise to rein in what they saw as wasteful natural resource use. Reclamation, the work of turning waterlogged land into dry, buildable real estate, required the redistribution of the dump's ash heaps. Construction required an even grade, but as the fair's Chief Sanitary Engineer pointed out, some areas immediately adjacent to the river were bare of ashes. In other places, thirty feet of fill and refuse lay atop the original surface of the marsh.[79] More than one thousand acres of meadowland required draining, filling, stabilizing, and grading.

In what fair leaders framed as a battle to reclaim the meadowland, the area's unstable terrain proved to be a formidable challenge. The meadows developed in the former inner harbor of Flushing Bay. Protected from tidal wash, sediment settled on the valley floor, eventually deep enough to support shallow-water cattails, salt marsh grasses, and reeds. Fine silt, clay, and decomposing organic material overlaid the dense root network of these plants. This fibrous crust, known as meadow mat, ranged from one to six feet in depth. Underneath lay original tidal deposits, which in the center of the marsh measured up to eighty feet deep. The gummy sediment decreased in depth as the underlying sand rose to meet the surface on the meadows' east and west margins. Due to the unstable, uneven ash layer, the nature of the mat, and the softness of underlying silt deposits, foundation problems plagued grading and construction.[80]

Consulting engineer Carlton W. Proctor addressed the challenges of this unstable terrain. A nationally renowned foundation engineer, Proctor was well positioned to assess the construction challenges of Flushing Meadows. Proctor and his partner Daniel Moran built the foundations and piers of some of the nation's most important bridges, including the George

Washington Bridge in New York Harbor. Proctor's work revealed that, the language of "conquering" the meadowlands notwithstanding, fair construction was not a juggernaut against natural systems. Instead, fair builders attempted to accommodate the natural processes of the environment, paying close attention to the marsh's hydrology. Proctor was part of the pioneering generation of soil engineers who first combined engineering with the theories of soil mechanics in the 1920s.[81] Collaborating with civil engineer Donald M. Burmister of Columbia University, he oversaw subsoil borings and calibrated a range of soil characteristics, including moisture content, grain-size distribution, and plasticity to assess the marsh's construction challenges.[82]

Based on his soil engineering research, Proctor declared it unwise to "fight the ground" at Flushing Meadows. The studies revealed plasticity to be the key to successful land making in the meadowland. This characteristic measures the stability of soil as it absorbs or loses water. Gravel or granular soils lose very little if any of their shear, or resistance to sliding, when water is added. Significantly plastic soils, however, can become putty and ultimately fluid-like when mixed with water. Proctor identified the "semi-liquid character of the silt and its tendency to flow laterally under unequal superimposed loads" as the cause of previous building failures in the meadowlands. As one of the project's engineers reported, remolding the meadow mat crust could trigger mud waves, and neither information nor precedent existed "to indicate how far this influence would extend."[83]

The nature of the meadowlands presented not only a problem but also a solution to grading and construction challenges. Tests revealed that while in some respects volatile, the meadow mat had considerable inherent strength due to the marsh's densely woven root system. Because of the work of Proctor and Burmister, the fair's builders realized that the mat was strongest in its natural state. Additional soil tests revealed that a uniform, balanced blanket of ash fill would cause the least possible disturbance to the underlying strata of mud and preserve the inherent structural integrity of the meadow mat building surface. Uniform surface tension would furthermore gradually press water out of underlying sediment, and the resulting settlement would beneficially reduce the meadow's plasticity. As a result, the construction department adapted building techniques to the meadows' challenging terrain to carefully control subsurface movements.[84] On Proctor's recommendation construction engineers dug through ash fill, meadow mat, and semisolid tidal deposits to anchor foundations for railroads, the dam, and buildings in underlying sand and gravel. Pile drivers delivered fifteen-thousand-pound blows to hammer fir and spruce pilings, some up to eighty or one hundred feet long, through the jelly-like muck. More than five hundred piles supported the reinforced concrete ring supporting the 4,650-ton Perisphere, the fair's enormous iconic orb. Not only did builders need to obtain fill, they needed to successfully hold shoreline fill in place to prevent the tide from undoing their work. Shoreline development wrought new city-building techniques. The boat basin bulkhead was designed to withstand the lateral movement of not just mud and silt in the bay bottom but the pressure and movement of the gravel and ash fill behind it.[85]

The grading of the ash dump and filling of the marsh continued a long history of land making in New York. For nearly three centuries landfill occurred in localized projects by private waterfront landowners, like the BARC, without oversight of planners or engineers. The comprehensive work at Flushing Meadows was unprecedented in its scientific studies and municipal oversight. Proctor proudly pointed to Flushing Meadows reclamation as a demonstration of the value and necessity of soil analysis before substructure design and the practical applications of the "relatively new engineering expedient" of soil mechanics.[86]

Like foundation engineers Proctor and Burmister, landscapers also faced challenges due to the physical makeup of the meadow mat. While the meadow mat supported a variety of plants, it differed greatly from upland mineral topsoil. Soil reclamation teams reported that the area's excavated material was "practically sterile since most bacteria [could not] tolerate such highly acid conditions."[87] The fair's builders had to manufacture new topsoil to ensure the growth of planned lawns and gardens. In the summer of 1936, soil engineers began a year of experiments to reclaim the fertility of excavated meadow mat.[88] Parks department workers applied hydrated lime to reduce acidity and eliminate harmful chemical toxins. Hydrated lime, combined with long-term weathering, encouraged physical and chemical breakdown of the mat. Mechanical aeration reduced remaining lumps of mat and improved soil structure to finish the conversion of meadow mat into planting soil. The final step was the application of humus of well-digested organic matter to encourage plant growth.[89]

On this new canvas the parks department planned a grand public park. Moses declared it would be an American Versailles with vast lawns, water basins, and tree-lined allées. By the 1930s, Major Gilmore D. Clarke had become America's foremost public works landscape architect due in large part to his work in greater New York, and he wielded substantial artistic influence over the fair. Clarke worked with landscape architect Francis Cormier, famous for her work on Westchester and Gold Coast estates, and architect Aymar Embury II, celebrated for both his country estates and

**6.2.** Gilmore D. Clarke's proposed plan for Flushing Meadows Park. *Source*: Gilmore D. Clarke papers, 15–1-808. Division of Rare and Manuscript Collections, Cornell University Library.

public works such as Orchard Beach. The team planned formal European-style gardens, a Japanese garden, a bird sanctuary, and facilities for active recreation, including golf and boating facilities, bridle and bicycle paths. Once the fair closed, the ten thousand trees planted at the site would constitute the basic landscape of Flushing Meadows Park. Clarke belonged to the cohort of landscape designers and engineers who collectively fostered a shared vision of park planning across Westchester, Long Island, and the city beginning in the 1920s. A park system was suggested for Queens in the early 1900s, but only with Moses, the New Deal, and the fair was a borough-wide system realized. Flushing Meadows constituted Queen's keystone park and the final triumph of a regional infrastructure plan circulated by planners through the hinterlands of greater New York.

The structural demands of the tidal deposits below the mat, although largely invisible, dictated foundation engineering for the fair. The city and meadows environment had long been connected, but the meadows environment was remade through new understandings of the meadow mat's soil mechanics. Fair builders reshaped the nature below the meadow mat in conjunction with the reconfiguration of the nature of the mat itself. Landscape architects then replaced the nature atop the mat. This physical transformation of the space marked the evolution of the marsh into a massive man-made artifact. Sewer construction and the reclamation of unstable marsh into profitable real estate proved the enmeshed and mutually supportive work of environmental and technological innovation in city building at Flushing Meadows. Shoreline construction required collaboration from sanitary engineers and landscape architects who viewed urban form and the coastal environment as systems in need of unification. Since the late nineteenth century, city leaders and planners debated how best to alter the environment to support expansionist political goals. Fair construction elevated the level at which government bodies pursued environmental engineering, leading to increasingly large-scale and centralized planning. The next step was to open this space to the greater metropolitan public.

*"All Roads Lead to the Fair"*

The city and fair corporation took on the work of not only building a fair site but building wholesale an urban infrastructure system. The intertwined goals of the New Deal—economic recovery, job creation, government protection of the people's well-being, civic uplift, and public works— transformed the American landscape between 1933 and 1942. In New York State, as across the nation, hundreds of highways, bridges, parks, airports, and municipal buildings were built as part of the New Deal. The city's five boroughs alone had more than 770 unique public works projects. The city and corporation agreed that the fair's success depended on improving accessibility to Queens: greater New York needed modern roads and bridges. The East River had to be bridged to connect Queens to both Manhattan and the Bronx on the mainland. To prepare for expected traffic from New England, the TBA eventually planned a second crossing east of Flushing Meadows.

Bridging the East River was the first step to making Queens, and the fair site, accessible to automobile traffic. A bridge across Blackwell's Island to link midtown Manhattan and Long Island City had long interested city builders, both professional planners working in the public sector and civic-minded capitalists and propertied New Yorkers privately working to shape the regional city. In the fall of 1896 one such benefactor-investor—William Steinway—took his Daimler motor yacht out to view the newly built caissons, the watertight chambers for underwater bridge construction work on Queens's shore.[90] This work and the press it received pleased Steinway, since coverage included references to North Beach and Steinway settlement real estate. Although Steinway never lived to see it, Queensboro Bridge construction began in earnest in 1903.[91] The bridge was immediately popular upon opening in 1906. The structure was the only automobile route off Long Island north of Brooklyn's three spans. As Steinway had predicted, the city looked to build additional bridges to Queens. City engineers had first proposed a triborough bridge between Queens, Manhattan, and the Bronx in 1916. The date the city finally began such a project, however, was October 25, 1929, Black Friday for Wall Street's stock market. In the depression that followed the crash, the city lacked the necessary savings, bonds, and corporate stock to continue construction projects. The bridge remained incomplete until early 1933, when Moses successfully forwarded legislation to create the TBA. When the Triborough opened in July 1936, it was the largest bridge complex in the world. The mammoth three-mile bridge sifted traffic from the three boroughs and two islands across sixteen miles of approach road and viaducts.[92] Moses relocated the city's parks department headquarters to 285-acre Randall's Island at the base of the bridge in the East River. This move underscored the bridge's keystone position in his jurisdictional empire and the centrality of waterfront redevelopment to his vision of a modern metropolis.

Bridge- and road-building schemes were proposed before Moses rose to power in city and state government. Moses's opportunistic resource gathering and eye for regionally scaled development, however, made possible the construction of a comprehensive, modern traffic network. Moses, who eventually oversaw the largest urban park and parkway program of the New Deal, boasted publicly of his crucial role as a liaison between the various agencies responsible for park and road building that spanned the city's jurisdictionally complex environs. The TBA, in Moses's words, "provided the warp on the metropolitan loom" that allowed him "to weave together the loose strands and frayed edges of New York's metropolitan arterial tapestry."[93] Rather than float a separate bond issue for a specific capital project, as was the traditional authority practice, the TBA was authorized to float bonds on the authority's general revenues. This funding innovation had far-reaching effects. Not only did the ability to refinance bonds and tolls turn the Triborough into a self-perpetuating funding source, refinancing extended the life of the TBA indefinitely since the authority could not go out of existence until all outstanding bonds were redeemed.[94] And this was not Moses's only source of funding: federal money also played a crucial role.

In the 1930s Moses's funding expanded dramatically despite the fiscal straits of the Great Depression. Even before the New Deal, Moses—as head of both New York State's Emergency Public Works Commission and President Hoover's Reconstruction Finance Corporation in New York—had secured funds for his road projects. When New Deal relief became available, both the city and Moses further benefited from first the Civil Works Administration (CWA) and then the Public Works Administration (PWA), which was eventually reorganized as the Works Progress Administration (WPA). Moses masterfully directed New Deal funds toward his public works agenda, although he downplayed its role by selectively announcing which projects received federal relief funds or workers. WPA workers contributed to the construction of the Bowery Bay and Tallman Island plants; CWA and WPA funds employed workers at Orchard Beach.[95] Built on Steinway's former Bowery Bay site, the forty-million-dollar LaGuardia Airport was the single most expensive project handled by the WPA nationwide.[96] Historian Mason Williams calculates that the city received fifty-eight dollars per capita in PWA grants and loans, far more than the national average of thirty-three dollars. When the WPA was organized, the city received its own unit independent from the state and effectively became the forty-ninth state for Depression-era relief. The parks department received the most CWA funds in the city; borough presidents received the second-highest portion to fund infrastructure such as streets and sewers. The parks department was also the largest beneficiary of PWA moneys, although all the agencies responsible for the city's physical property benefited.[97]

The New Deal was a major turning point in regional development because of unprecedented funds made available by the federal government but also because greater New York's planners and municipal leaders were particularly adept at working the bureaucracy. New Deal agencies favored "shovel ready" projects sponsored by a state or municipal agency on public property. Because of the work of the CRPNY, the RPA, and powerful park commissions such as the LISPC, the region was awash in proposed projects needing funding. Six months after the creation of the WPA, more than two hundred thousand New York City residents held WPA jobs, unlike the rest of the country, where relief work remained in planning stages or employed but a few thousand people. At its peak in January and February of 1936, the WPA's monthly outlay in the city was nearly twenty-three million dollars. By the end of 1937, LaGuardia could budget just 13 percent of municipal revenue for physical improvements, since the WPA and other federal funds covered the rest.[98] Moses oversaw the majority of workers and funds; between March 1934 and February 1935, for example, he directed the expenditure of more than thirty-two million dollars in relief funds. Moses approvingly reflected that after taking control of city parks and the TBA in addition to his state park and state relief work, "then at last it was possible to get something accomplished on a great scale, because little plans and accomplishments are of no use in the metropolitan community."[99]

The Westchester County Parks Commission (WCPC) developed an automobile-based regional transportation network in the 1920s. Moses had subsequently codified this park and parkway network across Long Island. Regional highway projects connecting to Fairfield County, Connecticut, via Westchester became the final mainland components of this regional system. In the 1930s the Henry Hudson Parkway and the expressway between the Grand Concourse and the Triborough, later renamed after its architect Major William Francis Deegan, eliminated the bottleneck of traffic in the south-central Bronx. Additional new Bronx roads linked to the WCPC's extensive arterial parkway system. In adjacent Fairfield, the easternmost corner of the city's suburbanizing environs, similar regional infrastructure was overdue.

Through the first decades of the twentieth century, Connecticut's legislature was reluctant to allocate funds to public works. The state government, headquartered in the north-central city of Hartford, was disinclined to fund projects associated with suburbanization in Fairfield County's metropolitan coastal corridor since it made up only 12 percent of the state. The pervasive tradition of New England town rule and lack of county jurisdictional and funding power meant no county-level organizations existed to fund highway improvements in relation to those under construction to the west in Westchester and the Bronx. The Fairfield County Planning Association (FCPA), formed in the twenties in the state's fastest growing county, advocated for regional planning. The FCPA argued that Fairfield should look to Westchester's advanced parkway system for a blueprint to accommodate commuters and turn metropolitan growth into an asset rather than a detriment.[100] In 1926, engineers from Connecticut's highway department consulted with Jay Downer of the WCPC to build a Connecticut extension of the Hutchinson River Parkway, but the project languished.

The opportunities for regional public works in southwestern Connecticut improved dramatically in the mid-1930s because of a break in Republican obstructionism in the state legislature. This break empowered Democrats to secure funding for the completion of Sherwood Island State Beach and approval for the construction of a parkway named in honor of Schuyler Merritt, a Republican congressman and an early supporter of the FCPA.[101] After a series of shelved reports, delayed funds, and both management and land speculation scandals, construction finally began in 1934 with the aid of newly available federal public works funds. The first eighteen miles of the Merritt, modeled on Westchester's Hutchinson and Bronx River parkways, opened in June 1938. When the second half of the "Gateway to New England" opened to traffic on September 2, 1940, 54,163 automobiles traveled the four-lane route, signaling its immediate popularity. The road-way received wide acclaim from architects and civil engineers for its 1,370 acres of landscaping and elaborate art deco overpasses.[102]

While Connecticut built a modern parkway through its coastal corridor communities, Moses used TBA funds to build additional expressways in New York City. New radial and circumferential highway routes provided a comprehensive citywide and intraregional circulation and transformed Queens. Queens was once dismissed as "an immense buffer State between Brooklyn and Nassau, challenging the advance of motorist armies from both borders."[103] New roads such as the Grand Central Parkway and its extension, however, made Queens accessible to greater New York's car-owning public. The state Department of Public Works constructed high-ways and bridges from plans prepared by the parks department, while the Queens Bureau of Highways oversaw local road development. As the borough improved its roads, bottlenecks developed where city boulevards met rural roadways. The right-of-way in Queens averaged seven-hundred-feet wide, encompassing seventy-feet-wide parkways. Across the county line, Nassau roads often measured only eighteen-feet wide in much nar-rower or nonexistent rights-of-way. The inconsistencies dumbfounded LISPC planners interested in facilitating intracounty access.[104] The LISPC built the Southern State Parkway in 1925–1933, the Northern State Park-way in 1931–1933, and the Wantagh State Parkway linking them in 1938.

Regional public works authorities built a system of roads and bridges that linked parks and made the sprawling metropolitan area a more defin-able region. The new roads of Long Island formed a continuous forty-three-mile chain of highways between the Triborough Bridge and Jones Beach on the South Shore. In the coming years, the Belt Parkway around Brooklyn and Queens opened to complete a full loop around western Long Island. City highways seamlessly linked Nassau parkways to those of Westches-ter via the Triborough. The fair corporation boasted "All roads lead to the Fair in '39." The parks department estimated Flushing Meadows Park to be

**6.3.** *Opposite*: Bird's-eye view of the fair. This rendering captures the comprehensive arterial road network underway in the 1930s in preparation for the fair. Rather than differentiate between WCPC, LISPC, or city-run parks and highways, promotional materials such as this map emphasized instead the region's connectivity. The round white Perisphere sits in the center of the park. Rikers Island, to the west of the mouth of Flushing Bay, has become an invisible part of the infrastructure and environmental reclamation required for the fair. *Source*: Map of the New York World's Fair and Approaches. Lionel Pincus and Princess Firyal Map Division, New York Public Library Digital Collections.

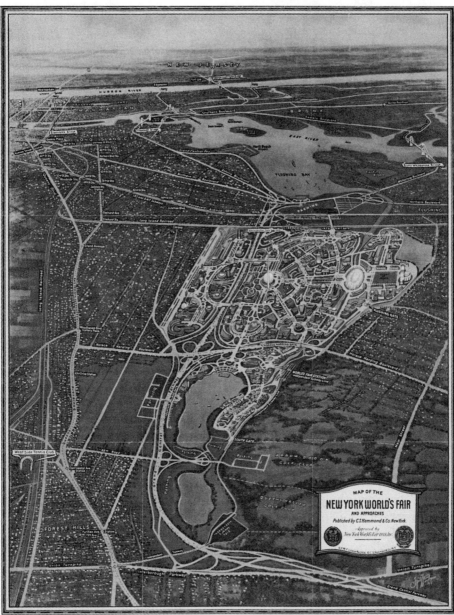

MAP OF THE
NEW YORK WORLD'S FAIR
AND APPROACHES
Published by C.S.Hammond & Co. New York

WORLD'S FAIR BUILDINGS
(PARTIAL LIST)

within a fifteen-minute drive of five million people and an hour drive of ten million along the "modern parkways leading to its doors."[105] In February 1937, Helen Keller noted the great changes underway from her home in Forest Hills. "The vast marsh . . . is being transformed into a beautiful parkway. . . . The World's Fair . . . will be almost at our door in 1939. Certainly our little house will no longer be the quiet spot only fourteen minutes by train from the Pennsylvania Station."[106]

The Bronx-Whitestone Bridge opened on April 29, 1939, the day before the fair. Moses declared it and its approaches "a logical and inevitable part of the great Belt Parkway program of the City."[107] The combination of local public works in Queens and regionally scaled public works that knit the city's coastal environs closer together marked a new chapter in the geography of greater New York. Public works and increased automobility tied spaces once considered hinterland into New York's sprawling networks of leisure, real estate, and commutation. The official poem of the World of Tomorrow captured the end of Long Island's insularity:

> Here on island (O connect here for all points of your travel)
> With many bridges extending: Triborough, Queensboro
> Brooklyn, Manhattan, Whitestone, iron harps suspended:
> Here at hub of island with many spokes converging:
> . . . . . . . . . . . . . . . . . . . . . . . . . . . . . . . . . . . . . . . . .
> This is the achievement: this is tomorrow.[108]

During the 1930s, New York's waterfront emerged from one hundred years of marginal use in sewer outfalls and dumps. The geographic scale and topographical and environmental challenges demanded innovation in soil, civil, and sanitary engineering and brought such work to the attention of regional planners. The fair's planners and civil engineers created a modern urban form characterized by arterial infrastructure, automobility, and remade nature, a vision mirrored in the World of Tomorrow.

## Visiting the Future in Queens

While the press and exhibit designers alike insisted on the groundbreaking futurism of the World of Tomorrow, greater New York was an exhibit of the promise of automobility and redevelopment of the urban-suburban threshold writ large. As Lewis Mumford summarized in the *New Yorker*, "if

**WHITESTONE'S BRIDGE AND WORLD'S FAIR CELEBRATIONS PARADE · RECEPTION · BALL WHITESTONE NAVAL ARMORY APRIL 29 1939**

**6.4.** This promotion for a local party in honor of the opening of both the Bronx-Whitestone Bridge and the New York World's Fair depicts the boroughs of the Bronx and Queens greeting each other. The Bronx-Whitestone Bridge, rushed to completion for the fair, became the easternmost bridge between Long Island and mainland New York. *Source*: Courtesy Queens Museum, 100.34.123WF39.

you combined Mr. Dreyfuss' clouds, which are fine, and Mr. Geddes' land-scape, which is marvelously good, you would have a pretty faithful model of the real world."[109] The piecemeal local roads that hindered Queens's previous development were a thing of the past. So, too, were the ramshackle junkyards and malodorous dump that had bothered suburban homeowners of greater Flushing. George McAneny, head of both the fair corporation and the RPA, used his position to advocate for comprehensive planning in Queens. The Queens fair stood apart from its predecessors in the scale of operations and the fact that permanent civic improvements were planned in advance and codified by legislation and contract.[110] The enabling legislation had included the power to zone nearby territory. The zoning banned billboards, private commercial parking fields, commercial side shows, and tourist camps, and it limited adjacent land use to low-density residential development across a mile radius from the fair.[111] New Deal–funded regional infrastructure bolstered just such patterns of development. Both the fair board of design and the region's planners presented a profound faith in the highway and the car as the keys to future prosperity. A car-centric present was at last realized through the road network that converged in Flushing Meadows.

In the 1930s, regional projects funded by state and federal appropriations outpaced the localism of county- and village-level building that characterized so much of the previous decade's growth. Westchester previously funded public works via county appropriations, bonds, and park revenue. During the Depression, however, home-rule parks programs became a liability; the county found itself unable to pay down its parkway debt or afford maintenance. In 1936, the county repeatedly turned parkway lights off to conserve funds.[112] In stark contrast, Moses sustained Long Island's parkways via special state appropriations. Home-rule public works became a burden, even for a wealthy county like Westchester. The Depression pushed its officials to reevaluate the advantages of state-backed and state-enforced planning powers. Westchester's county executive lamented, "If we had Moses to lead us out of the bulrushes, we could get the State to build parkways for us."[113] Support of the integrated public-sector offices of the New Deal and self-governing authorities echoed 1870s proannexation arguments in Westchester and 1890s proconsolidation arguments in Queens. The economic and political realities shifted the debate between localism and regionalism during the Depression. Planners and civic leaders debated without clear conclusion about whether an ever-larger authority was needed to balance development across the sprawling and jurisdictionally complex greater New York.

While Moses's regionalism was powerful and appealing, it was not without limits, as the afterlife of the fair site revealed. The World of Tomorrow
closed on October 26, 1940. The following day, 1,200 demolition workers swarmed the grounds. Nearly four hundred buildings needed razing
for Flushing Meadows Park's scheduled 1942 opening. But the debris of
dismantled buildings and the plowed-up pavement remained amid dying
landscaping into April 1941.[114] The fair corporation's first two million dollars in revenue had been earmarked for park construction, but the fair
failed to realize a profit—each dollar invested returned only forty cents by
the end of the 1940 season. Without profits to fund park work and with the
New Deal reserves funneled to the war effort, the park Moses envisioned
as the symbol of the modern metropolitan waterfront remained two-thirds
unfinished.[115]

No funds existed to complete Flushing Meadows Park, but the fifty-
nine million dollars invested in fair construction had already remade the
marsh environment. The fair's builders were selective, however, in what
qualities they valued in the environment, focusing on migrating birds,
for example, but not the larger network of relationships that made up the
meadows habitat. City and state officials pointed to the reappearance of
shorebirds along the fair site's beaches and lagoon as proof of its success.
Bird-watchers spied American egrets, little blue herons, and a variety of
dowitchers, sandpipers, and terns in the blueberry bushes and swamp aza-
lea along the fair's lakes. Following the Great Atlantic Flyway, millions of
migrating shorebirds, attracted to marshes like Flushing's despite the rise
and fall of Corona Dump, stopped in greater New York's estuaries.[116] Although the marshland available to migrating birds had in fact shrunk in
size because of fair development, the fair's builders claimed to have turned
an "unhealable" landscape into a wildlife sanctuary.[117]

The restructured nature of the fair site helped obscure the technological innovation, political influence, regulatory oversight, and public
moneys that motivated city officials and the fair corporation to reshape
the natural world. The remaking of northern Queens's coastal margin reflected the environmental goals of Progressive Era public works, including
the reclamation and "improvement" of marshland, sanitation advances,
and calls for modern public spaces in clean environments. Spaces seen
as "natural," such as a waterfront or a park, are often mediated by human
activities. Later in life Moses declared environmental preservation was a
primary objective of his career. On the upper East River and Flushing Bay,
Moses recalled, "we were busy . . . salvaging the environment."[118] Moses
also declared the LISPC work had been ahead of the times, an example of

"conservation in a broader sense, before the 'E's' of ecology, environmentalism, ecetics [aesthetics] became slogans of extremists and crusaders."[119] Some of this work had environmental benefits. Water pollution abatement was a clear achievement: while dissolved oxygen percentages declined through the first decades of the century, in the 1930s the values leveled off and slowly rebounded. The restructured meadowland underscored city builders' expectations of absolute control over the environment. The ecological problems that reclamation exacerbated or caused because of the removal of marshland, however, were overlooked or unacknowledged. This work took place before the rise of environmental planning and the field of ecology that together reevaluated the ecological processes and coastal flood protection provided by marshes.

The transformation of Flushing Meadows reveals the way in which regional planning developed in New York's environs was brought into the city. Urban growth altered a number of thresholds between the regional city and its environs: it reshaped the borderland between city and suburb and the environmental edge of the coast, and it deemphasized political boundaries through intrajurisdictional public works. From his office at the base of the Triborough Bridge in the East River, Moses coordinated the reclamation of East River islands. The river was no longer a boundary of separation but flowed through the center of the greater city. Moses's parks department and TBA both built recreation facilities in lands left over from parkway and bridge construction. The TBA declared that this work opened up distant edges of the city "hitherto neglected or inaccessible, restor[ing] sections which [had] become dumps and eyesores, and set[ting] a standard of usefulness and appearance."[120] As he had with the TBA, Moses envisioned the fair as an instrument of both urban growth and regional planning with which to make the periphery a centerpiece of the modern metropolis. The erasure of the gritty in-between spaces of the urban periphery, as captured in Flushing Meadow's ash heaps, made space for an expanded vision of suburban growth across the environs of greater New York.

# Epilogue

## *The Limits of the World of Tomorrow*

General Motor's Futurama celebrated automobility as the mark of American modernity. Designer Norman Bel Geddes dramatized congestion as America's essential problem and offered superhighways as the solution. Futurama featured vast seven-lane highways separated for travel at speeds of fifty, seventy-five, and one hundred miles per hour.[1] The comprehensive system knit sprawling suburbs and decentralized industrial nodes, structuring the metropolitan area of the future. Futurama did not explicitly sell automobiles. It sold the transportation infrastructure system on which GM's customers depended. For the ride's finale, the model's gradually increasing scale gave visitors the impression of descending from the air for a close-up of skyscrapers and GM vehicles, a frozen moment of traffic at a city corner. Exiting the ride with an "I Have Seen the Future" button in hand, visitors found themselves on a life-size replica of the intersection. One visitor described the ride's end as perpetuating a dream: "You stepped out from your chairs and found yourselves on the elevated sidewalks of the 'same' street corner you had just seen a moment ago . . . this part was continuous with the rest of the World's Fair and thence with the rest of the world."[2] A reporter for the *New York Herald Tribune* pithily captured Futurama's achievement: "General Motors has spent a small fortune to convince the American public that if it wishes to enjoy the full benefit of private enterprise in motor manufacturing it will have to rebuild its cities and its highways by public enterprise."[3]

Lewis Mumford disagreed that the decentralization on display in Futurama was good for New York or other American cities. Improved transportation networks were crucial to the regionally scaled development of the coastal corridor. Urban thinkers like Mumford and planners agreed on the need for regional connectivity and that parkways could be an essential part of this circulation.[4] But Mumford expressed concern that uncontrolled decentralization along highways would cause aimless drift. When an unhappy GM official questioned Mumford's criticism, Mumford explained that he did not inherently reject Futurama's expressways but rather questioned what he believed to be their wasteful and obsolete centripetal effects.[5] He had first publicized his disapproval of highways and decentralization in his criticism of the *Regional Plan of New York and Its Environs*. He likened Futurama's ideology to New York's 1811 grid and the regional plan, condemning them all for assuming indefinite sprawl: "both the planners of 1811 and those of 1931 believed in the Manifest Destiny of New York. One saw Manhattan, the other the metropolitan region, as a limited area, waiting to be filled up."[6] Mumford lamented that the fair offered visitors the conclusion that expansion was uncontested and that land outside the city was "merely tributary to New York."[7]

The World of Tomorrow assumed suburbanization to be an inevitable juggernaut, ignoring the dynamism of borderland residents and communities who negotiated growth. Despite the communitarian ideals of the fair's board of design, through exhibits like Futurama the fair ultimately bolstered rather than challenged a vision of unplanned decentralization. Local private-sector real estate interests would largely shape the regional city, not planning bodies like the Regional Plan Association, which demanded regionally balanced suburban land use. The fair's model town, the Town of Tomorrow, exemplified this narrow interpretation of regional development. Promotional material for the exhibit explicitly avowed that it was not "an actual 'Planned' community" and that it was not "intended to represent a model neighborhood plan."[8] The town filtered community planning ideology until only the single-family home on a suburban lot remained. The fifteen design homes encouraged private consumption with exhibits of home furnishings and appliances. The Town of Tomorrow confirmed for Mumford that the future of regional development was one of sprawl, "the usual suburban wilderness."[9]

Critics declared the fair catered exclusively to existing patterns of white, middle-class, suburban homeownership and failed to address the injustices this trend exacted on the city's racially and economically diverse residents.[10] The unparalleled migration of whites from city to suburbs in New Jersey and Westchester, Fairfield, and Nassau counties in the two decades after 1940 created a net negative of white population. Combined with nonwhite urban population growth, white suburbanization dramatically shifted New York's racial composition.[11] Segregation of the regional city had roots in the racism that shaped the East Bronx leisure network as a white-only space; postwar racialization of city and suburb would perpetuate spatial inequality with social, political, and cultural ramifications.

New York's World's Fair celebrated a suburbanized future for white

Americans. The official emblem of the Trylon was visible from the golf course of Pomonok County Club in Flushing. The fair literally cast a shadow over Long Island's elite country club landscape and portended the decline of the Gold Coast and the rise of the fair's vision of the modern metropolitan landscape: suburbia.[12] In the 1920s, the Committee on the Regional Plan of New York (CRPNY) celebrated the Gold Coast as an essential preserve of open space and had joined residents in speaking out against the Long Island State Park Commission's proposed Northern State Parkway. To the park commission's supporters, park protests highlighted the "undemocratic snobbery" of the millionaire colony.[13] The antidevelopment stance of estate villages, however, largely mitigated subdivisions and sprawl. Because of the North Shore estate district, Nassau was primarily white, and its population density remained as low as twelve persons per acre as late as 1930.[14] But as the Gold Coast landscape gave way to increased suburbanization and regional development, professional planners faced unanticipated challenges.

Beyond the World of Tomorrow, Queens and Nassau sat poised to become the locus of growth focused on single-family homes in car-dependent landscapes. By the mid-1930s, industry was still largely absent from Nassau, but agriculture had ceased to play an important part in the county's life.[15] The county was already primarily a place of residency. The nature of its development, however, was shifting from the estates of the few to the subdivisions of the many. In *The Great Gatsby*, F. Scott Fitzgerald judged Long Island Sound to be "the most domesticated body of salt water in the Western hemisphere," designating the Sound as "that great wet barnyard," a catchall for the enormous human population of greater New York.[16] A little over a decade later, as Long Islanders could testify, Fitzgerald's judgment proved prescient.

A series of taxation changes destabilized Gold Coast properties in the 1910s and 1920s and made possible the rapid subdivision of the North Shore. New national income and estate taxes created substantial assessments; these taxes further increased as the result of Congress's attempt to shore up the federal budget with steeper rates and lower exemptions for personal income taxes during the 1930s. The combined pressures of income tax modernization and the Depression strained the finances of even North Shore barons. By the mid-1930s, John E. Aldred's property taxes had risen to $25,000; annual upkeep costs had reached more than $100,000. During the Depression, Aldred lost his fortune and his mansion.[17] In the 1920s, wealthy barons successfully monopolized land, preventing subdivisions and blocking Moses's program of large public parks and beaches in their elite playground. At a dinner held in his honor in 1937, Moses expressed satisfaction that the power of estate owners had diminished, declaring "it would seem there is no one left to quarrel with."[18] Moses never acknowledged, however, that despite his deep dislike of Gold Coast privilege, by staving off development, those estates prevented intensive subdivision that would come to exact new costs.

By the 1940s, the era of Nassau's open development and low density had passed. While the city's population leveled off at 7.3 million during World War II, the region's suburban population exploded. Nassau's population grew from four hundred thousand in 1940 to 1.3 million by 1960; Westchester's rose from 570,000 to eight hundred thousand.[19] During the two decades following the war, Suffolk and Nassau were the nation's fastest growing counties. Long Island came to symbolize suburbanization. The fate of prominent newspaper publisher Frank A. Munsey's estate epitomized this transition. When Munsey died, he left his fortune and Long Island property to the Metropolitan Museum of Art, which decided to subdivide the land. By 1940, 1,456 people lived on what had been Munsey's six-hundred-acre country home.[20]

Because New York expanded continuously between 1840 and 1940, the location of its urban edges and the extent of its hinterland were constantly impermanent. In 1840, the city, confined to twenty-three-square-mile Manhattan, had housed a population of 312,000. Sixty years later, following immigration, annexation, and consolidation, it housed 3.4 million people across three hundred square miles.[21] But regional networks and populations always spanned the city's political boundaries. As city boroughs, satellite cities, and suburbanizing counties drew development and new residents, the thresholds between rural, urban, and suburban spaces shifted continuously. Transportation technology and infrastructure drove the integration of liminal spaces into urban networks. Greater New York's governments, with the support of real estate developers, reconstructed the relationship between core and periphery to encourage suburban automobility.[22] Looking back on the extent and density of suburbanization, it seems likely that regional parks foreclosed on even greater development. Nevertheless, regional parks introduced many urbanites to the suburbs of Nassau and Westchester, and parkways that provided access to large parks enabled the first wave of automobile-focused suburbanization that tied communities of the environs to urban networks of leisure, real estate, and public works.

The reach of development across the coastal corridor was clear to journalist Frederick Coburn by the early 1900s. Coburn declared urban growth cumulative in character, leading to the urbanization of the New England coast and the creation of a "500-mile city" with New York at its center. Connecticut manufacturers in cities like Bridgeport opened selling offices in Manhattan, while commuters made permanent homes from summer cottages and traversed the region daily. Such increased connectivity reshaped the intermediate spaces of the urban-suburban threshold. The new phenomenon of suburbia, Coburn concluded, was undoubtedly "the city of the future."[23] He was right. Suburbanization continued to accelerate in the 1920s, with newly introduced limited-access parkways that increased the speed of automobile travel and extended the scope of the region within commutation distance. The borough of Queens quadrupled in population between 1910 and 1940; the Bronx tripled, gaining more than 950,000 residents. Nassau's population doubled between 1920 and 1930, but postwar growth was even more impressive across New York's suburban counties.

The World of Tomorrow synthesized a cohesive message on the merits of mass suburbanization. A pamphlet distributed in the New York State fair building, funded in part by homebuilders and the Long Island Railroad, assured potential new homeowners that Long Island was "not only the breathing spot of the extremely well-to-do with great estates along the shore line, as has often been inferred, but is populated to a large degree by the man of moderate means."[24] A suburban future seemed assured. In fact, in just seven years, in the spring of 1947, the building firm of Levitt and Sons announced its two-thousand-home suburb Levittown for Nassau.

The making of greater New York is an inherently regional story, a history of the periphery and center developing in tandem. A number of lessons for regionally scaled urban history emerge from this history. First, nonprofessional planning and city-building patterns took place alongside professional and governmental development projects. Second, a corollary lesson is that through nonprofessional planning, property regimes and homeowners' mobilization of powerful private property rights shaped patterns of development and allowed the metropolis to grow. Third, local government emerged in this process as a significant player in the search to find an effective level of government to oversee growth. And finally, the building of the regional city is a reminder that urban expansion always reshapes the natural world while at the same time the environment influences the form and function of regional space. In the island and tideland setting of greater New York, this environmental change introduced longstanding new advantages and disadvantages to city building in the coastal corridor.

New suburbs and roads engendered regionally scaled environmental consequences. The multiplication of subdivisions created car-dependent landscapes of extensive highways, parking lots, drive-throughs, regional shopping malls, and sprawling low-density suburbs. The changed landscape introduced suburbanites to new ecological threats in their homes, communities, and bodies. Open space and wildlife habitat were lost when woods and pastures were cleared for homes; cleared hillsides and marshes encouraged flooding and landslide due to exacerbated runoff; and smog, pesticides, and the proliferation of septic tanks created new pollution concerns.[25]

The waterways and tides, islands, and estuary system of New York Harbor and the Sound were parameters that shaped the form and function of the regional city. City officials increased connectivity between the largest islands of Manhattan and Long Island and the mainland via docks and bulkheads to facilitate maritime transit and with bridges, tunnels, and arterial roadways. The remaining islands of the harbor were developed to support the urban system. New York's islands housed the indigent, insane, and incarcerated; took in the city's waste as dumps; or offered resorts and parkland retreats. Coastal change was inherent to and an essential focus of city building on the upper East River and Sound.

Planners and city officials claimed the coastal corridor's environment to be a determining factor in regional growth and planning. In practice, however, nature was often approached as a blank canvas awaiting transformation via human intervention. In the nineteenth century Andrew Haswell Green spoke of a greater New York defined by topographical features, arguing that a shared environment already tied urban core and edge together to achieve his expansionist political goals. In the following century, CRPNY planners echoed Green, stating that "the geographical, the geological and the climatic features are pre-existent to any plan for its improvement, and therefore they control the plan rather than are controllable by the plan."[26] Yet beyond such evocations, planners and bureaucrats evaluated the environment as to its future usefulness as potential developable real estate or a site of public recreation. From the 1890s channelization of the Harlem River to the 1930s reengineering of Flushing Meadows, planners expressed confidence in humans' ability to alter the environment. In parkland evaluations, nature was imbued with the cultural values of park use, location, and design, which determined the way planners and bureaucrats altered the natural world. With regard to both parks and property, nature was assumed to require human intervention and "improvement."

Aspects of the coastal environment improved because of government intervention to mitigate harbor pollution and to make the region's beaches cleaner for bathing. Yet geological and ecological features never dominated planning debates. Indeed, state and federal coastal wetland protections were not enacted until the late 1960s and 1970s. The ubiquity of landfill over intertidal marsh impaired wetlands. In 1914, 36.5 square miles of tidal marsh remained in Connecticut; by 1959, 45 percent of that total had been lost. In 1956, Sherwood Island's marsh became a gravel holding site and then a parking lot. In Queens and the Bronx, Moses described tidelands as worthless and unattractive; Clason Point's shore was but "barren salt marsh" with "an irregular and muddy shore line." He confidently claimed land making would improve coastal aesthetics.[27] The result of this approach, which Moses shared with professional planners and city officials, was the eradication of most of the region's remaining tidal wetlands by midcentury.[28]

Planned and unanticipated changes to the materiality of the coastline and the land-water interface reshaped greater New York's coastal environment. Dredging, such as that required for Harlem River channelization, affected habitats and destroyed offshore oyster reefs and eelgrass beds. Marsh landfill, like that required for Playland Beach, Orchard Beach, and the World's Fair, was particularly detrimental to the ecological health of New York Harbor and the Sound. Landfill covered over the ecosystem of the harbor and estuary bottom, harming oyster reefs and eelgrass beds that served as fish habitat.[29] The common reed often grew in place of cordgrass when salt marshes were dug up, a plant that decomposes into less nutritious matter than cordgrass and degrades the overall health of a tidal marsh. In the twenty-first century, the Army Corps of Engineers reported that the upper East River's intertidal ecosystem had suffered "extreme aquatic ecosystem habitat degradation due to coastal filling and shore hardening."[30]

Landfill in the coastal corridor additionally decreased the ability of the ecosystem to filer pollutants and store water. Landfill practices accelerated in the early twentieth century, decreasing the coast's capacity to filter and trap pollutants at the same time water pollution increased. Salt marshes can absorb pollutants that could otherwise contaminate adjacent waterbodies. In its natural state, a tidal marsh also soaks up and stores water, a natural form of flood control.[31] Landfill and man-made vertical structures, like William Steinway's bulkhead and lumber basin, ended the ability of tidelands to absorb storm surge and prevent flooding and exacerbated coastal erosion. By the mid-1900s the flat, expansive coastal environments that once functioned as flexible thresholds between land and water were transformed by primarily "hard" construction approaches. The remade littoral was a coastline of near-vertical structures of wood and concrete, metal and stone. The man-made shore reflected wave energy rather than absorbing it, leading not to stabilized shorelines but increased erosion around vertical structures such bulkheads. Shoreline hardening ended the natural coast's ability to absorb storm surge as well as regular tidal fluctuations.

Today, in the second decade of the twenty-first century, biologists, engineers, and urban planners question the long-term effects of coastal landfill and shoreline hardening. A series of punishing storms—from Hurricane Irene in August 2011 to Tropical Storm Lee in September 2011 to Superstorm Sandy in October 2012—pushed city officials to face the unintended consequences of a remade coast. The history of coastal change in greater New York reveals the unintended consequences engendered by a fundamental feature of American city building: the wholesale restructuring of the natural world. In building greater New York, property owners, recreationalists, and park designers who assumed they could control the coast's material nature in fact created new environmental challenges alongshore that New Yorkers still face.

The history of the rise of the regional city speaks to two fundamental and interrelated tensions of American life: the friction between private and public interests and the countervailing impulses to centralize or decentralize government. The rise of greater New York reveals how environment, residential and industrial decentralization, recreation, and public works tied the urban core and periphery together and gave shape to the region. It also reveals the geographic, socioeconomic, and political boundaries that continued to separate communities. The ambivalent and countervailing attitudes of people living or owning property at the rapidly expanding metropolis's edge also mattered. Local communities responded to regionalism in a variety of ways, at times supporting it, as was the case of Westchester's park commission, but at other moments subverting it and rejecting large-scale government intervention in regional growth. The tensions inherent to regional growth explain why by 1940, New York leaders could predict the degeneration of unplanned growth into sprawling suburban tracts while simultaneously celebrating the region's model peripheral public park network.

Greater New York's scale and complexity make the metropolis a unique, extreme example of American urban growth. Features such as the tension between private interests, centered on property, and public-sector planning, however, were also the essential characteristics of urbanization writ large. This tension became an acute point of conflict particularly at the turn of the twentieth century as city and suburb decentralized and regulatory government took on new, larger roles to structure urbanism.

Urban demand for recreational spaces threatened the property rights of landowners in greater New York from working-class Throgs Neck to elite Greens Farms. The threat grew when new regional commissions gained power on the periphery in the twentieth century. In response, people living or owning property at the rapidly expanding metropolis's edge used local government to erect new barriers to regional authority. Property owners discovered the powers of municipal regulations, and local governments could be used to pursue private-sector interests when faced with perceived threats to property rights. Home-rule autonomy complemented property owners' rights to exclude. It enabled residents to define who or what belonged inside the boundaries of not just private property but within municipal boundaries. Localism and privatism paradoxically served as two of the major forces that shaped regional development at the turn of the century.

Localism was an increasingly potent force in the twentieth century as urban officials and residents alike began pursuing grassroots representation and community-based special interests. Greater New York emerged during an era when city leaders disavowed fragmentation and special-interest politics and embraced technology as a fundamentally progressive force. In the postwar era, urbanites became increasingly suspicious of the idea of technology such as the automobile as a panacea. New Yorkers also grew increasingly wary of top-down bureaucracy and government intervention via master planning.[32] The shifting public assessment of Progressive Era regionalism and Moses's legacy continues to influence contemporary struggles over the boundaries between metropolitan planning and private property rights.

By the late 1920s, intense suburban localism disconnected new commuters from the social and economic problems of urban life, problems often best suited to a regional solution. In 1929, the authors of the *Regional Plan of New York and Its Environs* counseled local governments to think "of the totality rather than the parts." Small communities, planners urged, needed "to recognize that the great whole is a living thing, with a certain spirit of its own, a sort of anatomy, and something like a functional physiology."[33] But suburban development demonstrated yet again the effect of private-sector and local decisions over regional planners. The upper East River–Sound leisure corridor had early in the century demonstrated how shared interest could be mobilized as the basis of race and social distinction when whites blocked black access to leisure amenities. By the 1930s the Gold Coast again revealed how locals could refute regionalism and the public works and progressive reform associated with it. In 1933, a staggering sixty-three villages and 173 special districts governed Nassau, a tally that included estate villages. In the most jurisdictionally complicated part of the county, as many as twenty-four government units exercised authority over a single tract of 120 acres, or one government for every five acres of ground. Locals endorsed home-rule measures that fostered exclusion and privatization, preserving racial segregation and insisting on social, political, and economic separation in the region.[34] The suburbanization promoted by the 1939 fair, and in many other ways in New York and beyond, encouraged the privatization that transformed the nation and came to define postwar American living.

The contradictory impulses of centralization and decentralization form a second fundamental tension in metropolitan growth. The stunning scale of urbanization around New York Harbor heightened officials' desire to find the best level of government at which to shape growth. In the nineteenth century, New York City's administration of the South Bronx alternated between a strong central government with principal powers residing in the mayor's office and bodies like the Board of Commissioners of Central Park and home-rule government that diffused power across the city, such as the Department of Street Improvements for the Twenty-Third and Twenty-Fourth Wards. In the following decades, city leaders debated whether and how government might mediate between neighborhood issues and the large-scale authority needed to address the functions of such a sprawling and jurisdictionally complex area. Officials and private citizens questioned whether urbanization should be shaped from the top down, to unify and integrate multiple interests and promote a broad public interest, or from the bottom up, through home rule that considered the special interests of local communities. Progressive Era planner John Nolen boasted that because of the interplay of the local and the regional, "a new interrelationship between communities [was] born, and the city woke up to find itself a metropolis."[35] Regional expansion was not so simple, nor so congenial.

Drives to centralize and decentralize government were often simultaneous, and a clear answer as to which course was the best for greater New York's urbanism never appeared. As a result, multiple characterizations of the "region" existed during the hundred-year rise of the New York metropolis. Local custom could augment definitions of the region forwarded by central city officials; the concept was inherently changeable, and more than one definition existed at once. While metropolitan in scale, the work of developing the edge was frequently rooted in localized private-sector patterns. At times local actors subverted official visions of the region as a unit or rejected relations to a "center" that residents of the edge did not necessarily regard as a center. Notwithstanding the region's changeability and the diversity of landscapes and populations, the pillars by which greater New York's diverse city builders defined the city's hinterland remained consistent; what varied was whether planners, private citizens, politicians, or local communities defined the region by inclusion or rejection of the concept. The unpredictable dynamism of the region's multiple scales of locality and connectedness made the urban periphery a richly contested territory.

# Acknowledgments

This project would have been impossible without the existence of local archives and the work of local historians. My research led me to dozens of libraries and archives, small and large, scattered across greater New York. I am indebted to the staff of the New York Public Library Lionel Pincus and Princess Firyal Map Division, the New York City Municipal Library, the Westchester County Archives, the Bronx County Court, the Fairfield Museum and History Center, the La Guardia and Wagner Archives, the Office of the Town Clerk of the Town of North Hempstead, the Special Collections Department at Hofstra University, the former New York City Parks Library, the New Rochelle Public Library, and the historians of the Village of Sands Point, the Town of Oyster Bay, and the NYC Parks Olmsted Center.

The Graduate School of Arts and Science at Rutgers and the Department of History supplied funds for regional travel. I was also lucky to receive funding that made possible trips to the New York State Archives, the Rockefeller Archive Center, and the Lemelson Center for the Study of Invention and Innovation at the National Museum of American History. A Mellon Fellowship in Urban Landscape Studies at the Dumbarton Oaks Research Library and Collections, provided me with a wonderful and productive semester in Washington, DC. While my time there was primarily focused on different research, the time with the library's remarkable scholars and staff influenced this book for the better. Finally, thanks are due to the City University of New York and Queens College for the support offered to new faculty. Course releases made possible my time at Dumbarton Oaks and my participation in CUNY's Faculty Fellowship Publication Program.

I could not have asked for a better place to undertake my graduate work than the Department of History at Rutgers University. I am particularly thankful for the instruction of Dorothy Sue Cobble, Nancy Hewitt, Paul Israel, Jennifer Mittelstadt, and the late Susan Schrepfer, whom I consider to be model teachers and mentors. I was fortunate to complete my graduate work under the guidance of accomplished and supportive scholars. Paul

Clemens, Ann Fabian, Alison Isenberg, and Neil Maher supplied invaluable feedback on my work and generously shared their knowledge and time. A final year spent jointly at Rutgers and at Princeton University under Alison Isenberg's continued direction sped the completion of my initial work. Alison provided essential encouragement and critical feedback throughout my graduate experience. Her confidence in the merit of the earlier versions of this project and my ability to realize it proved vital. Elizabeth Blackmar's insightful comments on the first iteration of this project made it stronger and more complex than I had originally imagined it could be. Ann Fabian took on the Herculean task of teaching me the craft of writing. I am thankful for her patience and determination. Any accomplishments I can claim I owe directly to the mentorship of these scholars.

This project benefited from the feedback of a number of colleagues and friends. Thanks are due in particular to Adam Wolkoff for fielding questions on law cases. I also appreciate the thoughtful comments of the participants in the New York Metro Seminar in Environmental History as well as Judy Anderson, Dennis Halpin, Melissa Fuster, Kyle Shelton, and Adam Zalma. Thank you all for pushing me to clarify my ideas. A version of chapter 5 appeared in the *Journal of Urban History*, and the anonymous reviewer's comments led me to important revisions for the chapter. Thanks are also due to my fellow urban-environmental history enthusiast Melanie Kiechle. I am thankful for her continued friendship and brilliant advice.

This project stretched across my time at Rutgers, Towson University, and Queens College. At Towson, Ronn Pineo was an ideal mentor, welcoming me and supporting my research and teaching, including a very pungent field trip to a waste management plant. I have been similarly lucky at Queens College. The history department is full of friendly, wise, and ambitious scholars who make Queens a wonderful place to be. Thanks are particularly due to Joel Allen, Kristin Celello, and Peter Conolly-Smith for teaching me the CUNY ropes. As chair, Joel's insistence on the importance of research enabled me to take time off from teaching to dedicate crucial time to my research, for which I am grateful. At the University of Chicago Press, Timothy Gilfoyle and Timothy Mennel championed this project for a long time, and working with them both to finally see it published has been a pleasure. Timothy Gilfoyle has been generous with his time and comments, and I owe him particular thanks.

Researching and writing this book was a satisfying adventure. It was also an incredible challenge. For the friendship offered by good friends from my childhood, my years at Cornell University, and adventures in New York City and Washington, DC, I am thankful. I can never repay my extended family for their generosity in supplying me with the necessities of food, wine, and excellent company, but I will always try. In particular, my godfather Michael Murphy has been a steadfast champion and a frequent host, and I am very grateful for all he does for me and our family. Mike Kochanek has made me happy for a long time now, but his support during graduate school and my first years teaching was vital. His friendship, love, patience, and sense of fun are a gift I feel lucky for every day. My mother, Dr. Mary Murphy Schlichting, and my father, Dr. Kurt Schlichting, made my career as an academic possible. My twin sister Kerry Schlichting has played every role imaginable in my life, including editor, traveling companion, hostess, and sympathetic listener. As always, the thing I want most is to make my family proud. For her countless hours, smart conversation, enthusiasm, and crucial support whenever my nerve wavered, this work is dedicated to my mother, Mary Murphy.

# Notes

## Introduction

1.  Frederick Coburn, "The Five-Hundred Mile City," *World Today* 11 (December 1906), 1252.
2.  Matthew J. Bruccoli, "Explanatory Notes," in F. Scott Fizgerald, *The Great Gatsby* (1925; repr., New York: Scribners, 1992), 207.
3.  Andrew Needham and Allen Dieterich-Ward, "Beyond the Metropolis: Metropolitan Growth and Regional Transformation in Postwar America," *Journal of Urban History* 35, no. 7 (November 2009): 948.
4.  George E. Waring Jr., *Report on the Social Statistics of Cities, Part I: New England and Middle States, A: New England* (Washington, DC: Department of Interior Census Office, 1885), 531–32. On the significance of his report, see Thomas Bender, *The Unfinished City: New York and the Metropolitan Idea* (New York: New Press, 2001), 15.
5.  Walt Whitman, "Crossing Brooklyn Ferry," *Leaves of Grass, 1860: The 150th Anniversary Facsimile Edition*, ed., Jason Stacy (Iowa City: University of Iowa Press, 2009), 376–83, and Edward K. Spann, *The New Metropolis: New York City, 1840–1857* (New York: Columbia University Press, 1981), 402–3.
6.  Spann, *New Metropolis*, 411–12, and Ira Rosenwaike, *Population History of New York City* (Syracuse, NY: Syracuse University Press, 1972), 16, 58. Clifton Hood uses Rosenwaike to calculate the doubling of the population, a calculation that furthermore covers the greater city and not just Manhattan. See Hood, *722 Miles: The Building of the Subways and How They Transformed New York* (New York: Simon and Schuster, 1993), 35, 269n9.
7.  Waring, *Report on the Social Statistics of Cities*, 564–65. On mileage statistics for 1888, see Hood, *722 Miles*, 25.
8.  Peter Muller, "The Evolution of American Suburbs: A Geographical Interpretation," *Urbanism Past and Present*, no. 4 (Summer 1977): 1–10. On the dynamics of the urban-suburban edge, see Henry C. Binford, *The First Suburbs: Residential Communities on the Boston Periphery, 1815–1860* (Chicago: University of Chicago Press, 1985).
9.  *Report of the Board of Commissioners of the Department of Public Parks of the City of New York,[. . .] relating to improvements of portions of the counties of Westchester and New York, the improvement of Spuyten Duyvil Creek and Harlem River* [. . .]. (Albany, NY: Argus, 1871), 17–18.
10. New Rochelle, New York, Planning Board, *City Plan and Twenty-Year Program of Public Improvements for New Rochelle* (New York: Technical Advisory, 1929), 36.
11. Robert Fishman, "The Anti-Planners: The Contemporary Revolt against Planning and Its Significance for Planning History," in *Shaping an Urban World*, ed. Gordon E. Cherry (London: Mansell, 1980), 243–44, 251, and Kathryn J. Oberdeck, "From

Model Town to Edge City: Piety, Paternalism, and the Politics of Urban Planning in the United States," *Journal of Urban History* 26, no. 4 (May 2000): 508.

12. The literature on urban form in nineteenth-century New York is vast. Surveys that focus on nineteenth-century antecedents to planning include David C. Hammack, *Power and Society: Greater New York at the Turn of the Century* (New York: Russell Sage Foundation, 1982); David Schuyler, *The New Urban Landscape: The Redefinition of City Form in Nineteenth-Century America* (Baltimore: Johns Hopkins University Press, 1986); and David M. Scobey, *Empire City: The Making and Meaning of the New York City Landscape* (Philadelphia: Temple University Press, 2003).

13. On the evolution of park administration see Galen Cranz, *The Politics of Park Design: A History of Urban Parks in America* (Cambridge, MA: MIT Press, 1982): 157–81.

14. Revision of Caro's assessment began as early as 1989 with Joann P. Krieg, ed., *Robert Moses: Single-Minded Genius* (Interlaken, NY: Heart of the Lakes Publishing, 1989), and Jameson W. Doig, "Regional Conflict in the New York Metropolis: The Legend of Robert Moses and the Power of the Port Authority," *Urban Studies* 27, no. 2 (1990): 226. Joel Schwartz argues that Moses's projects aligned with conventional planning wisdom. See Schwartz, *The New York Approach: Robert Moses, Urban Liberals, and Redevelopment of the Inner City* (Columbus: Ohio University Press, 1993). The turning point in Moses revisionism was the 2008 publication of Hilary Ballon and Kenneth Jackson's *Robert Moses and the Modern City: The Transformation of New York* (New York: W. W. Norton), published in conjunction with a three-part exhibition hosted by the Museum of the City of New York, the Queens Museum of Art, and Columbia University's Miriam and Ira D. Wallach Art Gallery in 2007. Even so, Moses's Long Island projects, particularly his pre–World War II work, have not received significant reexamination beyond Owen Gutfreund's "Rebuilding New York in the Auto Age: Robert Moses and His Highways," Ballon and Jackson, *Robert Moses*, 86–93. For an example of the myths and broad characterizations that still haunt scholarship on Moses, see Tim Mennel's historiographic survey of urban renewal, "A Fight to Forget: Urban Renewal, Robert Moses, Jane Jacobs, and the Stories of Our Cities," *Journal of Urban History* 37, no. 4 (2011): 627–34.

15. Moses's environmental conservation is explored in Lawrence Kaplan and Carol P. Kaplan, *Between Ocean and City: The Transformation of Rockaway, New York* (New York: Columbia University Press, 2003), and Ted Steinberg, *Gotham Unbound: The Ecological History of Greater New York* (New York: Simon and Schuster, 2014).

16. Catherine McNeur, *Taming Manhattan: Environmental Battles and the Antebellum City* (Cambridge, MA: Harvard University Press, 2014), 181–85, and Michael Rawson, *Eden on the Charles: The Making of Boston* (Cambridge, MA: Harvard University Press, 2010), 272–76.

17. Daniel Schaffer, ed., *Two Centuries of American Planning* (Baltimore: Johns Hopkins University Press, 1988); Stanley K. Schultz, *Constructing Urban Culture: American Cities and City Planning,1800–1920* (Philadelphia: Temple University Press, 1989), and Jon Peterson, *The Birth of City Planning in the United States, 1840–1917* (Baltimore: Johns Hopkins University Press, 2003). For the evolution of the field, see Mary Corbin Sies and Christopher Silver, "Introduction: The History of Planning History," in *Planning the Twentieth-Century American City*, ed. Mary Corbin Sies and Christopher Silver (Baltimore: Johns Hopkins University Press, 1996), 1–36.

18. Scobey, *Empire City*, and McNeur, *Taming Manhattan*, 45–94.

19. *Report of the State Park Commission to the General Assembly for the Fiscal Year Ended September 30, 1914*, State of Connecticut Public Document no. 60 (Hartford, CT: Hartford Printing, 1914), 38.

20. Jack H. Archer, Donald L. Connors, and Kenneth Laurence, *The Public Trust Doctrine and the Management of America's Coasts* (Amherst: University of Massachusetts Press, 1994).

21. Robert C. Wood, *1400 Governments: The Political Economy of the New York Metropolitan Region* (Cambridge, MA: Harvard University Press, 1961), 1.

22. *The Graphic Regional Plan: Atlas and Description* [ . . . ] , vol. 1 of *Regional Plan of New York and Its Environs* (New York: Regional Plan of New York and Its Environs, 1929), 126. For a discussion of how the Hudson physically and politically divides the region, see Michael N. Danielson and Jameson W. Doig, *New York: The Politics of Urban Regional Development* (Berkeley: University of California Press, 1983), 58.

23. Eric H. Monkkonen, Robin L. Einhorn, Jon C. Teaford, Michael R. Fein, Keith D. Revell, and the contributors in Ballon and Jackson, *Robert Moses*, have examined government's role in shaping the urban form in the twentieth century. See Eric H. Monkkonen, *The Local State: Public Money and American Cities* (Stanford, CA: Stanford University Press, 1996); Robin L. Einhorn, *Property Rules: Political Economy in Chicago, 1833–1872* (Chicago: Chicago University Press, 1991); Michael R. Fein, *Paving the Way: New York Road Building and the American State, 1880–1956* (Lawrence: University Press of Kansas, 2008); and Keith D. Revell, *Building Gotham: Civic Culture and Public Policy in New York City, 1898–1938* (Baltimore: Johns Hopkins University Press, 2005).

24. Much of the literature on regionalism overlooks local responses to or effects on it, including Danielson and Doig, *New York*, and David A. Johnson, *Planning the Great Metropolis, The 1929 Regional Plan of New York and Its Environs* (London: E and FN Spon, 1996). An exception is Thomas J. Sugrue, "All Politics Is Local; The Persistence of Localism in Twentieth-Century America," in *The Democratic Experiment: New Directions in American Political History*, ed. Meg Jacobs, William J. Novak, and Julian E. Zelizer (Princeton, NJ: Princeton University Press, 2003). Jon C. Teaford's scholarship on home-rule government includes *City and Suburb: The Political Fragmentation of Metropolitan America, 1850–1970* (Baltimore: Johns Hopkins University Press, 1979); *The Unheralded Triumph: City Government in America 1870–1900* (Baltimore: Johns Hopkins University Press, 1984), and *Post-Suburbia: Government and Politics in the Edge Cities* (Baltimore: Johns Hopkins University Press, 1997).

25. Michael Rawson, *Eden on the Charles: The Making of Boston* (Cambridge, MA: Harvard University Press, 2010), 1, 3. *Eden on the Charles* and Christopher C. Sellers's *Crabgrass Crucible: Suburban Nature and the Rise of Environmentalism in Twentieth-Century America* (Chapel Hill: The University of North Carolina Press, 2013) represent the best recent scholarship uniting urban, suburban, and environmental history.

26. On the agency of nature see Rawson, *Eden on the Charles*, 19–20, and Linda Nash, "The Agency of Nature or the Nature of Agency?" *Environmental History* 10, no. 1 (January 2005): 67–69, and Ted Steinberg, "Down to Earth: Nature, Agency and Power in History," *American Historical Review* 107, no. 3 (June 2002): 798–820, cited in Rawson, *Eden on the Charles*, 296n25.

27. On the social emphasis of romanticism and the pastoral idea, see Schuyler, *New Urban Landscape*; Rawson, *Eden on the Charles*; Peter J. Schmitt, *Back to Nature: The*

*Arcadian Myth in Urban America* (New York: Oxford University Press, 1969); Thomas Bender, *Toward an Urban Vision: Ideas and Institutions in Nineteenth-Century America*. Lexington: University Press of Kentucky, 1975); and John R. Stilgoe, *Borderland Origins of the American Suburb, 1820–1939* (New Haven, CT: Yale University Press, 1990).

28. Roy Rosenzweig and Elizabeth Blackmar, *The Park and the People: A History of Central Park* (Ithaca, NY: Cornell University Press, 1992), 10.

29. Rawson, *Eden on the Charles*, 3.

30. Steinberg, *Gotham Unbound*, 394.

31. David Soll, *Empire of Water: An Environmental and Political History of the New York City Water Supply* (Ithaca, NY: Cornell University Press, 2013); Carol Sheriff, *The Artificial River: The Erie Canal and the Paradox of Progress, 1817–1862* (New York: Hill and Wang, 1996; and Elizabeth J. Pillsbury, "An American Bouillabaisse: The Ecology, Politics and Economics of Fishing around New York City, 1870-Present" (PhD diss., Columbia University, 2009). Wider environmental surveys of the city and New York State can be found in Mathew Gandy, *Concrete and Clay: Reworking Nature in New York City* (Cambridge, MA: MIT Press, 2002); David Stradling, *The Nature of New York: An Environmental History of the Empire State* (Ithaca: Cornell University Press, 2010); and Steinberg, *Gotham Unbound*.

32. Henry James, *The American Scene* (New York: Harper and Brother, 1907), 139–40.

*Chapter 1*

1. Phineas Taylor Barnum, *Struggles and Triumphs, or Forty Years' Recollections of P. T. Barnum* (New York: American News, 1871), 759–61.

2. William Steinway Diary, November 5, 1870, William Steinway Diary Project, 1861–1891, Smithsonian Institution National Museum of American History, http://americanhistory.si.edu/steinwaydiary (hereafter Steinway Diary).

3. *Illustrated Pamphlet on the Founding and Development of Steinway, N. Y.* (Long Island City: Steinway and Sons), 4, reproduced at Steinway Diary, http://americanhistory.si .edu/steinwaydiary/gallery/?category_id=1.

4. David Scobey, *Empire City: The Making and Meaning of the New York City Landscape* (Philadelphia: Temple University Press, 2003), 34–35.

5. Edward K. Spann, *The New Metropolis: New York City, 1840–1857* (New York: Columbia University Press, 1981), 404.

6. P. T. Barnum, "Present and Past," *Murray's Magazine* (London) 7 (January–June 1890): 103, and *Funny Stories Told by Phineas T. Barnum* (London: George Routledge and Sons, 1890), 361. On Barnum and the emergence of mass commercial entertainment in America, see Neil Harris, *Humbug: The Art of P.T. Barnum* (Boston: Little, Brown 1973); James W. Cook, *The Arts of Deception: Playing with Fraud in the Age of Barnum* (Cambridge, MA: Harvard University Press, 2001); and P. T. Barnum, *The Colossal P.T. Barnum Reader: Nothing Else Like It in the Universe*, ed. James W. Cook (Urbana: University of Illinois Press, 2005).

7. "Town of Steinway," *New York Times*, July 12, 1896, 25.

8. Barnum, *Struggles and Triumphs*, 559, and "Local Intelligence," *New York Times*, July 11, 1866, 2.

9. Barnum, *Struggles and Triumphs*, 762; see also *The Life of P. T. Barnum* (New York: Redfield, 1855), 401, and Charles Harvey Townshend, *The Commercial Interests of*

*Long Island Sound in General, and New Haven in Particular* (New Haven, CT: O. A. Dorman, 1883), 7.

10. Henry F. Walling, *Taintor's Route and City Guides, New York to the White Mountains via the Connecticut River* (New York: Taintor Brothers, 1867), 3, 8.

11. On the antebellum roots of company towns, see John S. Garner, *The Model Company Town: Urban Design Through Private Enterprise in Nineteenth-Century New England* (Amherst: University of Massachusetts Press, 1984), and John W. Reps, "The Towns the Companies Built," in *The Making of Urban America: A History of City Planning in the United States* (Princeton, NJ: Princeton University Press, 1965), 414–38.

12. Tin utensil manufacturer Florian Gorsjean built a company town at Woodhaven Village, and Conrad Poppenhusen, a manufacturer of rubber goods, incorporated College Point in 1870. See Vincent F. Seyfried, *300 Years of Long Island City: 1630–1930* ([New York?]: Edgian, 1984), 69. On philanthropist-sponsored housing in the Bronx, see Joel Schwartz, "Community Building on the Bronx Frontier: Morrisania, 1848–1875" (PhD diss., University of Chicago, 1972).

13. Barnum, *Struggles and Triumphs,* 550.

14. Henry W. Bellows, "Cities and Parks: With Special Reference to the New York Central Park," *Atlantic Monthly* 7 (1861): 416–29, quoted in Edwin G. Burrows and Mike Wallace, *Gotham: A History of New York City to 1898* (New York: Oxford University Press, 1999), 949.

15. Barnum, *Struggles and Triumphs,* 759.

16. Marc Linder and Lawrence S. Zacharias, *Of Cabbages and Kings County: Agriculture and the Formation of Modern Brooklyn* (Iowa City: University of Iowa, 1999); Tom Andersen, *This Fine Piece of Water: An Environmental History of Long Island Sound* (New Haven, CT: Yale University Press, 2002); and Marilyn E. Weigold, *The Long Island Sound: A History of Its People, Places, and Environment* (New York: New York University Press, 2004).

17. Map in Samuel Orcutt, *A History of the Old Town of Stratford and the City of Bridgeport, Connecticut* (New Haven, CT: Tuttle, Morehouse, and Taylor, 1886), vol. 2, between pages 996 and 997.

18. George Curtis Waldo Jr., *History of Bridgeport and Vicinity* (New York: S. J. Clarke, 1917), 1:37–38. For population see Campbell Gibson and Kay Jung, *Historical Census Statistics on Population Totals by Race, 1790 to 1990, and by Hispanic Origin, 1970 to 1990, for Large Cities and Other Urban Places in the United States* (Washington, DC: Population Division, US Census Bureau, 2005), table 7, http://purl.access.gpo.gov /GPO/LPS75196.

19. Gibson and Jung, *Historical Census Statistics.*

20. Ira Rosenwaike, *Population History of New York City* (Syracuse, NY: Syracuse University Press, 1972), 60, and Seyfried, *300 Years of Long Island City,* 84.

21. Edward Hagaman Hall, *A Volume Commemorating the Creation of the Second City of the World: By the Consolidation of the Communities Adjacent to New York Harbor* [ . . . ] (New York: Republic Press,1898), 16, quoted in Rosenwaike, *Population History,* 60.

22. John S. Billings, *Report on the Social Statistics of Cities in the United States at the Eleventh Census, 1890* (Washington, DC: Government Printing Office, 1895), 52, 54, 56.

23. *New York Evening Post* quoted in the *Long Island Star* (August 11, 1882), 2:5, reproduced in Seyfried, *300 Years of Long Island City,* 121; see also 114–18.

24. Seyfried, *300 Years of Long Island City,* 65, 107. For a history of marsh reclamation in

Hunter's Point, see Ted Steinberg, *Gotham Unbound: The Ecological History of Greater New York* (New York: Simon and Schuster, 2014), 140–42.

25.  "The Features of Greater New York," *New York Times*, July 12, 1896, 25. On the company's move, see *Illustrated Pamphlet on the Founding and Development of Steinway*, 22, reproduced at Steinway Diary. On labors' fight with Steinway and Sons, see Aaron Singer, *Labor Management Relations at Steinway and Sons, 1853–1896* (New York: Garland, 1986), 23–32, 38, 114–16; Richard K. Lieberman, *Steinway and Sons* (New Haven, CT: Yale University Press, 1995), 7–9, and Sean Wilentz, *Chants Democratic: New York City and the Rise of the American Working Class, 1788–1850*, 20th anniversary ed. (New York: Oxford University Press, 2004), 386–87, 371.

26.  David Tedone, ed., *A History of Connecticut's Coast: 400 Years of Coastal Industry and Development* (Hartford, CT: Coastal Area Management Program, 1982), 48.

27.  F. W. Beers, *Atlas of Long Island, New York: From Recent and Actual Surveys and Records* (New York: Beers, Comstock and Cline, 1873); Lionel Pincus and Princess Firyal Map Division, New York Public Library, *Map of Central Portions of the Cities of New York and Brooklyn* and *Map and Plan Showing the Street System in the 1st Ward of the Borough of Queens, Formerly Long Island City*, New York Public Library Digital Collections, http://digitalcollections.nypl.org/items/510d47e2-6338-a3d9-e040 -e00a18064a99. See also Seyfried, *300 Years of Long Island City*, 62, 70.

28.  Barnum, *Struggles and Triumphs*, 553; *Illustrated Pamphlet on the Founding and Development of Steinway*, 22, reproduced at Steinway Diary; and "Features of Greater New York."

29.  Steinberg, *Gotham Unbound*, 59. On "improvements" of wasteland, see also Jesse Goldstein, "*Terra Economica*: Waste and the Production of Enclosed Nature," *Antipode* 45 no. 2 (2013): 357–75.

30.  Steinberg, *Gotham Unbound*, 7, 75, 85, 128–9, 150.

31.  Oliver J. Dinius and Angela Vergara identify the two key elements of company towns to be "a single dominant industry with extensive company control over the daily life of a town." To this definition I add the company or individual executive's status as the dominant property owner. Like Dinius and Vergara, I employ a broader definition of company towns than Margaret Crawford, who argues that to classify as a model company town, a settlement must have the conceptual order of a comprehensive plan. See the introduction to *Company Towns in the Americas: Landscape Power and Working-Class Communities*, eds. Oliver J. Dinius and Angela Vergara (Athens: University of Georgia Press, 2011), 1–2, 12–13, 23, and Margaret Crawford, *Building the Workingman's Paradise: The Design of American Company Towns* (London: Verso, 1995), 23. Pullman, Illinois, set new standards for design and social control in the 1880s. See Stanley Buder, *Pullman: An Experiment in Industrial Order and Community Planning, 1880–1930* (New York: Oxford University Press, 1967), and Carl Smith, *Urban Disorder and the Shape of Belief: The Great Chicago Fire, the Haymarket Bomb, and the Model Town of Pullman*, rev. ed. (Chicago: University of Chicago Press; 2007).

32.  Two characteristic examples include William Cullen Bryant, "Can a City Be Planned?" *New York Evening Post*, March 16, 1868, 2, and "The Shaping of Towns," *American Architect and Building News* 2 (June 16, 1877), 186; (June 23, 1877), 195–96; and (June 30), 203–4, reproduced at John W. Reps, ed., *Urban Planning, 1794–1918: An International Anthology of Articles, Conference Papers, and Reports*, last modified

November 27, 2002, http://www.library.cornell.edu/Reps/DOCS/homepage.htm. Reps's website is a curated overview of nineteenth-century planning thought. For scholarship on city-building debates, see Scobey, *Empire City*; David Schuyler, *The New Urban Landscape: The Redefinition of City Form in Nineteenth-Century America* (Baltimore: Johns Hopkins University Press, 1986); and Thomas Bender, *The Unfinished City: New York and the Metropolitan Idea* (New York: New Press, 2001).

33.  Waldo, *History of Bridgeport*, 2:7; and "Bridgeport, Conn.," *New York Tribune*, July 18, 1872, 3.
34.  Barnum, *Life of P. T. Barnum,* 403.
35.  Joel Benton, "P.T. Barnum, Showman and Humorist," *Century Illustrated Magazine* 64, no. 4 (August 1902): 580, and "Bridgeport, Conn."
36.  "Iranistan, Bridgeport, Connecticut," *Frank Leslie's Illustrated Newspaper*, January 2, 1858, and "Iranistan, an Oriental Ville (Near Bridgeport, Connecticut)," lithograph (New York: Sarony and Major, ca.1852–1854), *Library of Congress Prints and Photographs Online Collection*, http://www.loc.gov/pictures/resource/pga.04090/. For nationwide coverage, see, for example, the Missouri *St. Louis Globe-Democrat*, August 24, 1880, 9; "The Home Residence of P. T. Barnum, the Showman, Is in Bridgeport, Conn.," North Carolina *Fayetteville Observer*, February 26, 1885; "Barnum's Burning: The Great Showman's Winter Quarters at Bridgeport in Flames," *Milwaukee Sentinel*, October 20, 1884, 2; and "Iranistan, Bridgeport, Connecticut." Waldo, *History of Bridgeport*, 2:7.
37.  Barnum, *Struggles and Triumphs*, 550.
38.  Mary K. Witkowski and Bruce Williams, *Bridgeport on the Sound* (Charleston, SC: Arcadia, 2001), 35; and "Barnum's Memory Honored," *New York Tribune*, July 5, 1893, 8.
39.  For the quote "nature as an organizer," see "P. T. Barnum as Mayor of Bridgeport, Conn.," *St. Louis Globe-Democrat* (Missouri), July 19, 1875, 3. See also "The Town of Bridgeport," *New York Times*, February 18, 1866, 5; "Bridgeport's Great Benefactor" *New York Times*, April 8, 1891, 2; and "P. T. Barnum Buried," *New York Tribune*, April 11, 1891, 7.
40.  Waldo, *History of Bridgeport*, 2:9, and Benton, "P.T. Barnum, Showman and Humorist," 586.
41.  The City of Bridgeport Town Clerk Land Records are only partially indexed for the nineteenth century. See *Bridgeport Town Records*, vol. 21 (Bridgeport, CT: Land Records, Office of the Town Clerk, n.d.).
42.  George Curtis Waldo Jr., *The Standard's History of Bridgeport* (Bridgeport, CT: Standard Association, 1897), 98; "Mr. Barnum's Life Story," and "Bequests by Barnum," *New York Times*, April 11, 1891, 1. See also H. G. Scofield, *Atlas of the City of Bridgeport Connecticut from Actual Surveys* (New York: J. B. Beers, 1876), and *Atlas of the City of Bridgeport, Conn, From Official Records and Actual Surveys* (Philadelphia: G. M. Hopkins, 1888).
43.  Map of Steinway & Sons Property, Long Island City New York, Box 040216, Folder 6, William Steinway Business Ventures Series, Steinway and Sons Collection, LaGuardia and Wagner Archives, Long Island City, New York (hereafter Steinway and Sons Collection). For acquisition details see Seyfried, *300 Years of Long Island City*, 70.
44.  "Bridgeport, Conn.," *New York Tribune*, July 18, 1872, 3, and Waldo, *Standard's History of Bridgeport*, 140.

45. For "my way," see 770–71; for "old fogys" and "we don't believe," see 758–59; for "looked upon me as a restless," see 772; and for "grasping farmers," see 770, Barnum, *Struggles and Triumphs*.

46. Steinway Diary, November 3 and 8, 1871. See also *Minute Book of the Astoria Homestead Co. Feb 2, 1869–May 9, 1896*, 2–3, Box 040196, Series 20, William Steinway's Business Ventures, Steinway Village, Steinway and Sons Collection, and Seyfried, *300 Years of Long Island City*, 73.

47. Steinway Diary, June 20, 1873.

48. On street openings, see Seyfried, *300 Years of Long Island City*, 71. Map of Steinway and Sons Property, Long Island City New York, Box 040216, Folder 6, William Steinway Business Ventures Series, Steinway and Sons Collection.

49. *Minute Book of the Astoria Homestead Co. Feb 2, 1869–May 9, 1896*, 2, 5, 28, Box 040196, Series 20, William Steinway's Business Ventures, Steinway Village, Steinway and Sons Collection.

50. *Ravenswood Improvement Co. Minute Book Apr. 21, 1892–Apr. 14, 1896*, 28–29; *Minute Book of the Astoria Homestead Co.*, 76, 88; *Minute Book of the Astoria Homestead Company June 20, 1896–Jan. 8, 1916*, 22; and "In the Matter of the Consolidation of the Astoria Homestead Company and the Ravenswood Improvement Company," *Ravenswood Improvement Co. Minute Book Apr. 21, 1892–Apr. 14, 1896*, 1–2, 12, and 63, Box 040197, Series 20, William Steinway's Business Ventures, Steinway Village, Steinway and Sons Collection. See also Map of Steinway and Sons Property, Long Island City New York, Box 040216, Folder 6, William Steinway Business Ventures Series, Steinway and Sons Collection.

51. Barnum, *Struggles and Triumphs*, 772.

52. *Bridgeport Town Records*, 166, 275, 421, 828, and 830. See also "Bridgeport, Conn.," *New York Tribune*, July 18, 1872, 3, and Barnum, *Struggles and Triumphs*, 389.

53. Barnum, *Struggles and Triumphs*, 415.

54. *Taintor's Route and City Guides, New York to Boston Via the Shoreline Route* (New York: Taintor Brothers, 1867), 9, and Thomas Henshaw, *The History of Winchester Firearms 1866–1992*, 6th ed. (Clinton, NJ: New Win, 1993), 8.

55. George E. Waring Jr., *Report on the Social Statistics of Cities, Part I: New England and Middle States, A: New England* (Washington, DC: Department of Interior Census Office, 1885), 388. On Hunter's Point see Seyfried, *300 Years of Long Island City*, 83–97,107, 112–13, 132. On Port Chester's status as a residential city, see "Cities of Westchester and Her County Seat," *New York Times*, May 3, 1903, 31. On Bridgeport's position, see "Waldemere—Seaside Residence of P. T. Barnum," *Horticulturist and Journal of Rural Art and Rural Taste* 27, no. 316 (October 1872), 288; "Bridgeport, Conn.," *New York Tribune*, July 18, 1872, 3; and "The Town of Bridgeport."

56. C. F. Theodore, William Steinway, and Albert Steinway, "Application for Use of Land under East River," March 3, 1871, Folder 38, Box 00203, Legal Papers Sub Series, William Steinway Business Ventures Series, Steinway and Sons Collection. On channelization approval, see Steinway Diary, March 28, 1873.

57. Steinway Diary, March 30, 1872, and "Town of Steinway," *New York Times*, July 12, 1896, 25

58. *Illustrated Pamphlet on the Founding and Development of Steinway*, 4, reproduced at Steinway Diary. Seyfried details the construction progress; see Seyfried, *300 Years of Long Island City*, 71–72. For worker statistics, see Singer, 92.

59. Steinway Diary, March 18,1896; October 15, 1892, and Lieberman, *Steinway and Sons*, 106, 116–17. On Daimler Motor, see also "The Daimler Motor," *Horseless Age: A Monthly Journal* 1, no. 1 (November 1895): 9, and *Illustrated Pamphlet on the Founding and Development of Steinway*, 26, reproduced at Steinway Diary.

60. Waldo, *The Standard's History of Bridgeport*, 54.

61. Steinway Diary, April 3, 1875. See also Singer, 96, and Lieberman, *Steinway and Sons*, 82–83, 105n15. Vincent F. Seyfried reconstructs Steinway's lines in *The New York and Queens County Railway and the Steinway Lines 1867–1939* (Hollis, NY: Vincent Seyfried, 1950).

62. *Eagle Almanac 1886–1892, Railroad Statistics*, cited in Seyfried, *New York and Queens County Railway*, 5– 6; Steinway Diary, March 23, 1892; March 24, 1892; March 30, 1892. See also Colin Hamilton Livingstone, *The Citizen Guide to Brooklyn and Long Island* (Brooklyn, NY: R. W. Wilson, 1893), 191, and *Documentary History of Railroad Companies*, Report of the Public Service Commission for the First District of the State of New York ([Albany, NY?]: [J. B. Lyon?], [1914?]), 5:543, 586–87, 1267–71.

63. "Long Island Territory," *New York Times*, May 24, 1896, 14.

64. "The Second East River Bridge," *New York Times*, July 1, 1875, 8, and Steinway Diary, May 11, 1887. Steinway attended dozens of meetings on bridge business. See for example Steinway Diary, November 20, 1886, and December 21, 1888. On project delays, see, for example, "Albany Working Methods: Subjects Now Before the Legislature," *New York Times*, April 6, 1878, 7; "Two Protests Entered: Arguments For and Against the Proposed East River Bridge," *New York Times*, July 8, 1887, 3; "The Long Island Bridge," *New York Times*, July 15, 1887, 3; "The Blackwell's Island Bridge," *New York Times*, March 10, 1890, 5; and "Bridge Schemes Blocked," *New York Times*, December 16, 1899, 5.

65. Steinway Diary, February 3, 1894. See also New York and Long Island Bridge Company, *Report of Board of Consulting Engineers, Appointed to Recommend a Plan for the New York and Long Island Bridge across the East River, at Blackwell's Island* (New York: Graphic Company, 1877); "Blackwell's Island Bridge," *New York Times*, May 28, 1887, 4; and "Blackwell's Island Bridge" *New York Times*, March 29, 1877, 8.

66. Steinway Diary, May 11, 1887.

67. Steinway played a large role in transit across all of New York City, not just Queens. See "New Companies Formed under the General Railroad Law during the Year Ending September 30, 1885," *Documents of the Assembly of the State of New York, One Hundred and Ninth Session, 1886*, vol. 3, nos. 24–26, pt. 1 (Albany, NY: Weed, Parsons, 1886), 390–91, and *Report of Committee on Docks, Adverse to Application of the New York and Long Island Railroad Company, for Permission to Join This City with Long Island by Railway Tunnels; Board of Alderman, January 29, 1889* (New York: Martin B. Brown, 1889), 20. For a history of the subway tunnel to Queens, see Clifton Hood, *722 Miles: The Building of the Subways and How They Transformed New York* (New York: Simon and Schuster, 1993), 163–65.

68. For "speedy consummation" and "vast benefit," see Steinway Diary, November 1, 1895; January 27, 1889; November 26, 1890; and December 7, 1890.

69. "Mr. Barnum's Life Story," *New York Times*, April 8, 1891, 1, and "Barnum as Mayor," *Daily Inter-Ocean*, April 23, 1875, 4.

70. Hood, *722 Miles*, 56–61.

71. Benton, "P.T. Barnum, Showman and Humorist," 586.

72. Queens County Clerk's Office, Title Deeds, Liber 432, 1874, pp. 81, 84, *Newtown Register*, July 22, 1880, cited in Singer, 95n52, and R. G. Dun and Company, March 18, June 23, 1878, R. G. Dun and Company credit report volumes, 1840–1895 (inclusive), Baker Library, Harvard University Graduate School of Business Administration, cited in Lieberman, *Steinway and Sons*, 82n16. For the 1870s depression and Steinway Village development, see D. W. Fostle, *The Steinway Saga: An American Dynasty* (New York; Scribner, 1995), 290–91.

73. Steinway Diary, August 6, 1872.

74. Smith, *Urban Disorder and the Shape of Belief*, 184–94, 207.

75. Henry William Blair, United States Congress, Senate Committee on Education and Labor, *Report of the Committee of the Senate upon the Relations between Labor and Capital and Testimony Taken by the Committee*, vol. 2 (Washington, DC: United States Government Printing Office, 1885), 1095.

76. *Illustrated Pamphlet on the Founding and Development of Steinway*, 22, reproduced at Steinway Diary.

77. For another company town that deviated from Pullman's standards, see Garner, *Model Company Town*, and Thomas A. McMullin, "Lost Alternative: The Urban Industrial Utopia of William D. Howland," *New England Quarterly* 55, no. 1 (March 1982): 25–38.

78. J. M. Kelsey, *History of Long Island City, New York* (New York: Long Island Star, 1896), 49–50. Scattered information about worker housing at Steinway is also available in Richard Pluntz, *A History of Housing in New York City: Dwelling Type and Social Change in the American Metropolis* (New York: Columbia University Press, 1990), 114. For census analysis, see Kerry Ehlinger, "Reinterpreting Steinway Village" (master's thesis, Columbia University, 1998), 43, 47.

79. *Long Island City Star*, March 17, 1890, and "Hauser und Bauplatze auf Steinway & Sons," circular, photo 04.003.0425, Steinway and Sons.

80. See "Bridgeport, Conn.," *New York Tribune*, July 18, 1872, 3, And Waldo, *Standard's History of Bridgeport*, 140.

81. For an example, see Palliser, Palliser, & Co., *Palliser's American Cottage Homes* (Bridgeport, CT: Palliser, Palliser & Co., 1878); Linda E. Smeins, *Building an American Identity: Pattern Book Homes and Communities, 1870–1900* (Walnut Creek, CA: AltaMira, 1999); and Daniel D. Reiff, *Houses from Books: Treatises, Pattern Books, and Catalogs in American Architecture, 1738–1950* (University Park: Penn State University Press, 2000).

82. Phineas Taylor Barnum, *How I Made Millions: or, The Secret of Success* (New York, Belford, Clark, 1884), 272.

83. Photo 04.002.1217, "Steinway Houses," and photo 04.002.0629, "These Houses are Possible the Ones on 20th Avenue . . . ," Steinway and Sons. See also Seyfried, *300 Years of Long Island City*, 72. On minimally designed company towns, see Alison K. Hoagland, *Mine Towns: Buildings for Workers in Michigan's Copper Country* (Minneapolis: University of Minnesota Press, 2010), xx.

84. Blair, *Report of the Committee of the Senate*, 1904–5; Benton, "P. T. Barnum Showman and Humorist," 586.

85. "Waldemere—Seaside Residence of P. T. Barnum," 288.

86. Mountain Grove Cemetery, *Lots about Lots; Or, The Great Fair, and What Preceded It: Sold for the Benefit of the Mountain Grove Cemetery* (Farmer Office Presses, 1879), 5. On mid-nineteenth-century rural cemeteries, see Thomas Bender, "The 'Rural'

Cemetery Movement: Urban Travail and the Appeal of Nature," *New England Quarterly* 47, no. 2 (June 1947), 196–211. For an environmental history of the movement, see Aaron Sach's *Arcadian America: The Death and Life of an Environmental Tradition* (New Haven, CT: Yale University Press, 2013).

87. On donated plots, see *Bangor Daily Whig and Courier* (Maine), no.145 (June 19, 1883), col. C, and "Bridgeport, Conn.," *New York Tribune*, July 18, 1872, 3.

88. The extension of Lafayette Street, Lambert Street (Warren Avenue), and Myrtle Avenue to the shore, and a dozen cross streets were laid out but never finished. Waldo, *History of Bridgeport*, 1:99. See also Waldo, *Standard's History of Bridgeport*, 81.

89. "Department of Parks," *Bridgeport Municipal Register* (1873), 109, Seaside Park, Bridgeport, Connecticut Subject File, Job Number 12021, Frederick Law Olmsted Documentary Editing Project, Special Collections, University Library, American University, Washington, DC (hereafter Frederick Law Olmsted Documentary Editing Project).

90. Barnum, *Struggles and Triumphs*, 760.

91. For a timeline of parcel acquisitions, see "Chronology of Seaside Park," *Seaside Park* and "Department of Parks," *Bridgeport Municipal Register* (1873), 110–14, 118–19, Seaside Park, Bridgeport, Connecticut Subject File, Job Number 12021, Frederick Law Olmsted Documentary Editing Project. See also Barnum, *Struggles and Triumphs*, 759–62, and "Waldemere—Seaside Residence of P. T. Barnum," 288.

92. *Bridgeport Standard*, October 7, 1865, quoted in *National Register of Historic Places, Barnum/Pallisser Historic District, National Register 82000995*, 8-3.

93. "Department of Parks," *Bridgeport Municipal Register* (1873), 117, Seaside Park, Bridgeport, Connecticut Subject File, Job Number 12021, Frederick Law Olmsted Documentary Editing Project. See also *The Daily Standard* for January 15, 1867, reproduced in Waldo, *History of Bridgeport and Vicinity*, 1:278.

94. Clipping, *Bridgeport Standard*, August 26, 1865, Seaside Park, Bridgeport, Connecticut Subject File, Job Number 12021, Frederick Law Olmsted Documentary Editing Project; Frederick Law Olmsted and John Charles Olmsted, *Beardsley Park Landscape Architects' Preliminary Report September 1884* (Boston: privately printed, 1884), 5. A topographical map by civil engineer Abner C. Thomas bearing the stamp of Olmsted, Vaux and Company is the only known record of the design for Seaside; no copies of Olmsted and Vaux's final plan survive. Abner C. Thomas, "Topographical Map of Sea Side Park," (New York: Olmsted, Vaux and Co., 1866), DL*275640.16, Division of Home and Community Life, Smithsonian Museum of American History, Washington, DC.

95. Olmsted and Olmsted, *Beardsley Park Landscape Architects' Preliminary Report September 1884*, 4. See also the introduction to *The Papers of Frederick Law Olmsted*, vol. 6, *The Years of Olmsted, Vaux, and Company 1865–1874*, ed. David Schuyler, Jane Turner Censer, Carolyn F. Hoffman, and Kenneth Hawkins (Baltimore: Johns Hopkins University Press, 1992), 28. On Viele, see Waldo, *Standard's History of Bridgeport*, 67. On the seawall, see "Department of Parks," *Bridgeport Municipal Register* (1873), 111, Seaside Park, Bridgeport, Connecticut Subject File, Job Number 12021, Frederick Law Olmsted Documentary Editing Project.

96. Sylvester Baxter, "Seaside Parks for the People," *New York Observer and Chronicle* 76, no. 27 (July 1898): 22; "Department of Parks"; and *Appleton's Hand-Book of American Travel: Northern and Eastern Tour* (New York: D. Appleton, 1870), 146–47.

97. "Fragments," *Park and Cemetery and Landscape Gardening* 10, no. 7 (September 1, 1900), 164.
98. "Features of Greater New York," *New York Times*, July 12, 1896, 25. On the distance to the trolley stop, see also Jeffrey Kroessler, "North Beach: The Rise and Decline of a Working-Class Resort," in *Long Island Studies: Evoking a Sense of Place*, ed. Joann P. Krieg (Interlaken, NY: Heart of the Lakes Publishing, 1988), 142. For the grant, see Theodore, Steinway, and Steinway, "Application."
99. See *Long Island City Star*, June 11, 1886, 1:6, cited in Kroessler, "North Beach," 144. For the railroad route, see Lieberman, *Steinway and Sons*, 105.
100. For the *World* quote, see Christopher Gray, "Where the Streets Smelled Like Beer," *New York Times*, March 22, 2012. For a history of the company, see George Ehret, *Twenty-Five Years of Brewing: With an Illustrated History of American Beer* (New York: Gast Lithograph and Engraving, 1881).
101. Theodore, Steinway, and Steinway, "Application.," See also Seyfried, *300 Years of Long Island City*, 73, and Steinway Diary, August 26th, Thursday, 1886. See also Kroessler, "North Beach," 143.
102. Map of Steinway and Sons holdings in Astoria (at that time part of Long Island City), circa 1874, From J. M. Kelsey's *History of Long Island City, New York* (New York: Lewis, 1896), reproduced at Steinway Diary.
103. "The Town of Bridgeport."
104. "Waldemere—Seaside Residence of P. T. Barnum," 288.
105. "'Waldemere,' Mr. Barnum's Residences at Bridgeport, Conn.," *Frank Leslie's Illustrated Newspaper*, August 29, 1874, 397.
106. "Necessity of Marsh Drainage," *Real Estate Record and Builders' Guide* (September 5, 1868), 1. For this article I am indebted to Steinberg, *Gotham Unbound*, 128n5. For marshes filled in northwestern Queens, see Seyfried, *300 Years of Long Island City*, 107.
107. Frederick Law Olmsted, "A Healthy Change in the Tone of the Human Heart (Suggestions to Cities)," *The Papers of Frederick Law Olmsted*, vol. 9, *The Early Boston Years 1882–1890*, ed. Charles E. Beveridge, Ethan Carr, Amanda Gagel, and Michael Shapiro (Baltimore: Johns Hopkins Press, 2013), 338.
108. Ira Rosenwaike, *Population History of New York City* (Syracuse, NY: Syracuse University Press, 1972), 59.
109. "Inventory and Appraisement of the Personal estate of William Steinway, April 5, 1898," cited in Fostle, *Steinway Saga*, 414–15, and Lieberman, *Steinway and Sons*, 106, 116–117. See also Seyfried, *The New York and Queens County Railway and the Steinway Lines*, 12.
110. "More Big Bridge Schemes," *New York Times*, February 6, 1892, 1, and Clifton Hood, *722 Miles: The Building of the Subways and How They Transformed New York* (New York: Simon and Schuster, 1993), 136, 148, 168.
111. For population, see Campbell Gibson and Kay Jung, *Historical Census Statistics on Population Totals by Race, 1790 to 1990, and by Hispanic origin, 1970 to 1990, for Large Cities and Other Urban Places in the United States* (Washington, DC: Population Division, US Census Bureau, 2005), http://purl.access.gpo.gov/GPO/LPS75196. On secession and annexation see Waldo, *Standard's History of Bridgeport*, 221–22; *History of Bridgeport*, 1:91, 93–96; Waring, *Report on the Social Statistics*, 388, and Thomas J. Farnham, *Fairfield: The Biography of a Community 1639–2000* (West Kennebunk, ME: Phoenix 2000), 181.

## Chapter 2

1. Andrew H. Green, "Communication of the Comptroller of the Park Relative to West-chester County, Harlem River, and Spuyten Duyvil Creek," in *Thirteenth Annual Report of the Board of Commissioners of the Central Park* [. . .], Board of Commissioners of Central Park (New York: Evening Post Steam Presses, 1870), 159. See also Andrew H. Green, *Public Improvements in the City of New York, Communication from Andrew H. Green to Wm. A. Booth, Esq., and Others. Sept. 28th, 1874* (New York: 1974), 9, 23.

2. Green, "Communication of the Comptroller," 153. See also *The Father of Greater New York: Official Report of the Presentation to Andrew Haswell Green of a Gold Medal* [. . .] (New York: Historical and Memorial Committee of the Mayor's Committee on the Celebration of Municipal Consolidation, 1899), 16, 29–30.

3. David C. Hammack, "Comprehensive Planning before the Comprehensive Plan: A New Look at the Nineteenth-Century American City," in *Two Centuries of American Planning*, ed. Daniel Schaffer (Baltimore: Johns Hopkins University Press, 1988), 146.

4. "Call it 'Trans-Harlem,'" *New York Times*, May 12, 1893, 11.

5. An 1870 headline quoted in Edwin G. Burrows and Mike Wallace, *Gotham: A History of New York City to 1898* (New York: Oxford University Press, 1999), 949.

6. William Bridges, "Remarks of the Commissioners for Laying Out Streets and Roads in the City of New York, Under the Act of April 3, 1807," *Map of the City of New York and Island of Manhattan with Explanatory Remarks and References* (New York: William Bridges, 1811), reproduced at John W. Reps, ed., *Urban Planning, 1794–1918: An International Anthology of Articles, Conference Papers, and Reports*, last modified November 27 2002), http://www.library.cornell.edu/Reps/DOCS/homepage.htm. See also Elizabeth Blackmar, *Manhattan for Rent, 1785–1850* (Ithaca, NY: Cornell University Press, 1989), 150, and Hilary Ballon, introduction to *The Greatest Grid: The Master Plan of Manhattan 1811–2011*, ed. Hilary Ballon (New York: Museum of the City of New York and Columbia University Press, 2012), 13–14.

7. William Cullen Bryant, "Can a City be Planned?" *New York Evening Post*, March 16, 186), 2.

8. On the innovative urbanism of civic leaders and planning visionaries, see David M. Scobey, *Empire City: The Making and Meaning of the New York City Landscape* (Philadelphia: Temple University Press, 2003).

9. Ballon, introduction to *Greatest Grid*, 17, 24, 27.

10. William Cauldwell, "Annexation," in *The Great North Side; Or, Borough of The Bronx*, ed. Bronx Board of Trade (New York: Knickerbocker Press, 1897), 22. On Morrisania's 1868 commission, see Frederick Law Olmsted and J. James R. Croes to William R. Martin, October 31, 1877, in Frederick Law Olmsted, *The Papers of Frederick Law Olmsted*, vol. 7, *Parks, Politics and Patronage, 1874–1882*, ed. Charles E. Beveridge, Carolyn F. Hoffman, Kenneth Hawkins, and Tina Hummel (Baltimore: Johns Hopkins University Press, 2007), 349n13.

11. Green, *Public Improvements*, 9; *Communication to the Commissioners of Central Park, Relative to the Improvement of the Sixth and Seventh Avenues* [. . .] *and Other Subjects* (New York: William Cullen Bryant, 1865), 36; and "Communication of the Comptroller," 148, 160–61. See also William R. Martin, "The Growth of New York," *Real Estate Record and Builder's Guide* 15, no. 368 (supplement, April 3, 1875), 91.

12. "Report of the Board of Commissioners of the Department of Public Parks of the City of New York, in Conformity 'With an Act of the Legislature, Passed April 15,

1871, Relating to Improvements of Portions of the Counties of Westchester and New York, the Improvement of Spuyten Duyvil Creek and Harlem River and to Facilities of Communication between Said Counties,'" in New York State, Senate Documents, vol. 4, no. 72 (Albany, NY: Argus, 1872), 2, 9, and *Twelfth Annual Report of the Board of Commissioners of the Central Park* [. . .] (New York: Evening Post Steam Presses, 1869), 57.

13.  *Thirteenth Annual Report of the Board of Commissioners of the Central Park* [. . .] (New York: Evening Post Steam Presses, 1870), 65.

14.  On comprehensive planning before planning professionalized, see Hammack, "Comprehensive Planning before the Comprehensive Plan," 139–44. On this topic and the use of the term *comprehensive*, see Jon Peterson, *The Birth of City Planning in the United States, 1840–1917* (Baltimore: Johns Hopkins University Press, 2003), 3.

15.  Green, *Public Improvements*, 9; *Communication to the Commissioners*, 36–7; "Surveying, Laying-Out and Monumenting the Twenty-Third and Twenty-Fourth Wards and Part of the Twelfth Ward of New York City, I," *Engineering News* 8 (February 12, 1881), 62; and "Report of the Board of Commissioners of the Department of Public Parks of the City of New York, in Conformity 'With an Act of the Legislature, Passed April 15, 1871,'" 19–20.

16.  *Father of Greater New York*, 35.

17.  *Report of the Board of Commissioners of the Department of Public Parks of the City of New York, Relating to Improvements of Portions of the Counties of Westchester and New York, and the Improvement of Spuyten Duyvil Creek* [. . .] (Albany, NY: Argus, 1871), 17–18.

18.  "Communication on the Subject of Water Supply," in ibid., 27–29.

19.  Green, "Communication of the Comptroller," 158.

20.  *Report of the Board of Commissioners of the Department of Public Parks of the City of New York, Relating to Improvements of Portions of the Counties of Westchester and New York, and the improvement of Spuyten Duyvil Creek*, 17–18.

21.  Green, "Communication of the Comptroller," 162, 159, 151–53. See also Martha Rogers, "The Bronx Parks System: A Faded Design" *Landscape* 27, no. 2 (1983): 13–21. For an overview of Green's power in the city, see Laura Wood Roper, *F.L.O.: A Biography of Frederick Law Olmsted* (Baltimore: Johns Hopkins University Press, 1983), 349.

22.  "The Yonkers Railway Scheme," *New York Times*, December 21, 1859, 4. For the assessment of Morrisania's grid as proof of New York City as an urbanizing juggernaut, see Richard Plunz, "Reading Bronx Housing, 1890–1940," in *Building a Borough: Architecture and Planning in the Bronx 1890–1940*, ed. Bronx Museum of the Arts (Bronx, NY: Bronx Museum of the Arts, 1986), 44.

23.  "Report of the Board of Commissioners of the Department of Public Parks of the City of New York, in Conformity 'With an Act of the Legislature, Passed April 15, 1871,'" 4, and Cauldwell, "Annexation," 7, 20–22. On the grid, see Edward K. Spann, "The Greatest Grid: The New York Plan of 1811," in *Two Centuries of American Planning*, ed. Daniel Schaffer (Baltimore: Johns Hopkins University Press, 1988), 26, and Hilary Ballon, introduction to *Greatest Grid*, 13.

24.  Evelyn Gonzalez, *The Bronx* (New York: Columbia University Press, 2004), 25, 21.

25.  "Advantages of the Great North Side," *New York Times*, March 24, 1895, 20.

26.  For an overview of the revisionism of Tweed and the machine, see Raymond A. Mohl, ed., *The Making of Urban America*, 2nd ed. (Wilmington, DE: Scholarly Resources, 1997), 340. For Corson's connections to Tammany and Tweed, see Joel

Schwartz, "Community Building on the Bronx Frontier: Morrisania, 1848–1875" (PhD diss., University of Chicago, 1972), 278–90.

27. Cauldwell, "Annexation," 24.

28. Neil S. Martin, "Westchester as an Evolving Suburb," in *Westchester County: The Past Hundred Years, 1883–1983*, ed. Marilyn E. Weigold (Valhalla, NY: Westchester County Historical Society, 1983), 93–94.

29. Outside of the districts to be annexed, 9,023 Westchester voters approved of annexation, while 2,643 voted against it. More than 55,300 New Yorkers voted for annexation and only 8,380 against it. Cauldwell, "Annexation," 25, and "Annexation of Towns in Westchester to New-York," *New York Times*, May 1, 1873, 5.

30. Cauldwell, "Annexation,"19.

31. Samuel Sloan, *Architectural Review and American Builder's Journal* 2 (1869), 7–8, quoted in Scobey, *Empire City*, 85. See also "City Railroad Traffic," *New York Times*, October 3, 1895, 4, and "Seek Another Driveway," *New York Times*, March 8, 2011, 9.

32. Frederick Law Olmsted and J. James R. Croes to William R. Martin, October 31, 1877, Olmsted, *Papers of Frederick Law Olmsted*, 7:349n13. Roper, *Biography of Frederick Law Olmsted*, 348–54.

33. Scobey, *Empire City*, 198–99.

34. Martin, "Growth of New York," 8.

35. For Green's conflicts with Olmsted and Martin, see Roper, *Biography of Frederick Law Olmsted*, 145–46, 154–55, 352–56; Scobey, *Empire City*, 262–63; and David Schuyler, *The New Urban Landscape: The Redefinition of City Form in Nineteenth-Century America* (Baltimore: Johns Hopkins University Press, 1986), 178.

36. Olmsted, Vaux, & Co., "Preliminary Report upon the Proposed Suburban Village at Riverside," reproduced in *Civilizing American Cities: Writings on City Landscapes*, ed. S. B. Sutton (New York: Da Capo, 1997), 295. See also Frederick Law Olmsted and J. James R. Croes, "Preliminary Report of the Landscape Architect and the Civil and Topographical Engineer, upon the Laying Out of the Twenty-third and Twenty-fourth Wards," (November 15, 1876), in Olmsted, *Papers of Frederick Law Olmsted*, 7:242–51. See also "The Future Suburbs of New-York," *New York Times*, February 25, 1877, 6.

37. Martin, "Growth of New York," 6. See also Olmsted and Croes, "Preliminary Report," 250n12.

38. Elizabeth Macdonald, "Suburban Vision to Urban Reality: The Evolution of Olmsted and Vaux's Brooklyn Parkway Neighborhoods," *Journal of Planning History* 4, no. 4 (2005): 295–321.

39. For Olmsted's large-scale planning opportunities in the 1870s, which included the Bronx and the Boston park system, see Olmsted, *Papers of Frederick Law Olmsted*, vol. 7.

40. Olmsted and Croes, "Preliminary Report," 247.

41. Frederick Law Olmsted and J. James R. Croes to William R. Martin, October 31, 1877, in Olmsted, *Papers of Frederick Law Olmsted*, 7:340.

42. Olmsted and Croes to Martin, and Bronx Board of Trade, *Great North Side*, 37.

43. Olmsted and Croes, "Preliminary Report," 244.

44. Frederick Law Olmsted, "Report to the Staten Island Improvement Commission of a Preliminary Scheme of Improvements (1871)," in *Landscape into Cityscape: Frederick Law Olmsted's Plans for a Greater New York*, ed. Albert Fein (New York: Van Nostrand Reinhold, 1981), 183.

45. Frederick Law Olmsted and J. James R. Croes, "Report of the Landscape Architect and the Civil and Topographical Engineering, Accompanying a Plan for Laying Out that Part of the Twenty-Fourth Ward Lying West of the Riverdale Road" (November 21, 1876), in Olmsted, *Papers of Frederick Law Olmsted*, 7:253. Olmsted's biographer Laura Roper concludes that the 1876 and 1877 reports "bear the stamp of Olmsted's thought and style and appear to have been written largely by him"; Roper, *Biography of Frederick Law Olmsted*, 355.
46. Olmsted and Croes, "Report of the Landscape Architect," 253.
47. Frederick Law Olmsted and J. James R. Croes to William R. Martin, October 31, 1877, in Olmsted, *Papers of Frederick Law Olmsted*, 7:347. See also Scobey, *Empire City*, 262.
48. Stanley K. Schultz, *Constructing Urban Culture: American Cities and City Planning, 1800–1920* (Philadelphia: Temple University Press, 1989), 3, and Stanley K. Shultz and Clay McShane, "To Engineer the Metropolis: Sewers, Sanitation, and City Planning in Late-Nineteenth-Century America," *Journal of American History* 65, no. 2 (September 1978): 389–411.
49. "Map of the 23rd and 24th Wards, New York, Compiled for an Index to Volumes of Important Maps," *Atlases of New York City Certified Copies of Important Maps Appertaining to the 23rd and 24th Wards, City of New York, Filed in the Register's Office at White Plains, County of Westchester, New York* (New York: E. Robinson, 1888), Lionel Pincus and Princess Firyal Map Division, New York Public Library, http://digitalgallery.nypl .org/nypldigital/id?1524169.
50. For an overview of the partisan politics of the parks commission in the 1870s, see Charles E. Beveridge, introduction to Olmsted, *Papers of Frederick Law Olmsted*, 7:3–5.
51. Introduction to Olmsted, *Papers of Frederick Law Olmsted*, vol. 6, *The Years of Olmsted, Vaux, and Company 1865–1874*, ed. David Schuyler, Jane Turner Censer, Carolyn F. Hoffman, and Kenneth Hawkins (Baltimore: Johns Hopkins University Press, 1992), 38. For the fight between Green and Martin, see Roper, *Biography of Frederick Law Olmsted*, 348–56.
52. Roper, *Biography of Frederick Law Olmsted*, 349.
53. Roper, 356, and Scobey, *Empire City*, 264.
54. Frederick Law Olmsted, "The Future of New York," *New York Daily Tribune*, December 28, 1879, 5. He expanded this critique in *A Consideration of the Justifying Value of a Public Park* (Boston: Tolman and White, 1881), 16. John C. Olmsted to Frederick Law Olmsted Jr., October 1, 1913, quoted in Schuyler, *New Urban Landscape*, 178.
55. "To Replan New York," *New York Times*, January 11, 1903, 6; "The Future Suburbs of New-York," *New York Times*, February 25, 1877, 6; "Street Plan for the Addition of 1895," *Real Estate Record and Guide*, December 25, 1897, 992; and "The Big Boulevard Scheme," *New York Times*, March 26, 1897, 4.
56. Bronx Board of Trade, *Great North Side*, 10.
57. US Census Office, *Report on the Social Statistics of Cities*, pt. 1 (Washington, DC: US Government Printing Office, 1886), 568, quoted in Ira Rosenwaike, *Population History of New York City* (Syracuse, NY: Syracuse University Press, 1972), 56–57.
58. Louis F. Haffen, "Department of Street Improvements," in Bronx Board of Trade, *Great North Side*, 39; "Advantages of the Great North Side," and "How Commuters Divide," *New York Times*, August 5, 1890, 8.
59. James D. McCabe Jr., *New York by Sunlight and Gaslight: A Work Descriptive of the Great American Metropolis* (Philadelphia: Douglass Brothers, 1881), 74, 82.

60. New York (State), *Report to the New York Legislature of the Commission to Select and Locate Lands for Public Parks in the Twenty-Third and Twenty-Fourth Wards of the City of New York* [ . . . ] (New York: M. B. Brown, 1884), 16, 20.

61. John Mullaly, *The New Parks beyond the Harlem* (New York: Record and Guide, 1887), 49.

62. Ann Schnitz and Robert Loeb, "'More Public Parks!': The First New York Environmental Movement," *Bronx County Historical Society Journal* 21, no. 2 (Fall 1984), 60.

63. Frederick Law Olmsted to John Charles Olmsted, October 29, 1890, in Olmsted, *Papers of Frederick Law Olmsted*, vol. 9, *The Last Great Projects 1890–1895*, ed. David Schuyler, Gregory Kaliss, and Jeffrey Schlossberg (Baltimore: Johns Hopkins University Press, 2015), 226.

64. Frederick Law Olmsted to John Charles Olmsted and Henry Sergeant Codman, July 9–11, 1892, in Olmsted, *Papers of Frederick Law Olmsted*, 9:540–41. See also Olmsted, "Future of New York," 5; Roper, *Biography of Frederick Law Olmsted*, 360–62, and Scobey, *Empire City*, 264–65.

65. "In the Annexed District," *New York Times*, June 14, 1891, 16, and "The New Street Commissioner," *New York Times*, November 2, 1890, 4.

66. "In the Annexed District."

67. Nate Gabriel, "Mapping Urban Space: The Production, Division and Reconfiguration of Natures and Economies," *City* 17, no. 3 (2013): 325–42.

68. See "Advantages of the Great North Side" and "James L. Wells," in *The Bronx and Its People: A History 1909–1927*, ed. James L. Wells, Louis F. Haffen, and Josiah Briggs (New York: Lewis Historical, 1927).

69. Haffen, "Department of Street Improvements," 41.

70. "North Side Men Rejoice," *New York Times*, March 27, 1896, 3.

71. "Advantages of the Great North Side," and Matthew P. Breen, *Thirty Years of New York Politics* (New York: privately printed, 1899), 726–29, quoted in Evelyn Gonzalez, "From Suburb to City: The Development of the Bronx, 1890–1940," Bronx Museum of the Arts, *Building a Borough*, 9.

72. "Louis J. Heintz is Dead," *New York Times*, March 13, 1893, 1, and "The New Street Commissioner," *New York Times*, November 2, 1890, 4.

73. "Plans Approved in the Main," *New York Times*, February 1, 1893, 9. On rising real estate valuations following Heintz's appointment, see "North Side Men Rejoice," and Simon Stevens, *Harlem River Ship Canal, Letter from Simon Stevens to the Commissioners of The Sinking Fund of The City Of New York* [ . . . ] (New York: C. G. Burgoyne, 1892), 19. For real estate values see "Addenda: The Growth of the North Side," Bronx Board of Trade, *Great North Side*, vi–vii.

74. Haffen, "Department of Street Improvements," 39.

75. "A Growing Borough," *New York Times*, May 30, 1897, 7, and "North Side Attractions," *New York Times*, March 31, 1895, 25.

76. Green, "Communication of the Comptroller," 151.

77. Daniel Van Pelt, *Leslie's History of the Greater New York*, vol. 1, *New York to the Consolidation* (New York: Arkell, 1898), 466.

78. "The Harlem River Canal," *Railroad Gazette*, December 28, 1894, 883, and Green, "Communication of the Comptroller," 151.

79. "The Harlem River Ship Canal," *Engineering News* 33, no. 25 (June 20, 1895): 399.

80. Fordham Morris, *Address of Mr. Fordham Morris* (New York: North Side Board of

Trade,1895), 8, italics in original. See also "The Harlem Ship Canal," reproduced in Stevens, *Harlem River Ship Canal*.

81. McCabe, *New York by Sunlight and Gaslight*, 83, and "Gen. Newton's Suggestions," *New York Times*, November 8, 1885, 7.

82. "Report of the Board of Commissioners of the Department of Public Parks of the City of New York, in Conformity 'With an Act of the Legislature, Passed April 15, 1871,'" 11–12.

83. Burrows and Wallace, *Gotham: A History*, 949.

84. "The Harlem Ship Canal," and "The Filling in of the Harlem," *New York Sun*, March 3, 1892, reproduced in Stevens, *Harlem River Ship Canal*, 28. See also "The Harlem Ship Canal: A Protest in the Interest of Railroads," *New York Times*, March 8, 1890, 5.

85. Stevens, *Harlem River Ship Canal*, 27; "The 'Annexed District,'" *New York Times*, January 4, 1891, 4; "Crossing the Harlem," *New York Times*, May 27, 1890, 4; and "The Harlem River Question," *New York Times*, February 8, 1891, 4.

86. "Harlem River Ship Canal."

87. The *New York Times* supported Stevens's proposal. See "The Harlem River," *New York Times*, March 3, 1892, 4; "The 'Annexed District'"; "The Filling in of the Harlem"; and "The Plan to Close the Harlem," *New York Times*, February 10, 1891, 8.

88. Green, "Communication of the Comptroller," 156.

89. "Disposes of the River and Harbor Bill," *New York Times*, August 17, 1890, 5.

90. Jacob W. Miller, "Relief for New York's Commercial Congestion," *New York Tribune*, September 29, 1907, C3. See also *Report of the Board of Commissioners of the Department of Public Parks of the City of New York, in Conformity with an Act of the Legislature Passed May 11, 1869, and an Act Passed May 19, 1870, Relating to Improvements of Portions of the Counties of Westchester and New York* [. . .] (Albany, NY: Argus, 1871), 6–8.

91. "The Harlem River Canal," *New York Times*, March 26, 1882, 2.

92. Seyfried, *300 Years of Long Island City*, 115–16, and "The Conquest of Hell Gate," US Army Corps of Engineers, New York District, http://www.nan.usace.army.mil /Portals/37/docs/history/hellgate.pdf. See also Ted Steinberg, *Gotham Unbound: The Ecological History of Greater New York* (New York: Simon and Schuster, 2014), 106–7, 126–27, 139, 142–43.

93. "Gen. Newton's Suggestions," 7.

94. Green, "Communication of the Comptroller," 158.

95. *Annual Report of the Department of Docks, City of New York*, vol. 21 (New York: Department of Docks, 1891), 107.

96. "Report of the Board of Commissioners of the Department of Public Parks of the City of New York, in Conformity 'With an Act of the Legislature, Passed April 15, 1871,'" 11–12. See also Joseph Elwood Betts, *The Sherwood-Elwood Connection: Story of the Transition of the Sherwood Elwood Farm to the Sherwood Island State Park* (Westport, CT: El-Lo Press, 2011), 40–56.

97. "Big Day in Gotham," *Chicago Daily Tribune*, June 18, 1895, 2. See also North Side Board of Trade, *Official Program of the Opening of the Harlem Ship Canal, June 17th, 1895* (New York: Freytag, 1895).

98. "Harlem Canal Dams Broken," *New York Times*, April 22, 1893, 9.

99. "The Harlem River Ship Canal," 400.

100. "The Harlem River Canal" See also Gina Pollara, "Transforming the Edge: Overview of Selected Plans and Projects," in *The New York Waterfront: Evolution and Building Culture of the Port and Harbor*, ed. Kevin Bone (New York: Monacelli, 1997), 165–66.

101. Van Pelt, *Leslie's History*, 466, and "Land and Water Parades," *New York Tribune*, May 26, 1895, 23.

102. Although the canal officially opened in 1895, it was not completed until 1938. See, for example, "Deeper Water Needed," *New York Tribune*, August 4, 1895, 4, and "The Harlem Ship Canal Has Become a Jest among Pilots Because of Neglect," *New York Tribune*, May 31, 1903, B8.

103. "Hudson Weds the Sound," *New York Times*, June 18, 1895, 1. Steinway Diary, June 17, Monday 1895.

104. "Harlem's Union of the Waters," *New York Herald*, June 18, 1895, 3.

105. "A Few Piscatorial Yarns," *New York Times*, April 7, 1901, 8.

106. "Where to Go A-Fishing," *New York Times*, June 21, 1891, 20. Marit Larson and Paul Mankiewicz, "Restoring Soundview Park: HEP Priority Restoration Site LI10," 2, Folder X-118 Soundview Park, Parks Library, New York City Department of Parks, New York, NY.

107. Joseph R. Bien and C. C. Vermeule, "Long Island Sound—Westchester North to Mt. Vernon—Queens South to Jamaica," in *Atlas of the Metropolitan District and Adjacent Country Comprising the Counties of New York, Kings, Richmond, Westchester* [...] (New York: J. Bien, 1891), plate 8; "Building a New City," *New York Tribune*, July 22, 1888, 11; US Coast and Geodetic Survey, *Harlem River*, Nautical Chart (1898), Historical Map and Chart Collection, Office of Coast Survey, NOAA. https://historicalcharts .noaa.gov/historicals/preview/image/3N274%e2%80%9398#searchInput, and US Geological Survey, *Harlem*, Topographical Quadrangle Map, 1:62500 scale (Reston, VA: US Geological Survey, 1967), USGS Historical Topographic Map Explorer, http://historicalmaps.arcgis.com/usgs/.

108. "Harlem's Union of the Waters," *New York Herald*, June 18, 1895, 3.

109. "Street Plan for the Addition of 1895," *Real Estate Record and Guide*, December 25, 1897, 992, "Board of Street Openings," *New York Times*, September 16, 1892, 10, and *Department of Street Improvements in the 23rd and 24th Wards Quarterly Report*, December 31, 1891, quoted in Gonzalez, *The Bronx*, 10.

110. "The Real Estate Field," *New York Times*, June 9, 1895, 23. See also *Father of Greater New York*, 36–40, and "The Trans-Harlem City," *New York Times*, March 28, 1896, 4.

111. Green, *Public Improvements*, 17.

112. *Father of Greater New York*, 35. Green repeats this argument in *Communication of Andrew H. Green to the Legislature of the State of New York, Copy of Act Creating Commission of Inquiry and Addresses of the President to the Commissioners* [...] ([New York?]: [1890?]), 18, 29.

113. Green, , 13, 8. See also John Foord, *The Life and Public Services of Andrew Haswell Green* (Garden City, NY: Doubleday, Page, 1913), 176.

114. "North Side Men Rejoice," and Charles B. Todd, *A Brief History of the City of New York* (New York: American Book, 1899), 286. On consolidation politics, see David C. Hammack, "Urbanization Policy: The Creation of Greater New York," in *Power and Society: Greater New York at the Turn of the Century* (New York: Russell Sage Foundation, 1982), 185–229, and Burrows and Wallace, *Gotham: A History*, 1222–33.

115. For "convinced that," see "The Features of Greater New York," *New York Times*, Jul 12, 1896, 25; for "manifest destiny," see North Side Board of Trade, *Official Program*.

116. US Census Office, *Report on the Social Statistics of Cities*, pt. 1 (Washington, DC: US Government Printing Office, 1886,) 568, quoted in Ira Rosenwaike, *Population History of New York City* (Syracuse, NY: Syracuse University Press, 1972), 56–57.

*Chapter 3*

1.  On the rise of country clubs, see Kenneth T. Jackson, *Crabgrass Frontier: The Suburbanization of the United States* (New York: Oxford University Press, 1985), 93, and Michael H. Ebner, *Creating Chicago's North Shore: A Suburban History* (Chicago: University of Chicago Press, 1988), 97.
2.  "Westchester Road Opening New Areas," *New York Times*, May 1, 1910, XX7. See also Roger Panetta, "Westchester, the American Suburb: A New Narrative," in *Westchester: The American Suburb*, ed. Roger Panetta (New York: Fordham University Press, 2006), 5–76.
3.  "Many Fine Seacoast Towns," *New York Tribune*, May 24, 1908, 1. On working-class leisure in polluted industrial neighborhoods, see Colin Fisher, *Urban Green: Nature, Recreation, and the Working Class in Industrial Chicago* (Chapel Hill: University of North Carolina Press, 2015), 6.
4.  "The Watering Places," *New York Herald*, July 30, 1860, 2.
5.  Jon Sterngass, *First Resorts: Pursuing Pleasure at Saratoga Springs, Newport, and Coney Island* (Baltimore: Johns Hopkins University Press, 2001).
6.  Evelyn Gonzalez, *The Bronx* (New York: Columbia University Press, 2006), 57.
7.  On diversity in Westchester, see George A. Lundberg, Mirra Komarovsky, and Mary Alice McInery, *Leisure, a Suburban Study* (New York: Columbia University Press, 1934). See also H. Paul Douglass, *The Suburban Trend* (New York: Century, 1925), 34, 89, 113, 119–21, cited in Elaine Lewinnek, *The Working Man's Reward: Chicago's Early Suburbs and the Roots of American Sprawl* (New York: Oxford University Press, 2014), 8.
8.  "Resorts along the Sound," *New York Times*, May 6, 1883, 6.
9.  "Features of Greater New York," 25. For the framing of the leisure corridor, I am indebted to Andrew W. Kahrl's work on the black recreational network along the Potomac River in *The Land Was Ours: African American Beaches from Jim Crow to the Sunbelt South* (Cambridge, MA: Harvard University Press, 2012).
10. "Many Fine Seacoast Towns," 12. For scenery as more than physical beauty, see Richard White, *Land Use, Environment, and Social Change: The Shaping of Island County, Washington* (Seattle: University of Washington Press, 1980), 145–48.
11. Leading revisionist scholarship on the socioeconomic and racial diversity of suburbia includes Becky M. Nicolaides, *My Blue Heaven: Life and Politics in the Working-Class Suburbs of Los Angeles, 1920–1965* (Chicago: University of Chicago Press, 2002); Andrew Weise, *Places of Their Own: African American Suburbanization in the Twentieth Century* (Chicago: University of Chicago Press, 2004); Richard Harris, *Unplanned Suburbs: Toronto's American Tragedy, 1900–1950* (Baltimore: Johns Hopkins University Press, 1996); and Richard Harris and Robert Lewis, "The Geography of North American Cities and Suburbs, 1900–1950: A New Synthesis," *Journal of Urban History* 27, no. 3 (March 2001): 262–92.
12. David Nasaw, *Going Out: The Rise and Fall of Public Amusements* (New York: Basic Books, 1993).
13. William Steinway Diary, August 29, 1886. The William Steinway Diary Project, 1861–1891, Smithsonian Institution National The William Steinway Diary Project, 1861–1891, Smithsonian Institution National Museum of American History, http:// americanhistory.si.edu/steinwaydiary (hereafter Steinway Diary). See also *Long*

*Island City Star*, June 11, 1886, 1:6, cited in Jeffrey Kroessler, "North Beach: The Rise and Decline of a Working-Class Resort," in *Evoking a Sense of Place: Long Island Studies*, ed. Joann P. Krieg (Interlaken, NY: Heart of the Lakes Publishing, 1988), 144.

14. "Many at Merritt Bake" *New York Tribune*, August 21, 1903, 5. See also "Clambakes are Popular," *New York Tribune*, July 31, 1902, 6, and "A Westchester Clambake," *New York Tribune*, July 30, 1902, 4.

15. Nasaw, *Going Out*, 4, 81, and Kathy Peiss, *Cheap Amusements: Working Women and Leisure in Turn-of-the-Century New York* (Philadelphia: Temple University Press, 1986), 42, 44. Contemporary studies of this change include Jesse Frederick Steiner, *Americans at Play: Recent Trends in Recreation and Leisure Time Activities* (New York: McGraw-Hill, 1933); Henry William Blair, United States Congress, Senate Committee on Education and Labor, *Report of the Committee of the Senate upon the Relations between Labor and Capital and Testimony Taken by the Committee* (Washington, DC: United States Government Printing Office, 1885); and Robert Coit Chapin, *The Standard of Living among Workingmen's Families in New York City* (New York: Charities Publication Committee, 1909). See also the subsequent reports by congress and the Department of Labor's Bureau of Labor Statistics surveyed in Joseph S. Zeisel, "The Workweek in American Industry 1850–1956," *Mass Leisure*, ed. Eric Larrabee and Rolf Meyersohn (Glencoe, IL: Free Press, 1958), 148. On New York in particular, see the New York State Bureau of Labor Statistics annual reports, cited in Peiss, *Cheap Amusements*, 42.

16. On Clason Point's trolley, see Aviation Volunteer Fire Co., *Aviation Volunteer Fire Co., Bronx, N.Y., 1923–1973: 50th Anniversary Featuring a Picture History of Clason Point and It's [sic] People* (New York: Aviation Volunteer Fire Co., 1973), Special Collections, Lehman College Library of the City University of New York (CUNY), The Bronx, New York. On trolleys see F. D. R., "Connecticut Revisited," *New York Times*, September 10, 1895, 4; "Old Boston Post Road," *New York Times*, June 21, 1896, 16, and "Now You Can 'Trolley' All the Way from New York to Boston," *New York Times*, December 4, 1904, SM7. For "the democracy of the trolley car," see *Trolley Exploring around New York City and Beyond* (New York: Brooklyn Eagle, 1908), 62, quoted in Panetta, "Westchester, the American Suburb," 38.

17. "Ads for Articles on North Beach Airport/Boardwalk [ . . . ]," Folder 2, Box 030024, Series Local Residents, Queens Local History Collection, LaGuardia and Wagner Archives, Long Island City, New York (hereafter Queens Local History Collection), and "Queens Plate No. 4," Atlas 148, vol. 10 (New York: Sanborn Map, 1914) Sanborn Maps, 1867–1970 Collection, New York Public Library, New York, NY. For Clason Point see George W. and Walter S. Bromley, *Annexed District: From Actual Surveys and Official Plans*, vol. 3 of *Atlas of the Borough of the Bronx* (Philadelphia: G. W. Bromley, 1927), plate 38.

18. "Secret Money Backs 'Hart's Coney,' says Broadway Merchant," *New York Evening Post*, July 25, 1925, 1.

19. John Kasson, *Amusing the Million: Coney Island at the Turn of the Century* (New York: Hill and Wang, 1978).

20. George Ehret, *Twenty-Five Years of Brewing: With an Illustrated History of American Beer* (New York: Gast Lithograph and Engraving, 1881); *A Day's Outing at Bowery Bay Beach: Containing History of Bowery Bay Beach (Illustrated), and of Old Newton, and Biographical Sketches of the Prominent People Living There* (New York, [1890?]); and

"Ads for Articles on North Beach Airport/Boardwalk." On Ehret's concession investments, see William Kells, interview by Jeffrey Kroessler, January 8, 1982, oral history
03.001.1.009,transcript, side A, 482, Queens Local History Collection.

21. Vincent F. Seyfried, and William Asadorian, *Old Queens, N.Y. in Early Photographs*
(New York: Dover, 1991), 146. On the fireworks, see Kells, interview by Jeffrey Kroessler, side B, 527.

22. "Sunday Walks in the German Quarter," *New York Times*, December 27, 1858, 8,
and "A German View of the Anti-Lager-bier on Sunday Memorial," *New York Times*,
July 23, 1859, 8. For midcentury critiques of German drinking culture on Manhattan,
see Catherine McNeur, *Taming Manhattan: Environmental Battles and the Antebellum
City* (Cambridge, MA: Harvard University Press, 2014), 215–16, and 296n56, and Roy
Rosenzweig and Elizabeth Blackmar, *The Park and the People: A History of Central Park*
(Ithaca, NY: Cornell University Press, 1992), 236, 233.

23. On the Bowery as a center of working-class commercialized sex, see Timothy J. Gilfoyle, *City of Eros: New York City, Prostitution, and the Commercialization of Sex, 1790–
1920.* New York: W. W. Norton, 1992), 119, 215; on middle-class fears of commercialized sexuality in leisure, see 38, 190, 196. For a critical analysis of the use of "vice" by
nineteenth-century reformers as well as historians, see Luis White, *The Comforts of
Home: Prostitution in Colonial Nairobi* (Chicago: University of Chicago Press, 2009), 7,
and "Prostitutes, Reformers and Historians," *Criminal Justice History* 6 (1985): 205–6.

24. "Sunday Beer and the Germans," *New York Times*, November 27, 1901, 8.

25. Theodore L. Cuyler, "The Sunday Saloon and the Germans," *New York Times*, December 21, 1901, 8.

26. Fisher, *Urban Green*, 44–47. On the nativism of mainstream Protestant reformers
in Chicago with regard to the German "continental Sunday," see also Thomas R.
Pegram, *Partisans and Progressives: Private Interest and Public Policy in Illinois, 1870–
1922* (Urbana, IL: University of Illinois Press, 1992), 105–12.

27. "A German Sunday," *New York Times*, September 28, 1873, 2. See also "Lager-Beer on
Sunday," *New York Times*, July 7, 1873, 2; Poultney Bigelow, "On Sunday, In This City,"
*New York Times*, October 23, 1895, 1; and A. R., "The German Laborer's Beer," *New
York Times*, October 22, 1903, 8.

28. Steinway Diary, June 20, 1886.

29. Kells, interview by Jeffrey Kroessler, side A, 45.

30. Steinway Diary, September 20, 1891.

31. Steinway Diary, June 6, 1886; July 2, 1886; and July 27, 1886.

32. Pool pass reproduced in "Ads for Articles on North Beach Airport/Boardwalk." On
the Geipel family, see Kroessler, "North Beach," 148.

33. "Features of Greater New York." See also "One-Day Trip to Greenwich," *New York
Times*, July 6, 1913, X4.

34. Peiss, *Cheap Amusements*, 12. Peiss draws on Chapin and the National Industrial Conference Board, *Family Budgets of American Wage Earners: A Critical Analysis*, Research
Report no. 41 (New York: Century, 1921), and Frank Hatch Streightoff, *The Standard
of Living among Industrial People of America* (Boston: Houghton Mifflin, 1911).

35. Chapin and the National Industrial Conference Board, *Family Budgets*, 281.

36. George Esdras Bevans, "How Working Men Spend Their Time" (PhD diss., Columbia
University, 1913). On recreational travel see Steiner, *Americans at Play*, 182, 193. On
spending money, see Louise Bolard More, *Wage-Earners' Budgets: A Study of Standards
and Cost of Living in New York City* (New York: Henry Holt, 1907), 161.

37.  Chapin and the National Industrial Conference Board, *Family Budgets*, 118, 211, 246.

38.  "Ads for Articles on North Beach Airport/Boardwalk." For total costs estimate, see Kroessler, "North Beach," 147.

39.  Bevans, 48; More, *Wage-Earners' Budgets* 142. Within the income range of $600–$1,000 or more, Chapin found the average annual expenditures were less than $8.50, a modest sum. Families with a $600 budget, the low end of the wage-earner spectrum, spent only $3.79; for those with budgets of $1,000, this expenditure reached $14.76; Chapin and the National Industrial Conference Board, *Family Budgets*, 210–11. See also Rosenzweig and Blackmar, *Park and the People*, 181, and Peiss, *Cheap Amusements*, 13.

40.  Chapin and the National Industrial Conference Board, *Family Budgets*, 210; for a summary of Chapin's work on this topic, see Rosenzweig and Blackmar, *Park and the People*, 181.

41.  Quoted in Peiss, *Cheap Amusements*, 53.

42.  Belle Linder Israels, "The Way of the Girl," *Survey* 22, (July 3, 1909): 489, quoted in Peiss, *Cheap Amusements*, 121.

43.  Kells, interview by Jeffrey Kroessler, side A, 667.

44.  "Rye Beach Woman Wishes She Was a Man at Trustees Meeting" and "Undressing in Cars," reproduced in Colin Dunne, John Dunne, Beth Griffin, Ted M. Levine, Robin Russell, and Arlene Weiss, *Rye in the Twenties* (New York: Arno, 1978).

45.  Kells, interview by Jeffrey Kroessler, side B, 070. See also Percy Loomis Sperr, "Bronx: Cornell Avenue–Newman Avenue," 1928, photograph, Bronx, Photographic views of New York City, 1870's–1970's, Irma and Paul Milstein Division of United States History, Local History and Genealogy, New York Public Library Digital Galleries, http://digitalgallery.nypl.org/nypldigital/id?700778f, and Kells, interview by Jeffrey Kroessler, side B, 101.

46.  Anne-Marie Cantwell and Diana diZerega Wall, *Unearthing Gotham: The Archaeology of New York City* (New Haven, CT: Yale University Press, 2001), 88.

47.  On New England, see Ellen Weiss, *City in the Woods: The Life and Design of an American Camp Meeting on Martha's Vineyard* (New York: Oxford University Press, 1987), and Dona Brown, *Inventing New England: Regional Tourism in the Nineteenth Century* (Washington, DC: Smithsonian Institution Press, 1995). On the Jersey Shore, see Charles E. Funnell, *By the Beautiful Sea: The Rise and High Times of That Great American Resort, Atlantic City* (New Brunswick, NJ: Rutgers University Press, 1983); Bryant Simon, *Boardwalk of Dreams: Atlantic City and the Fate of Urban America* (New York: Oxford University Press, 2006); and David Goldberg, *The Retreats of Reconstruction: Race, Leisure, and the Politics of Segregation at the Jersey Shore, 1865–1920* (New York: Fordham University Press, 2016). On the borscht belt and Catskills resorts, see David Stradling, *Making Mountains: New York City and the Catskills* (Seattle: University of Washington Press, 2007), 189–205.

48.  Marcia Dalphin, *Fifty Years of Rye 1904–1954* (Rye: City of Rye, 1955), 17. See also Maggie Walker, "A History of Rye Town Park," *Daily Item*, August 1972, 1, quoted in National Register of Historic Places, *Rye Town Park, Bathing Complex and Oakland Beach*, Rye, Westchester County, New York, *National Register* 3000252, 8-1. See also Eugene McGuire, "Rye Historical Society Walking Tour Script," unpublished script in author's possession, 10.

49.  "Rye Beach County Park," *New York Tribune*, March 10, 1901, B2; "Park Notes," *Park and Cemetery and Landscape Gardening*, February 1, 1901): 10–12; and Alvah P. French, *The History of Westchester County* (New York: Lewis Historical, 1925), 46.

50. "Oakland Beach Bathing Pavilion, Rye, New York," *Architects' and Builders' Magazine* 12 (1910–1911): 283; "Bungalows for Oakland Beach," *New York Tribune*, July 19, 1914; advertisements in *Rye Chronicle*, March 7, 1908), interior of front cover, 6, Knapp House Archives, Rye Historical Society, Rye, New York. See "Rye, Westchester, New York," roll 1664, page 1A–1B, and 8B, enumeration district 339, Bureau of the Census, *Fifteenth Census of the United States, 1930* (Washington, DC: National Archives and Records Administration, 1930), Ancestry.com (1997–2003).

51. Marjorie O'Shaughnessy, "Childhood Summers at Orchard Beach," *Bronx Historical Society Journal* 12, no. 1 (Spring 1975): 20.

52. Eugene O'Kane, interview by Tabitha Kirin, 309, August 18, 1987, side 2, transcript 36, The Bronx Institute Oral History Project, Special Collections, Lehman College Library of the City University of New York (CUNY), The Bronx, New York (hereafter Bronx Institute Oral History Project). See also *Annual Report of the Department of Parks of the Borough of the Bronx City of New York for the Year 1920* (New York: Clarence S. Nathan, 1921), 24.

53. New York Metropolitan Sewerage Commission, *Main Drainage and Sewage Disposal Works Proposed for New York City: Reports of Experts and Data Relating to the Harbor* (New York: Wynkoop Hallenbeck Crawford, 1914), 55.

54. New York Metropolitan Sewerage Commission, *Sewerage and Sewage Disposal in the Metropolitan District of New York and New Jersey: Report, April 30, 1910* (New York: Martin B. Brown, 1910), 278, and Gonzalez, *The Bronx*, 21–25, 31, 39–40, 62.

55. On urbanization and rapid transit see Gonzalez, *The Bronx*, 52–58, and Lloyd Ultan, *The Beautiful Bronx, 1920–1950* (New Rochelle, NY: Arlington House, 1979), 14. On population distribution, see Walter Laidlaw, Robert E. Chaddock, Neva Ruth Deardorff, and Cities Census Committee, New York, *Population of NYC, 1890–1930* (New York: Cities Census Committee, 1932), 34, 82; cited in Gonzalez, *The Bronx*, 6.

56. On the percentage of the borough in parkland, under agricultural use, or open, see New York Metropolitan Sewerage Commission, *Sewerage and Sewage Disposal in the Metropolitan District*, 266. For "unnecessary" quote, see "Residents of Unionport Impatient at Delay in Grading Tremont Avenue," *Bronx Home News*, August 16, 1907, 1.

57. Timothy Dwight, *Travels in New-England and New-York,* vol. 3 (London: William Baynes and Son, 1823), 504, 329.

58. "Regional Plan of New York and Its Environs," in Thomas Adams, Harold M. Lewis, and Theodore T. McCrosky, *Population, Land Values and Government*, vol. 2 of *Regional Survey of New York and Its Environs* (New York: Regional Plan of New York and Its Environs, 1929; repr., New York: Arno, 1974), 91–93, 97.

59. Adams, Lewis, and McCrosky, 73, fig. 18.

60. Ibid., 59.

61. Arthur Seifert, interview by Yoland Zick, 26, April 23, 1982, transcript, 16, Bronx Institute Oral History Project.

62. "East Bronx Street System," *New York Times*, August 3, 1898, 12.

63. Claude Hyman, "Oh Memories," in *The Bronx in the Innocent Years, 1890–1925*, ed. Lloyd Ultan and Gary Hermalyn (New York: Harper and Row, 1985).

64. New York (State), *Report to the New York Legislature of the Commission to Select and Locate Lands for Public Parks in the Twenty-Third and Twenty-Fourth Wards of the City of New York* [. . .] (New York: M. B. Brown, 1884), 61.

65. New York (State), *Report to the New York Legislature*, 117.

66. Christopher C. Sellers, *Crabgrass Crucible: Suburban Nature and the Rise of Environmentalism in Twentieth-Century America* (Chapel Hill: University of North Carolina Press, 2013), 20.

67. *Annual Report of the Board of Health of the Department of Health of the City of New York for the Year Ending December 31, 1908*, vol. 1 (New York: Martin B. Brown, 1909), 43; *Annual Report of the Board of Health of the Department of Health [. . .] December 31, 1909* (New York: Martin B. Brown, 1910), 26; and *Report of the Board of Health of the Department of Health of the City of New York for the Years 1910 and 1911* (New York: J.W. Pratt, Co., 1912), 27.

68. Anne Whiston Spirn, *The Granite Garden: Urban Nature and Human Design* (New York: Basic Books, 1984), 62.

69. Spirn, 51–52.

70. John C. Van Dyke, *The New New York: A Commentary on the Place and the People* (New York: Macmillan, 1909), 334. Comments on sound shore breezes were ubiquitous at the turn of the century. See, for example, William H. Avis, "The Sportsman Tourist: An Inexpensive Outing," *Forest and Stream* 58, no. 23, June 7, 1902, 442; Frederick Coburn, "The Five-Hundred Mile City," *World Today* 11, December 1906, 1254; and "By Boat and Trolley," *Washington Post*, June 8, 1902, 22.

71. John Mullaly, *The New Parks beyond the Harlem* (New York: Record and Guide, 1887), 19–20, 38.

72. Spirn, *Granite Garden*, 65.

73. "Editorial Article," *Hartford Daily Courant*, January 30, 1859, 2.

74. Phineas Taylor Barnum, *Struggles and Triumphs, or Forty Years' Recollections of P. T. Barnum* (New York: American News, 1871), 766–67, 771–72.

75. "The Beaches," *Rye Chronicle*, March 12, 1910, 1, Knapp House Archives, Rye Historical Society, Rye, New York.

76. Bromley and Bromley, *Atlas of the City of New York Borough of the Bronx*, plates 38, 41. On the vernacular landscape, see Daniel Campo's interpretation of J. B. Jackson's seminal *Discovering the Vernacular Landscape* (New Haven, CT: Yale University Press, 1984) in Campo, *The Accidental Playground: Brooklyn Waterfront Narratives of the Undesigned and Unplanned* (New York: Fordham University Press, 2013), 13–14.

77. Bill Twomey, "Childhood Memories of Edgewater Park," *Bronx County Historical Society Journal* 33, no. 1 (March 1996): 83–85.

78. *City of New York Office of the President of the Borough of the Bronx, Bureau of Engineer, General Map of the Borough of the Bronx, New York, April 1st 1935* (New York: Hagstrom Map Company, 1935).

79. Benjamin Waring, interview by Yolanda L. Zick, 130, September 17, 1983, transcript 5, Bronx Institute Oral History Project, 8–9. On Housing shortages in New York during and after World War I, see Robert M. Fogelson, *The Great Rent Wars: New York, 1917–1929* (New Haven, CT: Yale University Press, 2013), 17–39.

80. Waring, interview by Yolanda L. Zick, 3, and "Bronx Assembly District 8, Bronx, New York," roll T625_1142, page 10B; enumeration district 440, image 901, Bureau of the Census, *Fourteenth Census of the United States, 1930* (Washington, D.C.: National Archives and Records Administration, 1930) Ancestry.com.

81. Coburn, "Five-Hundred Mile City," 1259.

82. Becky M. Nicolaides includes owner-built bungalows as a type of informal housing, a

class of housing that she argues was built for and by working people. See Nicolaides, *My Blue Heaven*, 4, 29.

83. 19.8 Higgs Beach Tent 82, Bronx Regional and Community History Project, Photo Record, Bronx Institute Oral History Project. See also O'Shaughnessy, "Childhood Summers," 20.

84. Earl Ubell, "Bronx Boasts Country Hamlet of 257 Families," *New York Herald Tribune*, July 23, 1949, 9; "Jitterless Village Hides in the Bronx," *New York Times*, April 23, 1952, 3; and "Harding Park Folk Fight to Save Rural Life," *New York Herald Tribune*, January 10, 1953, 13.

85. Waring, interview by Yolanda L. Zick, 5.

86. On the similarities between small working-class suburban homes and Downing's designs, see Lewinnek, *Working Man's Reward*, 41.

87. 1.20—*Harding Park Rent Receipts*, Bronx Regional and Community History Project, Photo Record, Bronx Institute Oral History Project, and New York Committee on Slum Clearance, *Soundview: Slum Clearance Plan under Title I of the Housing Act of 1949 as Amended* (New York: Committee on Slum Clearance, 1959), 42. Elizabeth Blackmar documents the use of ground leases in Manhattan and the real estate hierarchy that emerged between 1780 and 1840 whereby one individual owned a plot of land, a second the lot's building, and a third the lease. See Blackmar, *Manhattan for Rent*, 28–95. Timothy J. Gilfoyle discusses this hierarchy in terms of prostitution and real estate. See Gilfoyle, *City of Eros*, 35.

88. Ubell, "Bronx Boasts Country Hamlet."

89. On long-term informal leases, see Clason Management Co. v. Altman, 40 A.D. 2d 635 (N.Y. App. Div. 1972). See also Alison K. Hoagland, *Mine Towns: Building for Workers in Michigan's Copper Country* (Minneapolis: University of Minnesota Press, 2010), 92.

90. Bromley and Bromley, *Atlas of the City of New York Borough of the Bronx*, plate 44. See also "Park of Edgewater," in John McNamara, *History in Asphalt: The Origin of Bronx Street and Place Names* (Harrison, NY: Harbor Hill Books, 1978), 179.

91. "Private House Sales," *New York Times*, April 14, 1923, 22; and Gary D. Hermalyn, "Silver Beach," *The Encyclopedia of New York City*, ed. Kenneth T. Jackson (New Haven, CT: Yale University Press, 1995), 1070–71. See also Kara Murphy Schlichting, "Rethinking the Bronx's 'Soundview Slums': The Intersecting Histories of Large-Scale Waterfront Redevelopment and Community-Scaled Planning in an Era of Urban Renewal," *Journal of Planning History* 16, no. 2 (May 2017): 112–38.

92. "Rye Westchester, New York," roll: 1664, pages 1A–1B, 8B, enumeration district, 339, Bureau of the Census, *Fifteenth Census of the United States, 1930* (Washington, DC: National Archives and Records Administration, 1930), Ancestry.com (1997–2003).

93. For example, Edmund C. Stanton's $100,000 home had a staff of five Irish, Scottish, and Welsh servants and caretakers. See "Rye, Westchester, New York," roll 1664, pages 10A, 11B, 12A, 12B, 13A, and 13B, enumeration district 339, Bureau of the Census, *Fifteenth Census of the United States, 1930* (Washington, DC: National Archives and Records Administration, 1930), Ancestry.com (1997–2003).

94. McGuire, "Rye Historical Society Walking Tour Script," 21.

95. "Rye Beachers Tell Woes," *New York Times*, June 25, 1905, 7.

96. "Judge Favors Paradise Park" (newspaper clipping), reproduced in Dunne et. al., *Rye in the Twenties*, and Kathryn W. Burke, *Playland* (Charleston, SC: Arcadia, 2008), 14.

97. "Rye Beach Woman Wishes She Was a Man at Trustees Meeting" and "Undressing in Cars," reproduced in Dunne et al., *Rye in the Twenties*.

98. *Bronx Record Times* (July 1923), cited in Burke, *Playland*. For "city slum" quote, see Leon N. Gillette and Gilmore D. Clarke, "Playland: An Amusement Park at Rye, Westchester County, N.Y.," *American Architect* 132, no. 2555 (October 20, 1928): 492.

99. Adams, Lewis, and McCrosky, *Population, Land Values and Government*, 91–93, 97.

100. "Fight New Yorkers at Election in Rye," *New York Times*, March 15, 1925, 6, and *Views of Rye* ([New York ?]: Blakeman Quintard Meyer, 1917), 8.

101. "Rye Starts Summer Bans on Buses and Boats," *Tribune*, May, 9 1925, Parks Department Clippings, Series 98, Westchester County Archives, Elmsford, New York (hereafter Parks Department Clippings).

102. For "only long enough to sleep" quote, see "Fight New Yorkers at Election in Rye." See also Burke, *Playland*, 107, and "'Blue Laws' Are Obsolete Owners of Parks Contend," *Rye Reporter*, May 20, 1925, Parks Department Clippings.

103. "Trustees Commerce Crusade to Keep Crowds from Rye Beach," reproduced in Dunne et al., *Rye in the Twenties*.

104. "Excursions from New York City," reproduced in Dunne et. al., *Rye in the Twenties*.

105. "Rye Refuses License to Amusement Park," *New York Times*, April 16, 1925, 16; "Bluest of Sundays," *New York Times*, May 8, 1925, 10; "Seek Writ to Halt Rye Sabbath Ban," *New York Times*, May 10, 1925, 3; and "Stop Everything, Why Not?" *White Plains Reporter*, May 11, 1925, Parks Department Clippings.

106. "Trustees Commerce Crusade to Keep Crowds from Rye Beach," reproduced in Dunne et. al., *Rye in the Twenties*; "Rye Tradesmen See Park Fight Victory," *New York Times*, May 21, 1925, 15. On Kelly, see "Rye, Westchester, New York," roll 1664, page, 9A, enumeration district 336, Bureau of the Census, *Fifteenth Census of the United States, 1930* (Washington, DC: National Archives and Records Administration, 1930) Ancestry.com (1997–2003).

107. "Rye Hearing Put Off," *New York Times*, May 13, 1925, 3; "Judge Favors Paradise Park," reproduced in Dunne et. al., *Rye in the Twenties*; "Injunction Halts Moves to Close Rye Beach Amusements on Sundays," *Statesman*, May 11, 1925, and "A Blessing as Yet Disguised," *Port Chester Daily Item*, June 11, 1928, Parks Department Clippings.

108. "Rye Starts Summer Bans on Buses and Boats"; "Rye Tradesmen See Park Fight Victory": "Seek Writ to Halt Rye Sabbath Ban"; "Charge Rye Uses Persecution in Fight on Parks," *Reporter*, July 15, 1925; "'Blue Laws' Can't Close Rye Parks, Judge Seeger Says," *New Rochelle Star*; and "Rye Parking Law Upset by Court in Parks Fight," *White Plains Register*, June 27, 1925, Parks Department Clippings.

109. Lundberg, Komarovsky, and McInery, *Leisure*, 30.

110. Bronx County Supreme Court, *In the Matter of Acquiring Title by the City of New York to Certain Lands* [. . .], Specialty Term for Trials, Supreme Court of the State of New York, Bronx County (Feb. 14, 1927), 243.

111. On Little Bridgeport, see Bess Liebenson, "Fairfield, Bridgeport, and the Trolley Shared," *New York Times*, July 7, 1996, CN17. On Little Danbury, see Beth Love, *Fairfield and Southport in Vintage Postcards* (Charleston, SC: Arcadia, 2000), 82. For Darcy quote, see "The Bungalow Spread," *Rye Chronicle*, May 4, 1912, 6, Knapp House Archives, Rye Historical Society, Rye, New York.

112. Coburn, "Five-Hundred Mile City," 1259.

113. *Report of the Westchester County Park Commission, April 30, 1925* ([White Plains?], 1925), 24.

114. Waring, interview by Yolanda L. Zick, 8–9; "Yacht Clubs on Cruising Routes," *Motor Boating* 44, no. 6 (December 1929), 18.

115. O'Kane, interview by Tabitha Kirin, side 2, 41–42. See also Robert Caro, *The Power Broker: Robert Moses and the Fall of New York* (New York: Knopf, 1974), 363.

116. *Tobin v. Hennessy* 129 Misc. 756, 223, N.Y.S. 618 WestLaw (N.Y. Supp. 1927). On yearly populations at Orchard Beach, see also *Department of Parks of the City of New York Annual Reports*. Parks Annual Reports, Press Releases Archives and Minutes, City of New York Parks and Recreation, http://www.nycgovparks.org/news/reports/archive.

117. Tobin v. Hennessy, 621.

118. *Annual Report of the Department of Parks of the Borough of the Bronx City of New York For the Year 1921* (New York: Herald Square, 1922), 7–8, and "Pelham Bay Park Squatters," *New York Times*, May 31, 1927, 20. See also Catherine A. Scott, *City Island and Orchard Beach* (Portsmouth, NH: Arcadia, 1999), 93.

119. Tobin v. Hennessy, 620.

120. Ibid., 619.

121. Ibid., 677. See also "Wants 'Squatters' Ousted," *New York Times*, May 31, 1927, 23, and New York Local Laws 1927, No. 10, adding to Greater New York Charter [Laws 1901, c. 466], § 612–e.

122. David Nasaw, *Going Out: The Rise and Fall of Public Amusements* (New York: Basic Books, 1993), and Bryant Simon, *Boardwalk of Dreams: Atlantic City and the Fate of Urban America* (Oxford: Oxford University Press, 2004), 16–18. See also Joel E. Black, "Space and Status in Chicago's Legal Landscapes," *Journal of Planning History* 12, no. 3 (July 2013): 239.

123. Bevans, 89, and Carol Groneman and David M. Reimers, "Immigration," *The Encyclopedia of New York City*, ed. Kenneth T. Jackson (New Haven, CT: Yale University Press, 1995), 582–83.

124. On blacks as "undesirables," see Victoria W. Wolcott, in *Race, Riots, and Roller Coasters: The Struggle over Segregated Recreation in America* (Philadelphia: University of Pennsylvania Press, 2012), 3, 9, 20. For regulations at parks, see Nasaw, *Going Out*, 86–88. Mixed-race swimming was a particular point of conflict. See Jeff Wiltse, *Contested Waters: A Social History of Swimming Pools in America* (Chapel Hill: University of North Carolina Press, 2007).

125. David Stradling, *Making Mountains: New York City and the Catskills* (Seattle: University of Washington Press, 2007), 185–6, 191, and Irwin Richman, *Borscht Belt Bungalows: Memories of Catskill Summers* (Philadelphia: Temple University Press, 2004), 2. On enforcement problems concerning the 1913 civil rights law, see Jeffrey Gurock, "The 1913 New York Civil Rights Act," *AJS Review* 1 (1976): 111–14.

126. On North Beach, see Kells, interview by Jeffrey Kroessler, side A, 152 and 379. For "no shore," see Bronx County Supreme Court, *In the Matter of Acquiring Title*, 22.

127. "Jail Escapes Feared as Harlem 'Coney' on Hart Island Rises," *New York Evening Post*, June 1, 1925), Old Fulton History, http://www.fultonhistory.com.

128. Ira Rosenwaike, *Population History of New York City* (Syracuse, NY: Syracuse University Press, 1972), 117, table 69, pp.141, 140.

129. "Negro Population Boosted 11,978 in Decade, Westchester Census Reveals," *New York Amsterdam News*, August 5, 1931, 6; Lundberg, Komarovsky, and McInery, *Leisure*, 51; and Lisa Keller, "Dreams Delivered: Following Diversity's Path in Westchester," in *Westchester: The American Suburb*, ed. Roger Panetta (New York: Fordham University Press, 2006), 330.

130. "Negro Coney Rises $500,000 in Price as City Delays Buying," *New York Evening Post*, July 18, 1925, 1.
131. Prison Association of New York, *A Study of the Conditions Which Have Accumulated under Many Administrations and Now Exist in the Prisons of Welfare Island* [. . .] (New York: Prison Association of New York, 1924), 38, and "Grab for Negro Vote Charged to Hylan in Hart's Island Coney," *New York Evening Post*, June 13, 1925, 1.
132. Bronx County Supreme Court, *In the Matter of Acquiring Title*, 235.
133. "'Hylan Folly' Loses $20,000 Hart's Tract, Now $160,000 'Coney,'" *New York Evening Post*, June 2, 1925, 1, and "New Hitch Develops in Harts 'Coney Ban," *New York Evening Post*, July 22, 1925, 8.
134. Prison Association of New York, *Study of the Conditions*, 40.
135. For the ideas of spatial policies underwriting racial inferiority and the subsequent transformation of African Americans into social outliers, see Black, "Space and Status in Chicago's Legal Landscapes," 227, 238–39. See also Wolcott, *Race, Riots, and Roller Coasters*, 4, 9. On the racial dimensions of environmental inequality, see Kahrl, *The Land Was Ours*; Andrew Hurley, *Environmental Inequalities: Class, Race and Industrial Pollution in Gary Indiana, 1945–1980* (Chapel Hill: University of North Carolina Press, 1995); and Robert Doyle Bullard, *Dumping in Dixie: Race, Class and Environmental Quality* (Boulder, CO: Westview Press, 1990).
136. "Says Hyland Held Up Hart's Island Sale," *New York Times*, June 9, 1925, 23.
137. "Jail Escapes Feared as Harlem 'Coney' on Hart Island Rises."
138. Committee of Finance, "Report of the Committee on Finance in Favor of Adopting an Ordinance Selecting a Site on Harts Island for the Use of the Department of Correction," in *The Municipal Assembly of the City of New York, Aldermanic Branch*, (June 9, 1925), 1015; "Report of the Committee on Finance in Favor of Adopting an Ordinance Selecting a Site on Harts Island for the Use of the Department of Correction," in *The Municipal Assembly of the City of New York, Aldermanic Branch* (June 16, 1925), 1058–59; and "Warn Park Builder to Obtain License," *New York Times*, June 4, 1925, 10.
139. "Unperturbed by Delay on Beach Permit, Riley Waits for Outcome on Pollution," *New York Amsterdam News*, July 9, 1930, 2. For neighborhood demographics, see "Bronx, Bronx, New York," roll 1476, page 7A, enumeration district 807, Bureau of the Census, *Fifteenth Census of the United States, 1930* (Washington, DC: National Archives and Records Administration, 1930) Ancestry.com.
140. Crolius v. Douglas Boat Club, 139 Misc. 29, 247, N.Y.S. 1 WestLaw (N.Y. Supp. 1930).
141. "Injunction Hits Throggs Neck Beach," *New York Amsterdam News*, September 3, 1930, 1.
142. "Negro Coney Rises," and *New York Legislative Documents One Hundred and Forty Fifth Session, 1922*, 20, nos. 95 to 108 incl. (Albany, NY: J. B. Lyon, 1922), 248–49.
143. Riley received an Order to Show Cause, which mandated he prove in court why the temporary injunction the order engendered should not be made permanent. George C. Crolius v. Elizer Reality Corp., Intrafraternity Council, and Solomon Riley, Special Term, Pt. 1, Sup. Ct. Bronx County, Nov. 12, 1931, Index No. 6545. See also Crolius v. Douglas Boat Club.
144. Jacob Grant, Aff. (Sept. 25, 1930), 2, and A. F. Harding, Aff. (Sept. 25, 1930), 1. George C. Crolius v. Elizer Reality Corp.
145. Andrew T. Williams, Closing Stmt., 1. George C. Crolius v. Elizer Reality Corp.

146. Crolius v. Douglas Boat Club, 3.

147. Ibid., 2.

148. "Riley May Receive Bath Beach Permit," *New York Amsterdam News*, July 23, 1930, 11, and "Riley Wins Permit for Bronx Resort," *New York Amsterdam News*, November 4, 1931, 1. In late October 1931, the Appellate Court issued a preemptory mandamus, a writ that legally insisted License Commissioner Geraghty issue a license. *Elizer Realty Company v. Geraghty*, 234 A.D. 682, 252, N.Y.S. 950 WestLaw (N.Y.A.D. 1 Dept. 1931).

149. "Prejudice Rears Its Ugly Heat [*sic*] at Throgs Neck," *New York Amsterdam News*, September 3, 1930, 1; "Riley Fights New Ban on His Beach," *New York Amsterdam News*, May 18, 1932, 1; and "Solomon Riley Dies in Bronx," *New York Age*, April 25, 1936. On the association of polluted coastlines with African Americans, see Kahrl, *The Land Was Ours*, 15.

150. "Negro Population Boosted 11,978 in Decade, Westchester Census Reveals," and Keller, "Dreams Delivered," 339.

151. J. A. Rogers, "Historic Westchester County Now the Home of Several Thousand Progressive Negros," *New York Amsterdam News*, December 18, 1929, 11. On minority domestic servant communities, see Andrew Weise, *Places of Their Own: African American Suburbanization in the Twentieth Century* (Chicago: University of Chicago Press, 2004), 26, 56–61, and Priscilla Murolo, "Domesticity and Its Discontents," in *Westchester: The American Suburb*, ed. Roger Panetta (New York: Fordham University Press, 2006), 356.

152. Wiese, *Places of Their Own*, 5, 51, 55.

153. "Westchester Scored on Aid to Negroes," *New York Times*, November 17, 1937, 19.

154. Phineas Taylor Barnum, *Struggles and Triumphs, or Forty Years' Recollections of P. T. Barnum* (New York: American News, 1871), 553, and *Illustrated Pamphlet on the Founding and Development of Steinway*, 22, reproduced at Steinway Diary.

155. Daniel C. Walsh, "The History of Waste Landfilling in New York City," *Ground Water* 29 (July–August 1991), 591.

156. "The Features of Greater New York," *New York Times*, July 12, 1896, 25.

157. "Fight on Rikers Island," *New York Times*, August 12, 1903, 2

158. "Typhoid from Oysters," *New York Tribune*, October 8, 1907, 11.

159. On total sewage output in New York Harbor, see New York Metropolitan Sewerage Commission, *Sewerage and Sewage Disposal in the Metropolitan District*, 76. At the turn of the century, American sewage treatment was in its infancy. See Martin V. Melosi, *The Sanitary City: Urban Infrastructure in America from Colonial Times to the Present* (Baltimore: Johns Hopkins University Press, 2000), 172.

160. Department of Health of the City of New York, *Annual Report of the Department of Health of the City of New York for the Calendar Year 1915* (1916; repr., n.p.: Forgotten Books, 2015), 45.

161. *Report of the Board of Commissioners of the Department of Public Parks of the City of New York: [ … ] Relating to Improvements of Portions of the Counties of Westchester and New York, and the Improvement of Spuyten Duyvil Creek and Harlem River, and to Facilities of Communication between Said Counties* (Albany, NY: Argus, 1871), 17–18.

162. New York Metropolitan Sewerage Commission, *Sewerage and Sewage Disposal in the Metropolitan District*, 45, 76, 227, 268, and "Kane's Park Resort Recalled by Oldtimer of Clason Point," Aviation Volunteer Fire Co, *Aviation Volunteer Fire Co., Bronx, N.Y.,*

*1923–1973: 50th Anniversary Featuring a Picture History of Clason Point and It's* [sic]
*People* (New York: Aviation Volunteer Fire Co., 1973), Special Collections, Lehman
College Library of the City University of New York (CUNY), The Bronx, New York;
Helena Orchuizzo, interview by Yolanda L. Zink, 17, March 26, 1982, Bronx Institute
Oral History Project. On the state of sewers in the 1960s, see F. X. Luts to M. Bender
(Jun. 8, 1967), 1, Folder X-118, Soundview Master Plans, Box X-118, New York City
Parks, Parks Library, New York.

163. Christopher C. Sellers, *Crabgrass Crucible: Suburban Nature and the Rise of Environmentalism in Twentieth-Century America* (Chapel Hill: University of North Carolina
Press, 2013), 11–38.

164. For "Inkwell" see Ultan, *Beautiful Bronx*, 150; and Vivian Cavilla, interview by Raymond C. Schloss, 241, May 23, 1983, Bronx Institute Oral History Project.

165. "Prohibition as Seen by a Maker of Beer," *New York Times*, December 30, 1917, XX4,
and "Where the Streets Smelled Like Beer," *New York Times*, March 22, 2012, RE6.

166. Kells, interview by Jeffrey Kroessler, side B, #112.

167. Aviation Volunteer Fire Co. and "Havoc by Wind at Clason's Point Amusement Park,"
*The Hartford Courant* Jun. 12, 1922, 1. For quote see Seifert, 38.

168. Percy Loomis Sperr, "Bronx: Stephens Avenue—Bronx River Avenue." photograph,
Bronx, Photographic Views of New York City, 1870's–1970's, Irma and Paul Milstein
Division of United States History, Local History and Genealogy, New York Public
Library Digital Galleries, March 25, 2011, http://digitalgallery.nypl.org/nypldigital/id
?701864f.

169. Clifton Hood, *722 Miles: The Building of the Subways and How They Transformed
New York* (New York: Simon and Schuster, 1993), 179, and Rosenwaike, *Population
History*, 133. Brooklyn and Staten Island experienced 56 and 84 percent growth,
respectively.

170. William C. Wright, "Biography of V. Everit Macy," chap. 16, p. 4, manuscript, Folder 1,
Series 187, V. Everit Macy Papers, Westchester County Archives, Elmsford, New York.

*Chapter 4*

1. Lee F. Hanmer, Thomas Adams, Frederick W. Loede Jr., Charles J. Storey, and
Frank b. Williams, *Public Recreation: A Study of Parks, Playgrounds, and Other Outdoor
Recreational Facilities*, Regional Survey of New York and Its Environs 5 (New York:
Regional Plan of New York and Its Environs, 1928), 88.

2. Jay Downer and James Owen, *Public Parks in Westchester County, 1925* (New York,
Lewis Historical, 1925), 22. On the influence of the automobile on urban planning,
see Norman T. Newton, *Design on the Land: The Development of Landscape Architec-
ture* (Cambridge, MA: Belknap Press, 1971), 563–64, and Clay McShane, *Down the
Asphalt Path: The Automobile and the American City* (New York: Columbia University
Press, 1994).

3. McShane, *Down the Asphalt Path*, 184.

4. Peter J. Schmitt, *Back to Nature: The Arcadian Myth in Urban America* (Baltimore:
Johns Hopkins University Press, 1969).

5. Lewis Mumford, *The Urban Prospect* (New York: Harcourt, Brace and World, 1968),
81, quoted in Hilary Ballon and Kenneth Jackson eds., *Robert Moses and the Modern
City: The Transformation of New York* (New York: W. W. Norton, 2007), 158.

6.  George A. Lundberg, Mirra Komarovsky, and Mary Alice McInery, *Leisure: A Suburban Study* (New York: Columbia University Press, 1934), 346, 352–55, italics in original.

7.  "Regional Park Planning," *New York Times*, January 11, 1926, 26.

8.  See Galen Cranz, *The Politics of Park Design: A History of Urban Parks in America* (Cambridge, MA: MIT Press, 1982); Roy Rosenzweig and Elizabeth Blackmar, *The Park and the People: A History of Central Park* (Ithaca, NY: Cornell University Press, 1992); Susan Currell, *The March of Spare Time: The Problem and Promise of Leisure in the Great Depression* (Philadelphia: University of Pennsylvania Press, 2005); Dominick Cavallo, *Muscles and Morals: Organized Playgrounds and Urban Reform, 1880–1920* (Philadelphia: University of Pennsylvania Press: 1981); and Gary S. Cross, "Crowds and Leisure: Thinking Comparatively Across the 20th Century," *Journal of Social History* 39, no. 3 (Spring 2006), 643.

9.  Jon Peterson, *The Birth of City Planning in the United States, 1840–1917* (Baltimore: Johns Hopkins University Press, 2003), 122, 152, 222, 324.

10. John Nolen, "March 27, 1930, to News Editor, for Immediate Release," Folder 8, Box 1, John Nolen Papers, 1890–1938, 1954–1960, Division of Rare and Manuscript Collections, Cornell University Library, Ithaca, New York.

11. Quoted in Lundberg, Komarovsky, and McInery, *Leisure*, 345.

12. Nathaniel Parker Willis, *Hurry-Graphs; Or Sketches of Scenery, Celebrities, and Society Taken from Life* (London: Henry G. Bohn, 1851), 83, and "Westchester County Proper" *New York Times*, May 31, 1874, 4.

13. *The Uptown News*, March 26, 1885, 2, cited in Ann Schnitz and Robert Loeb, "'More Public Parks!': The First New York Environmental Movement," *Bronx County Historical Society Journal* 21, no. 2 (Fall 1984): 57.

14. John Mullaly, *The New Parks beyond the Harlem* (New York: Record and Guide, 1887), viii. For a chronology of the New Parks Act debates, see Schnitz and Loeb, "'More Public Parks!,'" 51–66, and Martha Rogers, "The Bronx Park System, A Faded Design," *Landscape* 27, no. 2 (1983), 13–21.

15. Daniel Van Pelt, *Leslie's History of the Greater New York*, vol. 1, *New York to the Consolidation* (New York: Arkell, 1898), 546.

16. "To Rid Parks of Private Tenants," *New York Tribune*, June 1, 1908, C8.

17. "Hints for a Day's Outing," *New York Times*, July 20, 1890, 20.

18. On transit to the park, see Luis Pons, *Official History of Pelham Bay Park*, 8, Department of Parks and Recreation, "Bronx Parks, General" file, Parks Library, New York City Department of Parks, New York, NY (hereafter Parks Library). See also "Pelham Bay Park Popular," *New York Times*, August 14, 1910, X7. On the monorail, see "Monorail Car Falls on Its First Trip," *New York Times*, July 17, 1910, 1, and "City Island Road Sold," *New York Times*, July 10, 1914, 11.

19. Evelyn Gonzalez, *The Bronx* (New York: Columbia University Press, 2004), 57.

20. On this utopian vision for the Annexed District, see David Scobey, *Empire City: The Making and Meaning of the New York City Landscape* (Philadelphia: Temple University Press, 2003), 155, 249–55.

21. "The New Parks Bills," *New York Times*, April 3, 1885, 2, and "Hints for a Day's Outing," 20.

22. Fordham Morris, "Settlement and Early History," in *The Great North Side; Or, Borough of The Bronx*, ed. Bronx Board of Trade (New York: Knickerbocker Press, 1897), 1, and "Hints for a Day's Outing," 20.

23. Mullaly, *New Parks*, 78.

24. John C. Van Dyke, *The New New York: A Commentary on the Place and the People* (New York: Macmillan, 1909), 334.

25. Mullaly, *New Parks*, 16, 26, 76.

26. New York (State), *Report to the New York Legislature of the Commission to Select and Locate Lands for Public Parks in the Twenty-Third and Twenty-Fourth Wards of the City of New York* [. . .] (New York: M. B. Brown, 1884), 118.

27. Metropolitan Sewerage Commission of New York, *Supplementary Report on the Disposal of New York's Sewage: Critical Report of the New York Sewer Plan Commission on the Plans of Main Drainage and Sewage Disposal Proposed for New York* [. . .] *June 30, 1914* ([New York?], [1914?]), 19.

28. John Hooker Packard, *Sea-Air and Sea-Bathing* (Philadelphia: P. Blakiston, 1880), and George Oliver, "The Therapeutics of the Sea-Side: With Special Reference to the North-East Coast," *British Medical Journal* (November 19, 1870): 550–51, cited in Meghan Crnic and Cynthia Connolly, "'They Can't Help Getting Well Here': Seaside Hospitals for Children in the United States: 1872–1917," *Journal of the History of Childhood and Youth* 2 no. 2 (Spring 2009): 220.

29. Mullaly, *New Parks*, 42.

30. Ibid., 38, and "To Rid Parks of Private Tenants." On the Fresh Air Fund see Julia Guarneri, "Changing Strategies for Child Welfare, Enduring Beliefs About Childhood: The Fresh Air Fund, 1877–1926," *Journal of the Gilded Age and Progressive Era* 11, no. 1 (January 2012): 27–70.

31. "A Newport for Toilers," *New York Times*, March 12, 1889, 5, and "To Rid Parks of Private Tenants," *New York Tribune*, June 1, 1908, C8. See also Schnitz and Loeb, "'More Public Parks!,'" 51–66.

32. Carla Yanni, *The Architecture of Madness: Insane Asylums in the United States* (Minneapolis: University of Minnesota, 2007), 6, 58–59.

33. "A Newport for Toilers."

34. Parks board minutes reproduced in "Our Discussion on Centennial's Worthy of Note," October 26, 1983, Parks Library.

35. On lower Westchester population growth, see Michael Botwinick, "Introduction," *Westchester: The American Suburb*, ed. Roger Panetta (New York: Fordham University Press, 2006), vii, and "Regional Plan of New York and Its Environs," in Thomas Adams, Harold M. Lewis, and Theodore T. McCrosky, *Population, Land Values and Government*, vol. 2, *Regional Survey of New York and Its Environs* (New York: Regional Plan of New York and its Environs, 1929; repr., New York: Arno, 1974), 71.

36. Lundberg, Komarovsky, and McInery, *Leisure*, 29–30. On the number of clubs, see Michael R. Fein, *Paving the Way: New York Road Building and the American State, 1880–1956* (Lawrence: University Press of Kansas, 2008), 114.

37. For a Progressive-era history of the parkway see Hanmer et al., *Public Recreation*, 41. Numerous histories examine the design and construction of the Bronx River Parkway. While Newton, *Design on the Land*, is uncritical of the Bronx Parkway Commission's valuations of residents and land uses that predated the parkway, Barbara Troetel and Randall Mason offer more critical analysis of the commissioners' biases; see Troetel, "Suburban Transportation Redefined: America's First Parkway," *Westchester: The American Suburb*, ed. Roger Panetta (New York: Fordham University Press, 2006), 247–89, and Mason, *The Once and Future New York: Historic Preservation and the Modern City* (Minneapolis: University of Minnesota Press, 2009), 178.

38. While the city paid its share of the 1907 survey, it subsequently refused to authorize

funding for the purchase of land until 1913 and tried, without success, to halt its appropriations in 1922. See Mason, *Once and Future New York*, 199, and Clay McShane, *Down the Asphalt Path: The Automobile and the American City* (New York: Columbia University Press, 1994), 222.

39. *Bronx Parkway Commission, Final Report, December 31, 1925* (Bronxville, NY: Bronx Parkway Commission, 1925), 19.

40. Mason, *Once and Future New York*, 177–231.

41. Gilmore D. Clarke, "The Origins of the Parkway," February 1955, 2, and "The Design of the Motorway," May 22, 1962, 3, Folder, Manuscripts (arranged) Clarke—General, Box 4, Gilmore D. Clarke papers, ca.1920–1980, Special Collections, Columbia University Library, New York, NY.

42. "Reminiscences of Gilmore David Clarke: Oral History, 1959," 5, Special Collections, Columbia University Library, New York, NY.

43. Anne Whiston Spirn, "Constructing Nature: The Legacy of Frederick Law Olmsted," *Uncommon Ground: Rethinking the Human Place in Nature*, ed. William Cronon (New York: W. W. Norton, 1995), 92. See also Mason, *Once and Future New York*, 178, 181, 184.

44. *A General View of the Bronx River Parkway Reservation*, 1930, photograph of map, Parks Lantern Side Collection, Westchester County Park Commission Photograph Collection, Westchester County Archives, http://collections.westchestergov.com/cdm/singleitem/collection/pls/id/17/rec/2.

45. On Westchester's parkways, see Clarke, "The Design of the Motorway". On landscape designers' subsequent evaluations of the ecological benefits of rights-of-way, see Anne Whiston Spirn, *The Granite Garden: Urban Nature and Human Design* (New York: Basic Books, 1984), 72–73.

46. "Reminiscences of Gilmore David Clarke," 46.

47. *Report of the Bronx Parkway Commission* (Albany, NY: [J. B. Lyon?], 1912), 5, 9, 17–18.

48. On parkway traffic, see Troetel, "Suburban Transportation Redefined," 278; on Post Road traffic, see Michael R. Fein, *Paving the Way: New York Road Building and the American State, 1880–1956* (Lawrence: University Press of Kansas, 2008), 113.

49. "Downer Tells City Park Convention About This County," Parks Department Clippings, Series 98, Westchester County Archives, Elmsford, New York (hereafter Parks Department Clippings).

50. "More Good Advertising," *Tarrytown News*, May 19, 1927, and "The Opening of Playland," *Reporter*, May 25, 1928, Parks Department Clippings. See also Troetel, "Suburban Transportation Redefined," 249, 250.

51. Jon C. Teaford, *Post-Suburbia: Government and Politics in the Edge Cities* (Baltimore: Johns Hopkins University Press, 1997), 27; Edwin G. Michaelian, "Governing Westchester County," in *Westchester County: The Past Hundred Years, 1883–1983*, ed. Marilyn E. Weigold (Valhalla, NY: Westchester County Historical Society, 1983), 156–63, and Michael N. Danielson and Jameson W. Doig, *New York: The Politics of Urban Regional Development* (Berkeley: University of California Press, 1983), 79.

52. "How Westchester Got Her Parks," *New York Tribune*, April 8, 1926, Parks Department Clippings, and Jay Downer, "The Bronx River Parkway," in *Proceedings of the Ninth National Conference on City Planning Kansas City, Mo., May 7–9, 1917* (New York, 1918), 91. On the national prestige of the WCPC, see Marilyn E. Weigold, *Pioneering in Parks and Parkways: Westchester County, New York, 1895–1945*, Essays in Public Works History no. 9 (Chicago: Public Works Historical Society, 1980), 22–23.

53. *Report of the Bronx Parkway Commission* (Albany, NY: [J. B. Lyon?], 1912), 19, cited in Weigold, *Pioneering in Parks and Parkways*, 5.

54. Jay Downer and James Owen, *Public Parks in Westchester County, 1925* (New York, Lewis Historical, 1925), 24.

55. Downer quoted in Mann Hatton, "Westchester County Sees What's Coming and Is Putting 22 Million into Future," *Philadelphia Ledger*, May 8, 1925, Parks Department Clippings, and *Report of the Westchester County Park Commission, April 30, 1924* (Albany, NY, 1924), 45.

56. "More Good Advertising," *Tarrytown News*, May 19, 1927, and "The Opening of Playland," *Reporter*, May 25, 1928, Parks Department Clippings.

57. *Bronx Parkway Commission, Final Report, December 31, 1925* (Bronxville, NY: Bronx Parkway Commission, 1925), 78, and Downer and Owen, *Public Parks in Westchester*, 22.

58. *Report of the Westchester County Park Commission, 1924* ([White Plains, NY?]: The Commission, 1924), 76–7, and Downer and Owen, *Public Parks in Westchester*, 23.

59. *Report of the Westchester County Park Commission, 1928* ([White Plains, NY?]: The Commission, 1928), 10–11.

60. *Report of the Westchester County Park Commission, 1928*, 19.

61. Hanmer et al., *Public Recreation*, 42, 238.

62. Thomas Adams, "A Plan for Nassau County, Long Island, May 18, 1925," 4, Reel 88, New York (State), Governor Smith, Central Subject and Correspondence Files, 1919–1920, 1923–1928 New York State Archives, Albany, NY.

63. Hanmer et al., *Public Recreation*, 42, 238. See also David A. Johnson, *Planning the Great Metropolis: The 1929 Regional Plan of New York and Its Environs* (London: E and FN Spon, 1996), 137.

64. *Report of the Westchester County Park Commission, 1924*, 40.

65. *Report of the Westchester County Park Commission, 1932/1933* ([White Plains, NY?]: The Commission, 1933), 25, 28.

66. See "Swelling Sales Lead Westchester Realty Activity," *New York Times*, March 26, 1933, RE1.

67. *Bronx Parkway Commission, Final Report*, 56; *Report of the Westchester County Park Commission, 1924*, 40; and Hanmer et al., 87.

68. Olmsted, Vaux and Co., *Observations on the Progress of Improvements in Street Plans with Special Reference to the Parkway Proposed to Be Laid Out in Brooklyn, 1868* (Brooklyn: I. van Anden's Print, 1869), 23, cited in Clay McShane, *Down the Asphalt Path: The Automobile and the American City* (New York: Columbia University Press, 1994), 35.

69. "Home Building Features Westchester Development," *New York Times*, March 17, 1929, RE1.

70. John Nolen and Henry V. Hubbard, *Parkways and Land Values* (Cambridge: Cambridge University Press, 1937), 93, 99–100.

71. Ibid., 85–100, 123–24. For the quote, see Everit Macy, "Parks in the Modern Manner," *Survey*, July 1, 1930, 304.

72. *Bronx Parkway Commission, Final Report*, 25–27, and *Report of the Westchester County Park Commission, 1924*, 41–2.

73. "County Classes as Pioneer in Recreation Plan at Playland," March 20, 1928, and "Coney Island of the Future," *Rye Chronicle*, August 17, Parks Department Clippings.

See also Lundberg, Komarovsky, and McInery, *Leisure*, 78; "Regional Plan of New York and Its Environs," in *Graphic Regional Plan*, 394.

74. "Playland, How it Pays!," 42, Hufenland Pamphlet Collection, Westchester Historical Society Collection, Westchester County Archives, Elmsford, New York.

75. William C. Wright, "Biography of V. Everit Macy," chap. 16, p. 4, manuscript, Folder 1, Series 187, V. Everit Macy Papers, Westchester County Archives, Elmsford, New York.

76. *Report of the Westchester County Sanitary Sewer Commission from June 7, 1926 to November 30, 1929* (White Plains, NY, 1929), 42, and *Report of the Westchester County Sanitary Sewer Commission from November 20, 1929 to December 31, 1930* (White Plains, NY, 1930), 3. See also "Westchester Votes Rye Sewer Project," *New York Times*, (May 3, 1927), 13, and "Rye Village to Hear Improvement Plans," *New York Times*, July 18, 1926, 14.

77. Wright, "Biography of V. Everit Macy," 4.

78. Clarke, "Reminiscences of Gilmore David Clarke: Oral History, 1959," 5, Special Collections, Columbia University Library, New York, NY 52.

79. *National Association of Amusement Parks Bulletin*, Playland no. (August 15, 1929), 3.

80. Macy, "Parks in the Modern Manner," 303–4.

81. Leon N. Gillette and Gilmore D. Clarke, "Playland: An Amusement Park at Rye, Westchester County, N.Y.," *American Architect* 132, no. 2555 (October 20, 1928): 491.

82. "Playland, Utmost in Parks," *Amusement Park Management* 1, no. 4, Playland issue (March 1928), 127. See also "County Classes as Pioneer in Recreation Plan at Playland."

83. Gillette and Clarke, "Playland," 492.

84. "A Super-Park in the Making," *Amusement Park Management* 1, no. 4, Playland issue (March 1928): 134; Gillette and Gilmore, "Playland," 496; and Clarke, "Reminiscences of Gilmore David Clarke," 52.

85. *National Association of Amusement Parks Bulletin*, 3.

86. "A Super-Park in the Making."

87. Lundberg, Komarovsky, and McInery, *Leisure*, 79.

88. "Probe Discrimination at Playland," *New York Amsterdam News*, February 15, 1933, 3, and "New Charges Hurled at Playland," *New York Amsterdam News*, February 22, 1933, 2.

89. "Negroes Barred at Boat for Playland," *New York Amsterdam News*, August 19, 1931, 7.

90. "Probe Discrimination at Playland" and "Negroes Barred at Boat for Playland." The WCPC papers at the Westchester County Archives do not comment on the racial makeup of county park patrons and issues of racial discrimination at Playland.

91. "Negroes Allege Discrimination against Pupils," *Standard Star*, April 4, 1934, clipping, New Rochelle Public Library, New Rochelle, NY. See also Jeff Wiltse, *Contested Waters: A Social History of Swimming Pools in America* (Chapel Hill: University of North Carolina Press, 2007), and Victoria W. Wolcott, in *Race, Riots, and Roller Coasters: The Struggle over Segregated Recreation in America* (Philadelphia: University of Pennsylvania Press, 2012).

92. "Playland Manager Arrested," *New York Amsterdam News*, September 7, 1935, 1. For a retrospective on this desegregation work and the extent to which Westchester's shores remained largely restricted by the wealthy, see Paul D. Dennis Jr., "A Coast Off Limits to the Poor," *New York Times*, April 17, 1977, WC3.

93. Pauline Flippin, interview by Doris Rollins and Catherine Moore, October 1987, transcript, 3, Oral History Collection, New Rochelle Public Library, New Rochelle, NY.

94. For this idea of acting as a consumer in leisure and the claim to equal citizenship, see Wolcott, *Race, Riots, and Roller Coasters*, 3. See also "Playland Aide Seized for Ban on Negroes," *New York Times*, September 5, 1935, 24. The series of charges against Playland can be traced in the *New York Amsterdam News*: see "County Appeals Playland Cases," November 16, 1934, 16; "Start Crusades against Beaches," August 24, 1935, 1; and "Cashier Fined $150," September 28, 1935, 1. On the NAACP's success at Playland, see G. James Fleming, "New Rochelle NAACP Active All Year," *New York Amsterdam News*, November 14, 1936, 23, and "Right to Playland," *New York Amsterdam News*, July 4, 1936, 12.

95. "The Beautiful and the Damned," *Tarrytown News*, August 30, 1928, Parks Department Clippings.

96. "Westchester for Westchesterites!," *Mamaroneck Times*, June 11, 1928, Parks Department Clippings.

97. Frederick W. Sherman, "Exploiting Rye Beach," *Rye Chronicle*, November 26, 1927, Parks Department Clippings.

98. "Stop Everything—Why Not," *White Plains [Reporter?]*, May 11, 1925, Parks Department Clippings.

99. "Director of Playland Revokes Promises Made Rye People," reprinted in reproduced in Colin Dunne, John Dunne, Beth Griffin, Ted M. Levine, Robin Russell and Arlene Weiss, *Rye in the Twenties* (New York: Arno, 1978).

100. Charles Downing Lay, *A Park System for Long Island: A Report to the Nassau County Committee* (privately printed, 1925), 1, and "1929 Related Items," Folder 2, Box 3, Series 13915, Calendars and Minutes of Meetings of the State Council of Parks, 1925–1933, New York State Archives, Albany, NY (hereafter State Council of Parks). Other influential planners agreed. See "Notes by Thomas Adams: How We See Long Island," *A Plan for Nassau County, Long Island May 18, 1925*, 7, Reel 88, New York (State) Governor Smith, Central Subject and Correspondence Files, 1919–1920, 1923–1928, New York State Archives, Albany, NY.

101. *Population of States and Counties of the United States: 1790 to 1990*, ed. Richard L. Forstall (Washington, DC: Population Division, US Bureau of the Census, Department of Commerce, 1996), 111–15, https://www.census.gov/population/www /censusdata/PopulationofStatesandCountiesoftheUnitedStates1790-1990.pdf. On the housing boom of eastern Queens, see "History Promises to Repeat Itself in Sale of Corona-Flushing Subway Lots," *New York Times*, October 5, 1924, RE2.

102. Hanmer et al., *Public Recreation*, 64.

103. *First Annual Report of the Long Island State Park Commission [ . . . ] to the Governor and the Legislature of the State of New York, May 1925* (Albany, NY: Long Island State Park Commission,1925), 8, and *Second Annual Report of the Long Island State Park Commission [. . .] to the Governor and Legislature of the State of New York, December 1926* (Albany, NY: Long Island State Park Commission), and Hanmer et al., *Public Recreation*, 22.

104. *First Annual Report of the Long Island State Park Commission*, 15, 18. See also Hanmer et al., *Public Recreation*, 74–75, fig.44.

105. New York State Association, Committee on State Park Plan, *A State Park Plan for New York, With A Proposal For The New Park Bond Use* (New York: M. B. Brown, 1922),

5. On this report's significance, see Kathleen LaFrank, "Real and Ideal Landscapes along the Taconic State Parkway," in "Constructing Image, Identity, and Place," special issue, *Perspectives in Vernacular Architecture* 9, (2003): 248, and Robert Caro, *The Power Broker: Robert Moses and the Fall of New York* (New York: Knopf, 1974), 167.

106. Alfred E. Smith to Robert Moses, August 10, 1925, Reel 88, New York (State), Governor Smith, Central Subject and Correspondence Files, 1919–1920, 1923–1928 New York State Archives, Albany, NY.

107. New York State Association, *State Park Plan for New York*, 5. See also *New York State Council of Parks First Annual Report* [...] *to the Governor and Legislature of the State of New York*, Folder 4, Box 1, State Council of Parks, and "Suggested Letter from the Mayor-Elect to Goveror Lehman," 7, Folder Fiorello LaGuardia, Box 10, Series 1, Personal, Robert Moses Papers, New York Public Library Manuscripts and Archives Division, New York, NY (hereafter Robert Moses Papers).

108. "Statement of Mayor La Guardia as to City Park Bill," 1, Folder: Fiorello LaGuardia, Box 10, Series 1: Personal, Robert Moses Papers; *Memorandum on Proposed City Park and Parkway Extensions Prepared by the Metropolitan Conference of City and State Park Authorities* [...] (New York: Metropolitan Conference of City and State Park Authorities, 1925), Folder 4, Box 1, State Council of Parks; and also Metropolitan Conference on Parks, *Program For Extension of Parks and Parkways in the Metropolitan Region* [...] *as to Suggested Projects Within the City and to the Governor and Legislature of the State as to State Recommendations*, February 25, 1930, Folder 970, Box 108, Series E, Cultural Interests Park Ass. of New York City, 1926–1961, Record Group 2, Office of the Messrs. Rockefeller General Files, 1858–1981 (1879–1961), Rockefeller Family Archives, Rockefeller Archive Center, Sleepy Hollow, NY.

109. *Graphic Regional Plan*, 158, and *Memorandum on Proposed City Park and Parkway Extension*. For the quote, see "Park Boards Ready with City-Wide Plan," *New York Times*, January 10, 1926, E1, and Robert Dunbar Bromley, "The New Deal for the Parks Outlined by Their Director," *New York Times*, February 11, 1934, XX3.

110. Robert Moses, *The Expanding New York Waterfront* (New York: Triborough Bridge and Tunnel Authority, 1964), 1, and Alfred E. Smith to John D. Rockefeller, October 15, 1930, Family Records Series, Cultural Interests, F754 L.J. Parks (Long Island State Park Commission 1924–1961), Series 3 2E, Box 80, Folder 754, Rockefeller Archive Center.

111. *First Annual Report of the Long Island State Park Commission*, [...] *to the Governor and the Legislature of the State of New York, May 1925* (Albany, NY: Long Island State Park Commission,1925), 18.

112. Hanmer et al., *Public Recreation*, 64, and State Council of Parks, *New York State Parks*, Annual no. 2, Folder 2, Box 3, State Council of Parks.

113. *Memorandum on Proposed City Park and Parkway Extension*.

114. "Land for Buffer Parks," *New York Times*, December 11, 1925, 22. See also "Regional Park Planning," *New York Times*, January 11, 1926, 26, and "Park Boards Ready with City-Wide Plan," *New York Times*, January 10, 1926, E1.

115. *Graphic Regional Plan*, 125.

116. *First Annual Report of the Long Island State Park Commission*, 8, 15; *Second Annual Report of the Long Island State Park Commission* [...] *to the Governor and Legislature of the State of New York, December 1926* (Albany, NY: Long Island State Park Commission), 15; and Hanmer et al., *Public Recreation*, 22.

117. Hanmer et al, *Public Recreation*, 193–94.
118. *An Act to Amend Chapter 112 of 1924, Entitled "An Act to Provide for the Location, Erection, Acquisition and Improvement by the State of Parks [. . .] Creating a Commission Therefor, and Defining the Powers and Duties of Such Commission,"* "Related Items, 1925," Folder 4, Box 1, State Council of Parks.
119. Weigold, *Pioneering in Parks and Parkways*, 6; Marc R. Fasanella, "The Environmental Design of Jones Beach State Park: Aesthetic and Ecological Aspects of the Park's Architecture and Landscape" (PhD diss., New York University, 1991), 48; Eugene Lewis, *Public Entrepreneurship: Toward a Theory of Bureaucratic Political Power; The Organizational Lives of Hyman Rickover, J. Eager Hoover, and Robert Moses* (Bloomington: Indiana University Press, 1980), 177; and Caro, *Power Broker*, 175.
120. Robert Moses, "The Public Parks and Parkways of New York City and Long Island," April 8, 1936, Folder Robert Moses Park Work, Box 11, Series 1, Personal, Robert Moses Papers.
121. "Plan to Link Parks Furthered by Moses," *New York Times*, August 19, 1940, 17.
122. "New York Crowds Irk Westchester," *New York Times*, July 12, 1932, 11.
123. Robert Moses, "New York Reclaims its Waterfront," *New York Times*, March 7, 1948, SM 16. More recent revisionist scholarship on Moses has brought to light the master builder's long-standing commitment to coastal conservation. See Nicholas Dagen Bloom, *The Metropolitan Airport: JFK International and Modern New York* (Philadelphia: University of Pennsylvania Press, 2015), 115–17, 125, and Lawrence Kaplan and Carol P. Kaplan, *Between Ocean and City*, 79–81.
124. On the city's beach mileage, see Gina Pollara, "Transforming the Edge: Overview of Selected Plans and Projects," in *The New York Waterfront: Evolution and Building Culture of the Port and Harbor*, ed. Kevin Bone (New York: Monacelli, 1997), 176. On the city's coastal mileage, see Hanmer et al., *Public Recreation*, 177n1.
125. On the vernacular landscape, see Daniel Campo's interpretation of J. B. Jackson's *Discovering the Vernacular Landscape* (New Haven, CT: Yale University Press, 1984), 15, in Campo, *Accidental Playground*, 13–14.
126. "Park Camps to Go, Moses Announces," *New York Times*, March 7, 1934, 21.
127. "All Yacht Clubs, Except Morris, Have Vacated Property of City," *New York Times*, January 13, 1935, S6. On coastal land-use politics see John K. Walton, "Seaside Tourism and Environmental History," in *Common Ground: Integrating the Social and Environmental in History*, ed. Genevieve Massard-Guilbaud and Stephen Mosley (Newcastle upon Tyne: Cambridge Scholars, 2011), 71.
128. "Mayor Backs Moses on Orchard Beach," *New York Times*, April 10, 1934, 25. For the county court's "strained construction" comment, see "Moses Wins Again in Row over Camps," *New York Times*, June 12, 1934, 25.
129. L. H. Robbins, "Our Parks Creator Extraordinary," *New York Times*, March 20, 1938, 118. See also "Parks Come First, Moses Declares," *New York Times*, May 4, 1939, 23.
130. Department of Parks, *Memorandum of 1935 Budget Request of the Department of Parks*, Parks Library.
131. "New 'Jones Beach' Planned in Bronx," *New York Times*, February 28, 1934, 21.
132. Marshall Sprague, "Beach Treks Begin," *New York Times*, June 13, 1937, 145.
133. "New 'Jones Beach' Planned in Bronx," 21.
134. Moses, *Expanding New York Waterfront*, 23.
135. Daniel C. Walsh, "Urban Residential Refuse Composition and Generation Rates for

the 20th Century," *Environmental Science and Technology* 36, no. 22 (2002): 4938, and "The History of Waste Landfilling in New York City," *Ground Water Readers' Forum* 29 (July–August 1991): 591.

136. Aviation Volunteer Fire Co., *Aviation Volunteer Fire Co., Bronx, N.Y., 1923–1973: 50th Anniversary Featuring a Picture History of Clason Point and It's [sic] People* (New York: Aviation Volunteer Fire Co., 1973), Special Collections, Lehman College Library of the City University of New York (CUNY), The Bronx, New York.

137. For "something to complain about," see Moses, "New York Reclaims Its Waterfront," *New York Times*, March 7, 1948, SM 16. See also "Moses Acts to Halt Beach Pollution," *New York Times*, March 27, 1935, 14, and "Bronx Dumping to End," *New York Times*, May 23, 1935, 20.

138. For the quote, see Sprague, "Beach Treks Begin," *New York Times*, June 13, 1937, 145. See also "Moses Bars City Amusement Park in Bronx As Improper Enterprise for Municipal Board," *New York Times*, June 1, 1935, 17.

139. Moses, *Expanding New York Waterfront*, 23.

140. Bromley; "Bronx Park Needs Stressed on Tour" *New York Times*, July 24, 1931, 14; and "Moses Bars City Amusement Park in Bronx."

141. For quote, see "Pelham Bay Park," in Ballon and Jackson, *Robert Moses*, 162.

142. "Rowdyism Fought at Orchard Beach," *New York Times*, July 17, 1934, 17.

143. Robert Moses, "Hordes from the City," *Saturday Evening Post*, October 1931, 13–14, 92. See also letter from Moses to E. F. Chester, Chester-Polland Amusement Co., January 6, 1933, Folder, Box 4, State Council of Parks.

144. Lundberg, Komarovsky, and McInery, *Leisure*, 65.

145. "Park Camps to Go, Moses Announces," *New York Times*, March 7, 1934, 21.

146. "Pelham Bay Park," in Ballon and Jackson, *Robert Moses*, 161.

147. "Moses Plans Park Sports Fees to Support Recreation Centers," *New York Times*, March 2, 1936, 1, and "Extract from Affidavit of Sidney Solomon, Sworn Nov. 23, 1934 and Submitted in Support of Plaintiff's Motion for an Examination before Trial of James A. Sherry, Chief Clerk of the Department of Parks and W. Earle Andrews, General Superintendent of Dept of Parks of the City of New York," 34–35, Folder, New York City Park Department, Box 97, Series 6, New York City Dept. of Parks, Robert Moses Papers.

148. Lewis Mumford, "Report on Honolulu," in *City Development: Studies in Disintegration and Renewal* (New York: Harcourt, Brace, 1945), 133–34, quoted in Ballon and Jackson, *Robert Moses*, 159.

149. John Nolen, "Subject: The County Becoming the City Planning Unit," 1, Folder 8, Box 1, John Nolen Papers, 1890–1938, 1954–1960, Division of Rare and Manuscript Collections, Cornell University Library, Ithaca, New York.

150. "Coney Island of the Future," *Rye Chronicle*, August 17, Parks Department Clippings. See also Marilyn E. Weigold, *People and the Parks: A History of Parks and Recreation in Westchester County* (White Plains, NY: Westchester County Dept. of Parks, Recreation and Conservation, 1984), 34.

151. Moses, "Hordes from the City," *Saturday Evening Post*, October 1931, 90. See also David Stradling, *The Nature of New York: An Environmental History of the Empire State* (Ithaca, NY: Cornell University Press, 2010), 171.

152. Clarke, "Reminiscences of Gilmore David Clarke: Oral History, 1959," 53, Special Collections, Columbia University Library, New York, NY.

153. Letter from John M. Gaus to Robert Moses, January 22, 1934, Folder, Letters of Congratulation City Park Commissionership, Box 10, Series 1, Personal, Robert Moses Papers. On the singularity of his program, see Marta Gutman, "Equipping the Public Realm: Rethinking Robert Moses and Recreation," in Ballon and Jackson, *Robert Moses*, 73.
154. See Weigold, *Pioneering in Parks and Parkways*, 39.
155. Clarke, "Reminiscences of Gilmore David Clarke," 11, 276.
156. Johnson, *Planning the Great Metropolis*, 263.
157. Caro, *Power Broker*, 5–9, 339–46. Caro cites Moses's *Public Works: A Dangerous Trade* (New York: McGraw-Hill, 1970) as an example of the park commissioner's practice of taking credit for others' plans.

*Chapter 5*

1. Tiffany v. Town of Oyster Bay, 192 A.D. 126 WestLaw (N.Y. App. Div. 1920).
2. Tiffany v. Town of Oyster Bay, 104 Misc. 445 WestLaw (N.Y. Supp. 1918), and Frank B. Williams, "Foreshore and Rights in Land under Navigable Waters in the New York Region," in Lee F. Hanmer, Thomas Adams, Frederick W. Loede Jr., Charles J. Storey, and Frank B. Williams, *Public Recreation: A Study of Parks, Playgrounds, and Other Outdoor Recreational Facilities*, Regional Survey of New York and Its Environs 5 (New York: Regional Plan of New York and Its Environs, 1928), 202, 241.
3. "Justice Callaghan Dismisses Injunction, Mr. Tiffany Can Not Stop Bath House Building," *Oyster Bay Guardian*, June 16, 1916. On the court's reversal, see Tiffany v. Town of Oyster Bay 141 A.D. 720 WestLaw (N.Y. App. Div. 1910) at **912–913, and Tiffany v. Town of Oyster Bay, 209 N.Y. 1 WestLaw (N.Y. 1913).
4. "Court Decision Defied, Town Property Destroyed," *Oyster Bay Guardian*, June 9, 1916; "The Tiffany Case," *Oyster Bay Guardian*, June 16, 1916; and "Justice Callaghan Dismisses Injunction," Clippings, Town of Oyster Bay official historian's collection. John Hammond, email to author, Jan. 27, 2012. On Tiffany's nuisance claim, see Tiffany v. Town of Oyster Bay, 104 Misc. 445 WestLaw (N.Y. Supp. 1918).
5. "Militant," "Act for the Island," Folder 5, Box 7, Series D, William H. Burr Jr. Papers pertaining to Sherwood Island State Park, Fairfield Museum and History Center, Fairfield, Connecticut (hereafter William H. Burr Jr. Papers).
6. Anthony Baker, Robert B. MacKay, Carol A. Traynor, and Brendan Gill, eds., *Long Island Country Houses and Their Architects, 1860–1940* (New York: W. W. Norton, 1997).
7. F. Scott Fitzgerald, *The Great Gatsby* (1925; repr., New York: Scribners, 1992), 10.
8. Lewis Mumford, *The Urban Prospect* (New York: Harcourt, Brace and World, 1968), 81, quoted in "Jones Beach," in *Robert Moses and the Modern City: The Transformation of New York*, ed. Hillary Ballon and Kenneth Jackson (New York: W. W. Norton, 2007), 158.
9. Hanmer et al., *Public Recreation*, 71, 77.
10. Aldred, quoted in Helen Worden, "Where Are the Rich?," *New York World Telegram*, October 20, 1942, 17. This quote also appears in Robert B. MacKay, "Long Island Country Houses and Their Architects," *Long Island Historical Journal* 6, no. 2 (Spring 1994): 179.
11. Robert Caro, *The Power Broker: Robert Moses and the Fall of New York* (New York: Knopf, 1974), 146–50, 303–6, and David A. Johnson, *Planning the Great Metropolis:*

*The 1929 Regional Plan of New York and Its Environs* (London: E and FN Spon, 1996), 225–36.

12. *Williams v Gallatin* 128 N.E. 121, 123 (N.Y. 1920), and Molly Slevin, "This Tender and Delicate Business: The Public Trust Doctrine in American Law and Economic Policy, 1789–1920," (Harold Hyman & Stuart Bruchey eds., American Legal and Constitutional History: A Garland Series of Outstanding Dissertations, 1987), 11–12, cited in Michael Seth Benn, "Towards Environmental Entrepreneurship: Restoring the Public Trust Doctrine in New York," *University of Pennsylvania Law Review* 155 no. 1 (Nov. 2006), 205n33.

13. See Williams v. Gallatin, 128 N.E. 121, 123 (N.Y. 1920), cited in Michael Seth Benn, "Towards Environmental Entrepreneurship: Restoring the Public Trust Doctrine in New York," *University of Pennsylvania Law Review* 155, no. 1 (November 2006): 215n37. See also Jack H. Archer, Donald L. Connors, and Kenneth Laurence, *The Public Trust Doctrine and the Management of America's Coasts* (Amherst: University of Massachusetts Press, 1994).

14. *Map of the Shore Line of Nassau County, N.Y. Showing the Grants of Land under Water Made by the Commissioners of the Land Office [. . .] Prepared for the Commissioners of the Land Office under the Direction of Edward A. Bond, State Engineer and Surveyor 1901,* Series 119116, Maps of Grants of Lands under Water, [ca. 1777–1970], New York State Archives, Albany, NY.

15. Dennis P. Sobin, *Dynamics of Community Change: The Case of Long Island's Declining "Gold Coast"* (Port Washington, NY: Ira J. Friedman, 1968), 60.

16. "Village Law, Laws of the State of New York 1909, Chapter 64, Section 2," in *The Consolidated Laws of New York, Annotated, as Amended to the Close of the Regular and Extraordinary Sessions of the Legislature of 1917 . . . Book 63. Village Law* (Northport, NY: Edward Thompson Company, 1918), 11, and "Laws of the State of New York 1910, Chapter 258, Section 33," in *Laws of the State of New York, Passed at the One Hundred and Thirty-Third Session of the Legislature, Begun January Fifth, 1910, and Ended May Twenty-Seventh, 1910,* vol. 1 (Albany, NY: J.B. Lyon, 1910), 416. See also "Movement to Abolish Great Neck," *North Hempstead Record*, May 24, 1928, 1, Nassau County Museum Reference Library Collection, Special Collections Department, Hofstra University, Hempstead, NY. See also Sobin, *Dynamics of Community Change*, 99–103.

17. For a comprehensive list of estate village incorporation and dates, see Sobin, *Dynamics of Community Change*, 176, table 7.

18. "Movement to Abolish Great Neck."

19. Robert Moses to C. Chester Painter, March 5, 1929, "1929 Related Items," Folder 2, Box 3, State Council of Parks, Series 13915: Calendars and Minutes of Meetings of the State Council of Parks, 1925–1933, New York State Archives, Albany, NY.

20. Thomas H. Reed, *The Government of Nassau County: A Report Made to the Board of Supervisors* (Mineola, NY: Nassau County, 1934), 58. This challenges Robert C. Wood's conclusion that Long Island's special-purpose districts and units of general jurisdiction were "ineffective in the aggregate principally because its parts tend to cancel one another out"; Robert C. Wood, *1400 Governments: The Political Economy of the New York Metropolitan Region* (Cambridge, MA: Harvard University Press, 1961), 173.

21. "Laurelton Long Island Becomes an Incorporated Village," *Oyster Bay Guardian*,

July 16, 1926, 1, reproduced in Donald Glenn Nelson, "The Incorporated Village of Laurel Hollow: An Historical Essay," 6–7, History, Village of Laurel Hollow, 2013, http://laurelhollow.org/content/History/Home/:field=documents;/content/Documents/File/149.pdf.

22. Stephen Richard Higley, *Privilege, Power, and Place: The Geography of the American Upper Class* (Lanham, MD: Rowman and Littlefield, 1995), 37, and Jon C. Teaford, "Nassau County: A Pioneer of the Crabgrass Frontier," in *Nassau County: From Rural Hinterland to Suburban Metropolis*, ed. Joann P. Krieg and Natalie A. Naylor (Interlaken, NY: Empire State Books, 2000), 31.

23. "To Make Village 'Ideal,'" *New York Times*, December 28, 1931, 18, and "'Millionaire Village' to Be Taxless This Year," *New York Times*, May 1, 1936, 8.

24. "The Hearing Held," *Sands Point, New York*, clipping, vol. 3, Port Washington News Index, Port Washington Public Library, Port Washington, NY; "Certification of Consolidation of Village of Sands Point with Villages of Barker's Point and Motts Point into one Village by the Name of 'Sands Point,'" July 23, 1912, clipping, Town Clerk Records, Village of Sands Point, NY.

25. Quoted in Sobin, *Dynamics of Community Change*, 103.

26. "Saddle Rock Re-elects Mrs. Eldridge," *New York Times*, March 17, 1937, 18, and "Millionaire Village Born as Iced Drinks Clink," *New York Times*, July 30, 1931, 21.

27. Jon C. Teaford, *The Unheralded Triumph: City Government in America 1870–1900* (Baltimore: Johns Hopkins University Press, 1984), 106, 121.

28. "Inspect the Route of Proposed Parkway," *New York Times*, June 2, 1925, 2.

29. Quoted in Kate Van Bloem, *History of the Village of Lake Success* (Lake Success, NY: Incorporated Village of Lake Success, 1968), 31–32.

30. "Refuse to Renew Bar Beach Leases," *North Hempstead Record*, April 6, 1927, and "Squatters Take Over Old Roadway," *North Hempstead Record*, July 14, 1926, 1, Nassau County Museum Reference Library Collection, Special Collections Department, Hofstra University, Hempstead, NY. On legal and cultural localism, see David D. Troutt, "Localism and Segregation," *Journal of Affordable Housing* 16, no. 4 (Summer 2007): 325.

31. Helen Worden, "Where Are the Rich?," *New York World Telegram*, October 20, 1942, 17.

32. Troutt, "Localism and Segregation," 325.

33. "Planning Commissions Authorized for All Cities and Incorporated Villages in New York State," *American City* 9, no. 1 (July 1913): 79; "Article 16 of the Town Law Preempts Local Legislation," Brief for the New York State Builders Association, Inc., as Amicus Curiae, Kahmi v. Yorktown, 74 N.Y.2d 423 (N.Y. 1989), 22–23; and Lawrence M. Friedman, *A History of American Law*, 3rd ed. (New York: Simon and Schuster, 2005), 315–16, 554. See also Vincent Moore, "Politics, Planning, and Power in New York State: The Path from Theory to Reality," *Journal of the American Institute of Planners* 37 (March 1971): 66–77, and Troutt, "Localism and Segregation," 325–6.

34. "Wealthy Villagers Will Hold Election," *New York Times*, April 2, 1927, 17.

35. "Record of the Public Hearing Held at Town Hall, Manhasset New York, by the Town of North Hempstead on June 27, 1927, 1," Office of the Town Clerk, Town of North Hempstead, Manhasset, NY.

36. Thomas Carmody, quoted in Charles D. Newton, "Opinions to Commissioners of land Office," *New York Legislative Documents One Hundred and Fifty-Fourth Session*,

1921, vol. 21, nos. 51–61 inclusive (Albany, NY: J. B. Lyon, 1921), 307–8. On the rights implied in property leases, see Elizabeth Blackmar, *Manhattan for Rent, 1785–1850* (Ithaca, NY: Cornell University Press, 1989), 9.

37.    "Fight Hard for Beach," *New York Tribune*, May 9, 1912, 5, and *First Annual Report of the Long Island State Park Commission* [ . . . ] *to the Governor and Legislature of the State of New York, May 1925* (Albany, NY: Long Island State Park Commission, 1925), 37. See also J. Lance Mallamo, "Building the Roads to Greatness: Robert Moses and Long Island's State Parkways," in *Robert Moses Single-Minded Genius*, ed. Joann P. Krieg (Interlaken, NY: Heart of the Lakes Publishing, 1989), 160. On Moses's and the LISPC's effect on Long Island's road system, see also Owen Gutfreund, "Rebuilding New York in the Auto Age: Robert Moses and His Highways," Ballon and Jackson, *Robert Moses*, 86–93.

38.    Hanmer et al., *Public Recreation*, 77–78, fig. 44.

39.    Alfred E. Smith, "The Principles of State Park Planning as Applied to New York," in *1925 New York State Council of Parks First Annual Report* [ . . . ] *to the Governor and Legislature of the State of New York*, Folder 4, Box 1, Series 13915, Calendars and Minutes of Meetings of the State Council of Parks, 1925–1933, New York State Archives, Albany, NY.

40.    "Nation's Death List 300," *New York Times*, June 7, 1925, 1.

41.    "Heat Kills 21 Here; 93 are Overcome; May Last For Days," *New York Times*, June 6, 1925, 1. On Huntington, see Caro, *Power Broker*, 194.

42.    "Private Estates, Parks and Parkways," *New York Times*, June 7, 1925, E4; "Gov. Smith Calls an Extra Session to Act on Parks," *New York Times*, June 11, 1925, 1; "Says Public Must Have Shore Parks," *New York Times*, June 11, 1925, 1.

43.    "Governor Smith's Address to the Legislature on Parks," *New York Times*, June 23, 1925, 2. See also Silas Bent, "Governor Smith Outlines Park Battle," *New York Times*, June 14, 1925, XX3, and *1925 New York State Council of Parks First Annual Report* [ . . . ] *to the Governor and Legislature of the State of New York*, Folder 4, Box 1, Series 13915, Calendars and Minutes of Meetings of the State Council of Parks, 1925–1933, New York State Archives, Albany, NY.

44.    A version of this quote is often attributed to Smith. See Caro, *Power Broker*, 187, and Frank Graham, *Al Smith, American: An Informal Biography* (New York: G. P. Putnam's Sons, 1945), 140.

45.    "For Release in the Morning Papers of Tuesday, Jan 27, 1925," Reel 88, New York (State), Governor Smith, Central Subject and Correspondence Files, 1919–1920, 1923–1928, New York State Archives, Albany, NY. See also Robert Chiles, "Working-Class Conservationism in New York: Governor Alfred E. Smith and 'The Property of the People of the State,'" *Environmental History* 18, no. 1 (January 2013), 157–83.

46.    Moses, "Hordes from the City," *Saturday Evening Post*, October 1931, 92.

47.    Hanmer et al., *Public Recreation*, 77.

48.    For Smith quote, see "Governor Smith's Comment on Action of Legislative Leaders on the State Park Program on Long Island"; on state parks for the island, see *Second Annual Report of the Long Island State Park Commission* [ . . . ] *to the Governor and Legislature of the State of New York, December 1926* (Albany, NY: Long Island State Park Commission, 1926), and "Notes by Thomas Adams: How We See Long Island," *A Plan for Nassau County, Long Island May 18, 1925*, 7, all three items in Reel 88, New York (State), Governor Smith, Central Subject and Correspondence Files, 1919–1920, 1923–1928, New York State Archives, Albany, NY.

49. Karl Jacoby, *Crimes against Nature: Squatters, Poachers, Thieves, and the Hidden History of American Conservation* (Berkeley: University of California Press, 2003), 3, 6.

50. Michael Rawson, *Eden on the Charles: The Making of Boston* (Cambridge, MA: Harvard University Press, 2010), 171, and Michael H. Ebner, *Creating Chicago's North Shore: A Suburban History* (Chicago: University of Chicago Press, 1988), 105.

51. Teaford, "Nassau County: A Pioneer of the Crabgrass Frontier," in Krieg and Naylor, *Nassau County* 29–36; *Post-Suburbia: Government and Politics in the Edge Cities* (Baltimore: Johns Hopkins University Press, 1997), 19–29; and the review by Howard Gillette Jr., "In Assessing New Suburban Forms, Politics Matter," *Reviews in American History* 25, no. 3 (September 1997), 516.

52. "How Westchester Got Her Parks," *New York Herald Tribune*, April. 8, 1926, Parks Department Clippings, Series 98, Westchester County Archives, Elmsford, NY. For an overview of the Land Board controversy and the North State Parkway plan, see Caro, *Power Broker*, 192–98.

53. "Governor Smith's Address to the Legislature on Parks," *New York Times*, June 23, 1925.

54. Nassau County Committee, *The Nassau County Committee and State Parks on Long Island, with a Note on the Proposed Transfer of Hempstead's Town Lands to the State* (Glen Head, Long Island: The Nassau County Committee, 1925), 6, and Charles Downing Lay, *A Park System for Long Island: A Report to the Nassau County Committee* (privately printed, 1925), 1.

55. Lay, *A Park System for Long Island*, 8, 11.

56. Ibid., 7.

57. Ibid., 8.

58. David A. Johnson, *Planning the Great Metropolis: The 1929 Regional Plan of New York and Its Environs* (London: E and FN Spon, 1996), 167.

59. *First Annual Report of the Long Island State Park Commission [ . . . ] to the Governor and Legislature of the State of New York, May 1925* (Albany, NY: Long Island State Park Commission, 1925), 23. See also Johnson, *Planning the Great Metropolis*, 227.

60. Sam Bass Warner, *The Private City: Philadelphia in Three Periods of Its Growth*, rev. ed. (Philadelphia: University of Pennsylvania Press, 1987), 213.

61. Henry M. Earle, "A Protest against the Construction of a Certain Proposed Boulevard in a Section of the North Shore of Long Island," 3, Reel 88, New York (State), Governor Smith, Central Subject and Correspondence Files, 1919–1920, 1923–1928, New York State Archives, Albany, NY.

62. Thomas Adams, "A Plan for Nassau County, Long Island, May 18, 1925," Reel 88, New York (State), Governor Smith, Central Subject and Correspondence Files, 1919–1920, 1923–1928 New York State Archives, Albany, NY, quoted in Johnson, *Planning the Great Metropolis*, 232.

63. "Regional Plan of New York and Its Environs," in Thomas Adams, Harold M. Lewis, and Theodore T. McCrosky, *Population, Land Values and Government*, vol. 2 of *Regional Survey of New York and Its Environs* (New York: Regional Plan of New York and Its Environs, 1929; repr., New York: Arno, 1974), 58–59.

64. Regional Plan Association, "Requirements for Residential Land in the Region Analyzed by Counties," *Information Bulletin* no. 19 (June 18, 1934), cited in Reed, *Government of Nassau County*, 2n3.

65. *The Graphic Regional Plan: Atlas and Description [ . . . ]*, vol. 1 of *Regional Plan of New York and Its Environs* (New York: Regional Plan of New York and Its Environs, 1929), 362–363. For "public misfortune," 378.

66. *Graphic Regional Plan*, 281–82, 377–80, 406.

67. Ibid., 378.

68. Ibid., and Melissa A. Smith, "Russell Sage Foundation," *The Encyclopedia of New York City*, ed. Kenneth T. Jackson (New Haven, CT: Yale University Press, 1995), 1029.

69. *Graphic Regional Plan*, 282, 377, and Hanmer et al. *Public Recreation*, 94.

70. Moses, "Hordes from the City," *Saturday Evening Post*, October 1931, 90. For an over-view of Jones Beach, see Ballon and Jackson, *Robert Moses*, 158–60. Caro famously ac-cused Moses of deliberately building bridges too low for buses. While this argument has since been refuted, it remains a powerful myth about Moses's attitude toward the public. "Jacob Riis Park," and Kenneth T. Jackson, "Robert Moses and the Rise of New York: The Power Broker in Perspective," in Ballon and Jackson, *Robert Moses*, 169, 70.

71. Michael Simpson, "Meliorist versus Insurgent Planners and the Problems of New York, 1921–1941," *Journal of American Studies* 16 no. 2 (August 1982): 207, and Michael K. Heiman, *The Quiet Evolution: Power, Planning and Profits in New York State* (New York: Praeger, 1988), 63.

72. "Governor Smith's Address to the Legislature on Parks," *New York Times*, June 23, 1925. See also Johnson, *Planning the Great Metropolis*, 234–35.

73. Herbert Evison, ed., *A State Park Anthology* (Washington, DC: National Conference On State Parks, 1930), 195.

74. Albert M. Turner, "Report of Field Secretary," *State of Connecticut Public Document No. 60, Report of the State Park Commission to the General Assembly for the Fiscal Year Ended September 30, 1914* (Hartford: Hartford Printing Company, 1914).

75. Turner, "Report of Field Secretary," (1914), 17.

76. Ibid., 30–31.

77. Graham v. Walker, 78 Conn., 130, 133–34, 61 A. 98, 99 (1905);, see Jack H. Archer, Donald L. Connors, and Kenneth Laurence, *The Public Trust Doctrine and the Manage-ment of America's Coasts* (Amherst: University of Massachusetts Press, 1994), 104.

78. Town of Orange v. Resnick, 94 Conn. 573 (Conn. 1920), cited in Frank B. Williams, "Foreshore and Rights in Land under Navigable Waters in the New York Region," in Hanmer et al., *Public Recreation*, 204.

79. Turner, "Report of Field Secretary," 37–8.

80. Turner's 1917 "Report of the Connecticut State Park Commission to the Governor" was reproduced as Turner, "The Price of Procrastination," in Evison, *State Park An-thology*, 61.

81. Turner, "Report of Field Secretary," 23.

82. Turner, "Planning a System of State Parks for Connecticut," in Evison, *State Park Anthology*, 82.

83. This percentage reflected the 995,475 people living in the coastal lowlands of the state in 1910. Turner, "Report of Field Secretary," 17.

84. See Evison, *State Park Anthology*. A scholarly overview of state parks is available in Norman T. Newton's classic and comprehensive *Design on the Land: The Development of Landscape Architecture* (Cambridge, MA: Harvard University Press, 1971), 555–75. Newton, however does not distinguish between state and county park commissions with regard to funding and home rule.

85. Turner, "Report of Field Secretary," 31.

86. Woody Klein, *Westport, Connecticut: The Story of a New England Town's Rise to Promi-nence* (Westport, CT: Greenwood, 2000), 135.

87. Three deeds were filed between George W. Gair and the estate of Henrietta A.

Jennings, Ester Sherwood Minor, and Dorothy Sherwood Minor July 12, 1921, *Town of Westport Land Records*, Book 38, 94– 95, *Connecticut Town Portal- Land and Indexed Records* (2007–2013), https://connecticut-townclerks-records.com/. See also "George W. Gair Resents William H. Burr's Criticisms and Returns the Attack," March 21, 1929, William H. Burr Jr. Papers, MS B112, Folder 5, Box 7, Series D, William H. Burr Jr. Papers.

88. See the 1917 *Westporter-Herald* article cited in Joseph Elwood Betts's *The Sherwood-Elwood Connection: Story of the Transition of the Sherwood Elwood Farm to the Sherwood Island State Park* (Westport, CT: El-Lo, 2011), 6.

89. Turner, "Report of Field Secretary on Acquisition of Land," in *Report of the State Park and Forest Commission to the Governor for the Fiscal Term Ended June 30, 1924*, State of Connecticut Public Document No. 60 (Hartford, CT: Hartford Printing, 1924), 12, and "State Should Put Up Road Signs Leading to Island and Mark Off Parking Spaces Says Reader," clipping, Folder 5, Box 7, William H. Burr Jr. Papers

90. "Official Would Have Town Take Over Sherwood Island," Folder 5, Box 7, William H. Burr Jr. Papers.

91. "Proposed State Park Deal May Be Blocked by Residents," clipping, Folder 3, Box 7, William H. Burr Jr. Papers.

92. Nicolas K. Blomley's scholarship addresses the dynamics of private-public boundaries. See, for example, "From 'What?' to 'So What?': Law and Geography in Retrospect" in *Law and Geography*, ed. Jane Holder and Carolyn Harrison (Oxford: Oxford University Press, 2003), 17–33; "The Borrowed View: Privacy, Propriety, and the Entanglements of Property," *Law and Social Inquiry* 30, no. 4 (October 2005): 617–61; and "Flowers in the Bathtub: Boundary Crossings at the Public-Private Divide," *Geoforum* 36, no. 3 (May 2005): 281–96.

93. "Mean Water, Mean Crowds," Folder 5, Box 7, William H. Burr Jr. Papers.

94. Thomas J. Sugrue, *The Origins of the Urban Crisis: Race and Inequality in Postwar Detroit* (Princeton, NJ: Princeton University Press, 1996), 210–11.

95. James S. Duncan and Nancy G. Duncan, *Landscapes of Privilege: The Politics of the Aesthetic in an American Suburb* (New York: Routledge, 2004), 60.

96. William H. Burr to Edward E. Bradley, May 1, 1916, Folder 1, Box 7, William H. Burr Jr. Papers.

97. Turner, "Report of Field Secretary on Acquisition of Land," (1924), 20.

98. Turner, "Report of Field Secretary on Acquisition of Land" in *Report of the State Park Commission to the General Assembly for the Fiscal Year Ended June 30, 1926*, State of Connecticut Public Document No. 60 (Hartford, CT: Hartford Printing, 1926), 12; "Westport—On the Sound," *New York Times*, July 12, 1925, W5, and Sherwood Island Park Association, properties map of Sherwood Island, Folder 3, Box 7, William H. Burr Jr. Papers.

99. "Official Would Have Town Take Over Sherwood Island," clipping, Folder 5, Box 7, William H. Burr Jr. Papers.

100. "Not a Mean Property Owner" and "Beaches and Public," clippings, Folder 5, Box 7, William H. Burr Jr. Papers.

101. "Bridgeport Sunday Post 2/24/1929, 3 of 3," clipping, Folder 5, Box 7, William H. Burr Jr. Papers.

102. Copy of resolutions reprinted in "Town Meeting Called Oct. 27th 1924," Folder 3, Box 7, William H. Burr Jr. Papers.

103. "Westport Dredged Creek to Isolate Park," *Hartford Courant*, August 4, 1929, A4.

104. For "Gair's ditch," see "Bridgeport Editor Sees Nothing but Ulterior Motive in Town Dredging at Sherwood's Island," clipping, Folder 5, Box 7, William H. Burr Jr. Papers. See also Jamie Eves, "Beyond the Dry Highway and the No Trespass Sign: Albert Milford Turner, William H. Burr, Elsie Hill, and the Creation of Connecticut's First Oceanfront State Parks, 1914–1942," *Connecticut League of History Organizations Bulletin* 57, no. 2 (May 2004): 12.

105. "The Man of Sherwood's Island," *Bridgeport Telegram*, May 13, 1932, Folder 5, Box 7, Series D, William H. Burr Jr. Papers pertaining to Sherwood Island State Park, "Burr Family Papers, 1752–1940," Fairfield Museum and History Center, Fairfield, Connecticut.

106. For "whether the beauties," see clipping, *Bridgeport Sunday Post*, February 24, 1929, Folder 5, Box 7, William H. Burr Jr. Papers. For "shut up," "throw him out," and "we want to vote," see "Westport Votes to Delay Park Action to 1932," *Bridgeport Telegram*, February 10, 1931, Folder 5, Box 7, William H. Burr Jr. Papers.

107. "Public Beach for Fairfield County Asked," *Hartford Courant*, April 24, 1931, 3; see also "Politics: Maneuvering for Political Advantage in 1932," *Hartford Courant*, May 24, 1932, A8.

108. Deborah Wing Ray and Gloria P. Stewart, *Norwalk: Being an Historical Account of That Connecticut Town* (Canaan, NH: Phoenix, 1979), 161–63.

109. *Report of the State Park Commission to the Governor for the Two Fiscal Years Ending September, 30, 1916* (Hartford: Hartford Printing, 1916), 10.

110. *Report of the State Park Commission to the Governor for the Fiscal Term Ending June 30, 1924* (Hartford: Hartford Printing, 1914), 7.

111. Ibid., 21.

112. "Report of the Treasurer," *State of Connecticut Public Document No 55–60, Report of the State Park and Forest Commission to the Governor for the Fiscal Term Ended June 30, 1926* (Hartford: Hartford Printing, 1926), 57.

113. Turner, "Report of the Field Secretary," *Report of the State Park Commission to the Governor for the Fiscal Term Ended June 20, 1920*, State of Connecticut Public Document No. 60 (Hartford, CT: Hartford Printing, 1920), 14; "Report of Field Secretary on Acquisition of Land," (1924), 12.

114. William H. Burr to Mr. Deyo, January 10, 1923, Folder 2, 1920–1923 SISP Acquisition, William H. Burr Jr. Papers.

115. "Current Comment: The Fairfield Shore Park," *Bridgeport Times-Star*, reprinted in the *Hartford Courant*, May 15, 1931, 14, and "The Fairfield Park Commission," *Hartford Courant*, May 12, 1931, 12.

116. "Sherwood Island and What It Means to Connecticut-Fairfield County, the Town of Westport," Connecticut Forestry Association and the Fairfield County Planning Association, 1925, Folder 3, Box 7, William H. Burr Jr. Papers. See also "Public Parks along the Shore," *Hartford Courant*, April 30, 1931, 14; "A Veto Called For," *Hartford Courant*, May 19, 1931, 12; and "Sherwood Island and Roton Point," *Bridgeport Post*, clipping, Folder 5, Box 7, William H. Burr Jr. Papers.

117. Robert M. Ross to William H. Burr, January 7, 1931, Folder 4, 1930–1937 SISP Acquisition, William H. Burr Jr. Papers.

118. For "policy of inaction," see "State Provides No Accommodations Not Even the Most Primitive," and for editorials against isolationism, see "Officials Violate Law, Levitt Holds," "Sherwood Island Park Dispute Is Aired at Capitol," and "Showdown on Sherwood Island," all in Folder 5, Box 7, William H. Burr Jr. Papers.

119. Through the nineteenth and twentieth centuries, county government was stripped of its limited powers as new state agencies took over county functions. In October 1960, county government ceased to exist. See Rosaline Levensen, *County Government in Connecticut: Its History and Demise* (Storrs: Institute of Public Service, Extended and Continuing Education, University of Connecticut, 1966).
120. "Officials Violate Law, Levitt Holds," Folder 5, Box 7, William H. Burr Jr. Papers.
121. For "guard" quote, see "Militant," Folder 5, Box 7, William H. Burr Jr. Papers. For "common herd" quote, see A. Jonstone, Letter to the Editor, September 1, 1924, clipping, Folder 3, Box 7, William H. Burr Jr. Papers. For the public-private divide on beaches, see Carol M. Rose, *Property and Persuasion: Essays on the History, Theory, and Rhetoric of Ownership* (Oxford: Westview, 1994), 115, and Richard K. Norton, Lorelle A. Meadows, and Guy A. Meadows, "Drawling Lines in Law Books and on Sandy Beaches: Marking Ordinary High Water on Michigan's Great Lakes Shorelines under the Public Trust Doctrine," *Coastal Management* 39 (2011), 143.
122. The statistics for the Bronx used by the *Regional Plan of New York and Its Environs* were from 1927 rather than 1928; Hanmer et al., *Public Recreation*, 64. On the ideal ratio of country and city parks, see Hanmer et al., *Public Recreation*, 127.
123. Hanmer et al., *Public Recreation*, 241.
124. "State Park Plan before Committee," *Hartford Courant*, March 27, 1925, 2. On the region as a concept, see Thomas Adams, Harold M. Lewis, and Theodore T. McCrosky, *Population, Land Values, and Government*, vol. 2 of *Regional Survey of New York and Its Environs* (New York: Regional Plan of New York and Its Environs 1929; repr., New York: Arno, 1974), 204, quoted in Keith D. Revell, *Building Gotham: Civic Culture and Public Policy in New York City, 1898–1938* (Baltimore: Johns Hopkins University Press, 2005), 11n18.
125. Turner, "Report of Field Secretary on Acquisition of Land," (1924), 11.
126. "Sherwood Island and What It Means," 9.
127. Ibid., 4–5.
128. "Press Release in the Morning Papers of Monday, May 7," Reel 88, New York (State), Governor Smith, Central Subject and Correspondence Files, 1919–1920, 1923–1928 New York State Archives, Albany, NY. See also "Gov. Smith Thanks DeForest for Land," *New York Times*, May 7, 1928, 9, and *Report of the Long Island State Park Commission for the Year 1930* (Albany, NY: Long Island State Park Commission, 1930), 10. See also "Conference Held by Former Governor A.E. Smith on a Proposed Settlement of the Northern State Parkway Controversy," December 1929, and "Joint Statement by Robert Moses, President of the LISPC and Greenville Clarke, Counsel for the Nassau County Citizen's Committee," December 18, 1929, Folder 2, Box 3, Series 13915, Calendars and Minutes of Meetings of the State Council of Parks, 1925–1933, New York State Archives, Albany, NY. On the resulting detour, see Caro, *Power Broker*, 185, 277–80, 301.
129. Robert Moses, *Public Works: A Dangerous Trade* (New York: McGraw-Hill, 1970), 135, and Johnson, *Planning the Great Metropolis*, 225–36.
130. *Report of the Long Island State Park Commission for the Year 1930*.
131. *Twelfth Biennial Report of the State Park and Forest Commission to the Governor for the Fiscal Term Ended June 30, 1936* (Harford: Hartford Printing, 1936), 8.
132. Ibid., 11; *Thirteenth Biennial Report of the State Park and Forest Commission to the Governor for the Fiscal Term Ended June 30, 1938* (Harford: Hartford Printing, 1938), 11, 23–25; and "Shore Park in Fairfield County Sure," *Hartford Courant*, May 12,

1932, 1, and "Sherwood Island Bill for 435,000 Signed by Cross," *Bridgeport Post*, April 29, 1937, accessed at Friends of Sherwood Island State Park, "The 23-Year War," http://friendsofsherwoodisland.org/history/the-23-year-war/. For an overview of the restrictions the SPFC had to untangle, see Albert M. Turner, "Report of the Field Secretary," in *State of Connecticut Public Document No. 60, Thirteenth Biennial Report of the State Park and Forest Commission to the Governor for the Fiscal Term Ended June 30, 1938* (Harford: Hartford Printing, 1938), 10.

133. Charles Downing Lay, "The Raid on Hempstead Town!," Folder 63, Box 2, Series 7, Writings, Charles Downing Lay Papers, 1898–1956, 4477, Division of Rare and Manuscript Collections, Cornell University Library, Ithaca, NY.

134. William H. Burr to Devo, January 10, 1923, Folder 2, 1920–1923 SISP Acquisition, William H. Burr Jr. Papers. See also W. H. Burr, "Effort for Parks Urged," Folder 5, Box 7, William H. Burr Jr. Papers.

## Chapter 6

1. F. Scott Fitzgerald, *The Great Gatsby* (1925; repr., New York: Scribners, 1992), 27.

2. *New York World's Fair Preview April 29–30–May 1st* (New York: Junior League of the City of New York, 1939), Special Collections, World's Fairs Collection, Archives Center, National Museum of American History, Washington, DC (hereafter Special Collections, World's Fairs Collection).

3. Fitzgerald, *Great Gatsby*, 27.

4. "World's Fair Set-up Completed," *World's Fair Bulletin* 1, no. 1 (August 1936): 1; and *1939 Dawn of a New Day* ([New York?]: [1939?]), Special Collections, World's Fairs Collection.

5. For "the Fair itself" see *Your World of Tomorrow* (Rogers Kellogg Stillson, 1939), Folder 3, "Trylon and Perisphere-Democracity," Subseries B (2) Theme Center, Series 3, NYWF Records, Edward J. Orth Memorial Archives of the New York World's Fair, Archives Center, National Museum of American History, Washington, DC (hereafter Edward J. Orth). For "in its scope," see *Official Guide Book New York World's Fair, The World of Tomorrow* (New York: Exposition Publications, 1939), 25, Folder 1, Box 15, Collection 60, World's Expositions 1851–1965, National Museum of American History Archives Center, National Museum of American History, Washington, DC (hereafter World's Expositions 1851–1965).

6. Barbara Blumberg, *The New Deal and the Unemployed: The View from New York City* (Lewisburg, PA: Bucknell University Press, 1979), 36, 45.

7. Metropolitan Conference on Parks, *Program for Extension of Parks and Parkways in the Metropolitan Region* [. . .] *as to Suggested Projects within the City and to the Governor and Legislature of the State as to State Recommendations* (New York, 1930), Folder 970, Box 108, Series E, Cultural Interests Park Ass. of New York City, 1926–1961, Record Group 2, Office of the Messrs. Rockefeller General Files, 1858–1981(1879–1961), Rockefeller Family Archives, Rockefeller Archive Center, Sleepy Hollow, NY.

8. "Men Behind the New York World's Fair 1939," *World's Fair Bulletin* 1, no. 1 (August 1936), 1, and Frank Monaghan, *New York World's Fair 1939: "The Fairs of the Past and the Fair of the Future,"* rev. ed. (Chicago: Encyclopaedia Britannica, 1939), Special Collection, World's Fairs Collection.

9. On the overlap of New Deal, the World's Fair, and the Regional Plan Association, see Michael K. Heiman, *The Quiet Evolution: Power, Planning and Profits in New York State* (New York: Praeger, 1988), 63.

10. "Robert Moses Park Work," Folder, Robert Moses Park Work, Box 11, Series 1, Personal, Robert Moses Papers, and *The Triborough Bridge Authority Fifth Anniversary*, 6.

11. Frederick Newell, "What May Be Accomplished by Reclamation," *Annals of American Academy of Political and Social Science* 33 (1909), 174, 178. For Moses's uses of the term, see Robert Moses, "Backyard Conservation," *Forum and Century* 99, no. 1 (January 1939), 56.

12. Tom Daniels, *When City and Country Collide: Managing Growth on the Metropolitan Fringe* (Washington, DC: Island, 1999), 19.

13. John Peale Bishop, "World's Fair Notes," *Kenyon Review* 1, no. 3 (Summer 1939): 239.

14. "Building the 'World of Tomorrow,'" *American Builder and Building Age* 60, no. 6 (June 1938): 44, in Series 9, Publications Relating to the World's Fair, American Builder 1937-American City 1959, Edward J. Orth. For an overview of industrial design at the fair, see Jeffrey L. Meikle, *Twentieth Century Limited: Industrial Design in America 1925–1939* (Philadelphia: Temple University Press, 1979), 39–67, and Helen A. Harrison, ed., *Dawn of A New Day: The New York World's Fair 1939–40* (Flushing, NY: Queens Museum, 1980), 4.

15. "New York Looks at Tomorrow," *World's Fair News* 1, no. 5 (1936): 5, Folder 2, Box 16, Collection 60, World's Expositions 1851–1965, National Museum of American History Archives Center, National Museum of American History, Washington, DC (hereafter World's Expositions 1851–1965).

16. *The City*, directed by Ralph Steiner and Willard Van Dyke (1939; American Documentary Films, Inc.), film, https://archive.org/details/CityTheP1939.

17. "Theme Focal Exhibit," February 25, 1939; "Trylon and Perisphere, Democracity Stat Sheet," May 30, 1940; and "Visit the Perisphere and See Your 'World of Tomorrow,'" Folder 3, "Trylon and Perisphere-Democracity," Subseries B(2), Theme Center, Series 3, New York World's Fair Records, Edward J. Orth.

18. Norman Bel Geddes, *Magic Motorways* (New York: Random House, 1940), 36.

19. Bel Geddes, 261.

20. *Your World of Tomorrow*, Rogers Kellogg Stillson, Inc., 1939, Folder 3, "Trylon and Perisphere-Democracity," Subseries B (2) Theme Center, Series 3, NYWF Records, Edward J. Orth.

21. Bel Geddes, *Magic Motorways*, 239.

22. Pearl E. Levison, "World of Tomorrow," in *The Official Poem of the New York World's Fair, 1939, and Other Prize Winning Poems* (New York: Academy of American Poets, 1939), 5, Special Collections, World's Fairs Collection.

23. "Rehabilitation of Flushing Bay," *Flushing Meadow Improvement* 2, no. 4 (June 1937): 27.

24. Walter I. Willis, *Queens Borough; Being a Descriptive and Illustrated Book of the Borough of Queens, City of Greater New York* [. . .] (Brooklyn: Brooklyn Eagle, 1913), 62. For letters to Congress, see *61st Congress, 3rd Session, House of Rep Doc. no. 13333, Flushing Bay New York*, from House Documents, 61st Congress 3rd Session, Dec. 5 1910–March 4, 1911, vol. 20 (Washington DC: Government Printing Office, 1911), 8–9.

25. "M. J. Degnon Dies," *New York Times*, April 23, 1925, 21, and "Michael J. Degnon Built New York City's Subway" advertisement, *New York Times*, September 21, 1924,

RE3. For an overview of Degnon's businesses, see Vincent F. Seyfried, *Corona: From Farmland to City Suburb, 1650–1935* ([New York?]: Edgian, [1986?]), 66–70.

26. "Brooklyn's Plan for Ash Removal," *New York Times*, March 10, 1907, 6.

27. "Corona-Flushing Lots Sell for Over $1,000,000" *New York Times*, November 16, 1924, RE2, and *Queensborough* [monthly magazine of Queens Chamber of Commerce], May 16, 1924, cited in Seyfried, *Corona*, 318. Ted Steinberg offers a history of the company and dump in *Gotham Unbound: The Ecological History of Greater New York* (New York: Simon and Schuster, 2014), 208–21.

28. Francis Cormier, "Flushing Meadow Park: The Ultimate Development of the World's Fair Site," *Landscape Architecture* 29 (July 1, 1939): 162–70; Rebecca B. Rankin, ed., *New York Advancing: The Result of Five Years of Progressive Administration in the City of New York, F.H. La Guardia, Mayor* [. . .] (New York: Publishers Printing Company, 1939), 246–48, Folder, 1939–40 Theme Center and Surrounding Area, New York City Building, Subseries B(2), Theme Center, Series 3, NYWF Records, Edward J. Orth.

29. For "man-made glacier," see "Permanent Plan for Park After Fair," *Flushing Meadow Improvement* 1, no. 3 (Dec. 1936): 2–4. For "dismal scene" see Fitzgerald, 28.

30. Cormier, "Flushing Meadow Park," 168.

31. Benjamin Miller, *Fat of the Land: Garbage of New York, The Last Two Hundred Years* (New York: Basic Books, 2000), 118.

32. Helen Adams Keller, "Diary of Helen Adams Keller, February, 1937," February 13, 1937, 194, in *Helen Keller's Journal 1936–1937* (London: Michael Joseph, 1938), 296. *North American Women's Letters and Diaries*, Colonial–1950 (Alexandria, VA: Alexander Street Press, 2001–), June 4, 2013. On waterfowl protections, see Ann Vileisis, *Discovering the Unknown Landscape: A History of America's Wetlands* (Washington, DC: Island, 1997), 151, 156, 163.

33. Charles Downing Lay, "Tidal Marshes," *Landscape Architecture* 2, no. 3 (April 1912), 101–2. Other landscape designers spoke in similar terms of wetland beauty, including Frederick Law Olmsted, but they remained a minority; see Vileisis, *Discovering the Unknown Landscape*, 148.

34. Commission on Street Cleaning and Waste Disposal, *Report of the Commission on Street Cleaning and Waste Disposal, the City of New York, 1907* (New York: Commission on Street Cleaning and Waste Disposal, 1907), 101, 141.

35. "Fight on Rikers Island," *New York Times*, August 12, 1903, 2; "North Beach Unique as an Amusement Resort," *New York Times*, July 2, 1905, X5, and Commission on Street Cleaning and Waste Disposal, *Report of the Commission*, 101.

36. "Public Hearing No. 2," February 15, 1895, Municipal Archives 90-SWL-49, quoted in Miller, *Fat of the Land*, 78–79, 311.

37. "Riker's Island Odors Decreasing," *New York Times*, August 16, 1894, 9; "To Abate the Dumping Nuisance," *New York Times*, August 15, 1894, 5. See also Miller, *Fat of the Land*, 79–80.

38. George Soper, "Great Sanitation Questions That Now Confront the City" and "Appeals to Connolly in Letters Opposing Ash Contract Renewal," *Daily Star, Queens Borough*, December 15, 1921, 10; "Flatlands Protests against Another Ash Removal Plant," *Brooklyn Daily Eagle*, June 9, 1922, 5; and "War Declared upon Rats," *New York Times*, November 4, 1920, 13.

39. "Indict Queens Dump Owners for Nuisance," *Brooklyn Daily Eagle*, November 25, 1931, L.1.

40. "Smoke Nuisance Charged," *New York Times*, September 14, 1923, 19.

41. "Dogs Killing Off Rats at Dump, Supervisor Says," *Daily Star, Queens Borough*, October 2, 1925, 6, and "History of 'Mt. Corona,'" *Flushing Meadow Improvement* 2, no. 1 (March 1937): 15. This article appeared as "The Rise and Fall of 'Mt. Corona,'" in *American City* 52 (August 1937): 64–65.

42. "Civic Workers Sincere in Effort to Stop Corona Meadows Smoke Nuisance," *Newton Register*, February 2, 1924, 5, and "Indict Queens Dump Owners for Nuisance," *Brooklyn Daily Eagle*, November 25, 1931, L.1. On Tammany politics, see "Hand of Tammany in Corona Dump Fight?," *Daily Star, Queens Borough*, December 14, 1923, 1, and "This Dump Seems to Be an Official Pet," *Daily Star, Queens Borough*, June 14, 1929, 1. On the costs of the BARC contract and its termination, see "Oppose Contract for Ash Removal," *New York Times*, January 26, 1933, 24; "Ash Removal Suit Ordered to Trial," *New York Times* May 5, 1933, 10; "City Ash Removal Halted by Court," *New York Times*, December 30, 1933, 14; and Miller, *Fat of the Land*, 130, 134, 178–80.

43. "History of 'Mt. Corona,'" 15. See also *Official Guide Book New York World's Fair, The World of Tomorrow,* (New York: Exposition Publications, 1939), 28, Folder 1, Box 15, Collection 60, World's Expositions 1851–1965. See Francis Cormier, "Flushing Meadow Park: The Ultimate Development of the World's Fair Site," *Landscape Architecture* 29 (July 1, 1939):166–82, and "Site Virgin Land," *World's Fair Bulletin* 1, no. 1 (August 1936): 11. On Halleran's controversial role, see "Halleran Attacks Lockwood Awards," *New York Times*, June 15, 1936, 13; "Windels Defends 'Low' Land Awards," *New York Times*, June 16, 1936, 23; and "Halleran's Profits at Fair Site Shown," *New York Times*, June 17, 1936, 11. On real estate profits motivating fair work, see Steinberg, *Gotham Unbound*, 216.

44. "Corona-Flushing Lots Sell for over $1,000,000," *New York Times*, November 16, 1924, RE2, and Rebecca B. Rankin, ed., *New York Advancing: The Result of Five Years of Progressive Administration in the City of New York, F.H. La Guardia, Mayor* [. . .] (New York: Publishers Printing Company, 1939), xxii–xxiii, Folder, 1939–40, Theme Center and Surrounding Area, New York City Building, Subseries B(2), Theme Center, Series 3, NYWF Records, Edward J. Orth.

45. For "the last great open space," see L. H. Robbins, "It Begins to Look Like a Great Fair," *New York Times*, May 1, 1939, 119.

46. Robert Moses, "From Dump to Glory," *Saturday Evening Post*, January 15, 1938, 2.

47. Robert Moses, *A Half-Century of Achievement: Triborough Bridge and Tunnel Authority Links the Five Boroughs, Parks, Beaches, and Suburbs* (New York: Triborough Bridge and Tunnel Authority, 1948), 4. This quote also appears in "Moses Asks Steps to End Pollution," *New York Times*, May 24, 1937, 21.

48. Meyer Berger, "At the World's Fair," *New York Times*, July 21, 1939, 10.

49. Moses, "From Dump to Glory," 72. See also "'Beauty for Ashes': A Statement by the Executive Committee of the New York World's Fair 1964–1965," June 3, 1967, Folder 230, Box 8, Malcolm Wilson Papers, Rockefeller Archive Center, Sleepy Hollow, NY, and "Parkways and Expressways in Brooklyn and Queens," in *Robert Moses and the Modern City: The Transformation of New York*, ed. Hilary Ballon and Kenneth T. Jackson (New York: W. W. Norton, 2007), 221. For Moses's critique of the golf courses, see "Plan Golf Courses in Corona Meadows," *New York Times*, July 22, 1929, 35, and Moses, "From Dump to Glory," 13.

50. "North Queens Sewage Disposal Problem," *Flushing Meadow Improvement* 3, no. 6 (October 1938): 2, 7–8, Box Cue 1941–Family Circle 1939, Subseries 1, Magazines, 1937–1948, Series 9, Publications Relating to the World's Fair, Edward J. Orth.

51. George Elighton, "World's Fairs: From Little Egypt to Robert Moses," *Harper's Magazine*, July 1960, 27, Folder 10, Box 15, World's Expositions 1851–1965.

52. For "romantic saga" see *Souvenir View Book: New York World's Fair the World of Tomorrow* (New York: Tichnor Brothers, 1939), Special Collections, World's Fairs Collection. On organization, see "Fair to Cost Millions," *World's Fair Bulletin* 1, no. 1 (August 1936): 9, and "4 Agencies Speed Bay Improvement," 1, no. 3, *World's Fair World*, 14, Folder 16, Box 15 [filed with flat oversize box], Alfred E. Cohn Papers, 1900-(1920–1954)-1980, Rockefeller Archive Center, Sleepy Hollow, NY.

53. George Soper, "Great Sanitation Questions That Now Confront the City," *New York Times*, May 7, 1933, XX3.

54. Willis, *Queens Borough*, 3, 59. See also "Purifying Flushing Bay," *Flushing Meadow Improvement* 3, no. 3 (January 1938): 29, Box Cue 1941–Family Circle 1939, Subseries 1, Magazines, 1937–1948, Series 9, Publications Relating to the World's Fair, Edward J. Orth.

55. Benjamin Eisner, "Modern Sanitation Protects New York World's Fair," *Municipal Sanitation* 10 (April 1939): 173.

56. Clarence Perry, *The Rebuilding of Blighted Areas: A Study of the Neighborhood Unit in Replanning and Plot Assemblage* (New York: Regional Plan Association, 1933), 17. See also New York Metropolitan Sewerage Commission, *Sewerage and Sewage Disposal in the Metropolitan District of New York and New Jersey: Report, April 30, 1910* (New York: Martin B. Brown, 1910), 279.

57. Richard Harris, "Industry and Residence: The Decentralization of New York City, 1900–1940," *Journal of Historical Geography* 19, no. 2 (1993): 180.

58. "Walker Deplores Queens Conditions," *New York Times*, October 31, 1928, 32.

59. Perry, *Rebuilding of Blighted Areas*, 13. On critiques on unplanned growth in the outer boroughs, see Harris, "Industry and Residence," 180–84. Harris also references Perry.

60. "Modifying Sewer Sections Produces Savings in Construction Costs of Flushing Meadow Park Storm Water Trunk Sewer," *Flushing Meadow Improvement* 2, no. 3 (May 1937): 20. On local assessments, see Peter Marcuse, "The Grid as City Plan: New York City and Laissez-Faire Planning in the Nineteenth Century," *Planning Perspectives* 2, no. 3 (1987): 299, 304.

61. "Sewage Disposal Plants to Be Constructed by the Dept. of Sanitation" and "Purifying Flushing Bay," *Flushing Meadow Improvement* 1, no. 2 (November 1936): 6 and 29, respectively. For a summary of the sewage infrastructure constructed for the fair, see *New York World's Fair Bulletin*, Construction Edition, October 1936, 3, Folder 41, New York World's Fair 1939, Box 2, Series 6, Architecture, Planning, etc., Housing, Parks, Committees and Other, Charles Downing Lay Papers, 1898–1956, Division of Rare and Manuscript Collections, Cornell University Library, Ithaca, NY. On the area's drainage see New York Metropolitan Sewerage Commission, *Main Drainage and Sewage Disposal Works Proposed for New York City: Reports of Experts and Data Relating to the Harbor* (New York: Wynkoop Hallenbeck Crawford, 1914), 60–61.

62. Interstate Sanitation Commission, *Highlights of Water Pollution Abatement Activities (1936–1969)* (New York: The Commission, [197–]), 1; Tri-State Treaty Commission,

*Final Report of the Research and Engineering Committee* (1931), 4; *Historical Reports and Other Documents Issued by the IEC*, Interstate Environmental Commission, 2015, http://www.iec-nynjct.org/archive.htm; and New York Metropolitan Sewerage Commission, *Sewage Disposal Works Proposed for New York City*, 64. See also Dennis J. Suszkowski, "Sewage Pollution in New York Harbor: A Historical Perspective" (MS thesis, State University of New York at Stony Brook, 1973), 11, 54.

63.  "North Queens Sewage Disposal Problem" and "The Tallman's Island Sewage Treatment Works," *Flushing Meadow Improvement* 3 no. 4 (April 1938): 25–27; and "North Beach Screen Plant," *Flushing Meadow Improvement* 3, no. 1 (September 1937): 17. For estimates of fair daily sewage, see "Sewage Disposal Plants to Be Constructed by the Dept. of Sanitation," *Flushing Meadow Improvement* 1 no. 2 (November 1936): 6.

64.  "Purifying Flushing Bay," 29. For the sewage infrastructure constructed for the fair, see *New York World's Fair Bulletin*, Construction Edition, October 1936, 3, Folder 41, New York World's Fair 1939, Box 2, Series 6, Architecture, Planning, etc., Housing, Parks, Committees and Other, Charles Downing Lay Papers, 1898–1956, Division of Rare and Manuscript Collections, Cornell University Library, Ithaca, NY. On the area's drainage, see New York Metropolitan Sewerage Commission, *Main Drainage and Sewage Disposal Works Proposed for New York City: Reports of Experts and Data Relating to the Harbor* (New York: Wynkoop Hallenbeck Crawford, 1914), 60–61.

65.  "Sewage Disposal Plants to Be Constructed by the Dept. of Sanitation," *Flushing Meadow Improvement* 1, no. 2 (November 1936): 6.

66.  "Mammoth Sewers Drain Park Area," *Flushing Meadow Improvement* 1, no. 2 (November 1936): 3; "Park Plans for Sewers, Water and Lights," *Flushing Meadow Improvement* 1, no. 4 (January 1937): 37; and "Description of Drainage for Area Surrounding World's Fair Site," *Flushing Meadow Improvement* 2, no. 1 (March 1937): 23. On the area's drainage, see New York City Metropolitan Sewerage Commission, *Preliminary Reports on the Disposal of New York's Sewage* (New York: Evening Post Job Printing Office, 1911), 10.

67.  Marit Larson and Paul Mankiewicz, "Restoring Soundview Park, HEP Priority Restoration Site L110," *Tidal Exchange*, Spring 2006, 1.

68.  Eisner, "Modern Sanitation Protects New York World's Fair," *Municipal Sanitation* 10 (April 1939): 173.

69.  Susan Strasser, *Waste and Want: A Social History of Trash* (New York: Metropolitan Books, 1999), 15, 18, and Steinberg, *Gotham Unbound*, 241.

70.  George Soper, "Great Sanitation Questions That Now Confront the City," *New York Times*, May 7, 1933, XX3. For the ocean dumping of various boroughs in the early twentieth century, see Julie Sze, *Noxious New York: The Racial Politics of Urban Health and Environmental Justice* (Cambridge, MA: MIT Press, 2007), 59, 216n6.

71.  On incinerators, see "Delay to July 1 Given on Garbage," *New York Times*, December 5, 1933, 5. On dumping at Soundview and Rikers, see New York City Department of Parks, *The Reclamation of Park Areas by Sanitation Fill and Synthetic Topsoil* (New York: Comet, 1950), 1.

72.  For Riker's Island waste statistics, see "Fair Site Menaced by Rikers Island," *New York Times*, December 20, 1935, 27; "The New Plan for Riker's Island," *Flushing Meadow Improvement* 3, no. 1 (September 1937): 18; "End of Vesuvius," *Flushing Meadow Improvement* 1, no. 1 (October 1936): 4; and George Soper, "Disposal of Waste an Urgent Problem," *New York Times*, March 18, 1934, XX2. For an overview of landfill, see

Daniel C. Walsh, "The History of Waste Landfilling in New York City" *Ground Water Readers' Forum* 29 (July–August 1991): 591.

73. Robert Moses to Fiorello LaGuardia, November 1, 1935, 1–2, Folder New York Parks Dept. 1935, Box 97, Series 6, New York City Dept. of Parks, Robert Moses Papers, New York Public Library Manuscripts and Archives Division, New York, NY (hereafter Robert Moses Papers). For "notorious East River Vesuvius," see "End of Vesuvius," *Flushing Meadow Improvement* 1, no. 1 (October 1936): 4–5.

74. Memorandum to the mayor from Frances Lehrich, February 16, 1939, and Robert Moses to Edmund L. Palmieri, November 3, 1939, Folder 04, 1938–1940, Folder Parks, Dept. of, Soundview Park, Series Departmental Correspondence, Fiorello H. LaGuardia Collection, LaGuardia and Wagner Archives, Long Island City, New York.

75. "The New Plan for Riker's Island," *Flushing Meadow Improvement* 3, no. 1 (September 1937): 20, and "Fair Site Menaced by Rikers Island," *New York Times*, December 20, 1935, 27.

76. "End of Vesuvius," *Flushing Meadow Improvement* 1, no. 1 (October 1936): 4–5.

77. "Sewage Disposal Plants to Be Constructed by the Dept. of Sanitation," *Flushing Meadow Improvement* 1, no. 2 (November 1936): 6.

78. "Modifying Sewer Sections Produces Savings in Construction Costs of Flushing Meadow Park Storm Water Trunk Sewer," *Flushing Meadow Improvement* 2, no. 3 (May 1937): 21.

79. Benjamin Eisner, "Water Problems at the New York World's Fair," *Journal of the American Water Works Association* 31, no. 2 (February 1939): 190. On the conservationist approach to marshes, see Vileisis, *Discovering the Unknown Landscape*, 111, 113, 198. On the control of nature in the science of ecology and the conservation movement, see Donald Worster, *Nature's Economy: A History of Ecological Ideas* (Cambridge: Cambridge University Press, 1977), 113–14, cited in Steinberg, *Gotham Unbound*, 220.

80. Donald M. Burmister, "Laboratory Investigations of Soils at Flushing Meadow Park," *Transactions of the American Society of Civil Engineers* 107, no. 2 (January 1942): 187.

81. Karl Terzaghi, *Origin and Functions of Soil Mechanics*, Transactions of the American Society of Civil Engineers, paper no. 2619 (Cambridge, MA: Harvard University, 1953), 669, cited in Ben H. Fatherree, "The Rise of Geotechnology through 1936," in *The History of Geotechnical Engineering at the Waterways Experiment Station 1932–2000* (Vicksburg, MS: US Army Engineer Research and Development Center, 2006).

82. Burmister, "Laboratory Investigations," 188.

83. For "semi-liquid character" see Carlton S. Proctor, "Address before the Columbia Engineering Association and Engineering Faculty Annual Dinner at the Columbia Club [ . . . ]," *Flushing Meadow Improvement* 2, no. 2 (April 1937): 16. For "to indicate how far," see "Mammoth Sewers Drain Park Area" *Flushing Meadow Improvement* 1, no. 2 (November 1936): 3.

84. Burmister, "Laboratory Investigations," 198–99.

85. Proctor, "Address," 16, and "Life Beings at 8:00 A.M.," *Flushing Meadow Improvement* 2, no. 1 (March 1937): 2–3. On the Perisphere, see Henry Welles Durham, "Construction of the New York World's Fair," *Military Engineer* 31, no. 179 (September–October 1939), 361, and Gohnny Johns, "International Phases of the New York World's Fair," *World Affairs* 101, no. 2 (June 1938): 92. On the bulkhead, see "Flushing Bay Boat Basin" and "History of Dredging in Flushing Bay," *Flushing Meadow Improvement* 2, no. 2 (April 1937): 13 and 8, 10, respectively.

86. Proctor, "Address," 18.

87. "Top Soil Reclamation for Flushing Meadow Park," *Flushing Meadow Improvement* 1, no. 4 (January 1937): 23, 29.

88. Frank B. Kemp, "The New York World's Fair," *Technical Engineering News* 17/18 (April 1937): 73. This article exists in nearly the exact same form as "Top Soil Reclamation for Flushing Meadow Park," 23–29. On excavated mat, see Allyn R. Jennings, "Landscape Features of New York's World's Fair of 1939," *Parks and Recreation* 22, no. 8 (April 1939): 6.

89. Kemp, "New York World's Fair," 73, and "Top Soil Reclamation for Flushing Meadow Park," 28–29.

90. "The New East River Bridge Begun," *New York Times*, May 8, 1895, 16, and William Steinway Diary, September 8, 1895, The William Steinway Diary Project, 1861–1891, Smithsonian Institution National Museum of American History, http://americanhistory.si.edu/steinwaydiary (hereafter Steinway Diary).

91. Steinway, September 29, 1870, Steinway Diary; "Features of Greater New York" *New York Times*, July 12, 1896, 25; and "Queensboro Bridge Opens to Traffic," *New York Times*, March 31, 1909, 2.

92. Victor H. Bernstein, "New Triumph of Engineering," *New York Times*, July 5, 1936, XX1, and *The Triborough Bridge Authority Fifth Anniversary July 11, 1941* (New York, 1941), 13, 22, Box 2, Series 12, New York Tourism, Edward J. Orth.

93. For "metropolitan loom," see *Triborough Bridge Authority Fifth Anniversary*, 6. For "arterial tapestry," see Robert Moses, "New York Opens Bottlenecks," *New York Times*, October 13, 1940, 151.

94. Eugene Lewis, *Public Entrepreneurship: Toward a Theory of Bureaucratic Political Power; The Organizational Lives of Hyman Rickover, J. Eager Hoover, and Robert Moses* (Bloomington: Indiana University Press, 1980), 206–11.

95. "New York State," The Living New Deal, https://livingnewdeal.org/new-york/.

96. Barbara Blumberg, *The New Deal and the Unemployed: The View from New York City* (Lewisburg, PA: Bucknell University Press, 1979), 132, 144.

97. "Moses's Many Projects Are All Tied Together," *New York Times*, February 10, 1935, E11, and Robert Moses, *Public Works: A Dangerous Trade* (New York: McGraw-Hill, 1970), 687. For a contemporary, laudatory overview of Moses's New Deal–funded public works, see "Robert (Or-I'll-Resign) Moses," *Fortune* 17, no. 6 (1938): 71–79, 126–28. See also Robert Caro, *The Power Broker: Robert Moses and the Fall of New York* (New York: Knopf, 1974), 345, and Mason B. Williams, *City of Ambition: FDR, La Guardia, and the Making of Modern New York* (New York: W. W. Norton, 2013), 151–55.

98. Blumberg, *New Deal and the Unemployed*, 48, 74, 125–26.

99. Robert Moses, "Speech of Robert Moses at Jamaica, Queens County," October 31, 1934, 6, Folder 1, Box 10, Series 1, Personal, Robert Moses Papers.

100. *Pledge of the Membership of the Fairfield County Planning Association*, Folder 3, Box 7, Series D, William H. Burr Jr. Papers Pertaining to Sherwood Island State Park, Burr Family Papers, 1752–1940, Fairfield Museum and History Center, Fairfield, CT.

101. "Planning for Fairfield County," *Hartford Courant*, August 15, 1933, 8. On the politics of the Merritt, see Matthew Dalbey, *Regional Visionaries and Metropolitan Boosters: Decentralization, Regional Planning, and Parkways during the Interwar Years* (Boston: Kluwer Academic, 2002), 78–79. For early scandals, see "Connecticut Plans New

Scenic Highway," *New York Times*, April 23, 1926, 3. See also Bruce Radde, *The Merritt Parkway* (New Haven, CT: Yale University Press, 1996).

102. Larry Larned, *Traveling the Merritt Parkway* (Charleston, SC: Arcadia, 1998), 14, 34.

103. James Wechsler, "To Queens the Fair's a Coming-Out Party," *New York Times*, September 11, 1938, 148. See also "Recent Highway Developments in New York City and Their Relation to Flushing Meadow Park," *Flushing Meadow Improvement* 3, no. 5 (July 1938): 21.

104. "Statement by Governor Smith Accompanying Map Showing Proposed Solution of the Highway Traffic Problem of Long Island," *First Annual Report of the Long Island State Park Commission to the Governor and the Legislature of the State of New York, May 1925* (Albany, NY: Long Island State Park Commission, 1925), 60–63.

105. For "all roads," see "Recent Highway Developments," and *Official Guide Book New York World's Fair, The World of Tomorrow* (New York: Exposition Publications, 1939), 15, Folder 1, Box 15, Collection 60, World's Expositions 1851–1965. For population estimates, see "Flushing Meadow Park Site of Early Colonization," *Flushing Meadow Improvement* 2, no. 5 (July 1937): 16. For "modern parkways," see "Plan of New York State World's Fair Commission for a Water Amphitheatre on the Meadow," *Flushing Meadow Improvement* 1, no. 2 (Nov. 1936): 12.

106. Keller, "Diary of Helen Adams Keller," February, 1937," February 13, 1937, 194, in *Helen Keller's Journal 1936–1937* (London: Michael Joseph, 1938), 296.

107. "Recent Highway Developments," 21, Moses, *Public Works*, 191, and Harold Sweeny, and Triborough Bridge Authority, *Bronx-Whitestone Bridge* (New York: Triborough Bridge Authority, 1939), 1.

108. Pearl E. Levison, "World of Tomorrow," in *The Official Poem of the New York World's Fair, 1939, and Other Prize Winning Poems* (New York: Academy of American Poets, 1939), 8, Special Collections, World's Fairs Collection.

109. Lewis Mumford, "The Sky Line in Flushing: Genuine Bootleg," *New Yorker*, July 29, 1939, 38.

110. Moses makes this point in "From Dump to Glory," *Saturday Evening Post*, January 15, 1938, 13.

111. "Oct. Progress of New York's World's Fair," *World's Fair News* 1, no. 5 (November 1936), Folder 2, Box 16, World's Expositions 1851–1965. See also, "World's Fair Zoning" *Flushing Meadows Improvement* 1, no. 4 (January 1937): 3.

112. On reduced budgets, see *Report of the Westchester County Park Commission, 1932/1933* ([White Plains, NY?]: The Commission, 1923), 7. See also Michael R. Fein, *Paving the Way: New York Road Building and the American State, 1880–1956* (Lawrence: University Press of Kansas, 2008), 115, 167–73, and Marilyn E. Weigold, *Pioneering in Parks and Parkways: Westchester County, New York, 1895–1945*, Essays in Public Works History no. 9 (Chicago: Public Works Historical Society, 1980), 31.

113. "Westchester Told State Ignores It," *New York Times*, October 23, 1939, 43. See also "Bronx River Road Is Facing Closing," *New York Times*, March 14, 1942, 17, and in Fein, *Paving the Way*, 168n101.

114. Eugene Kinkead, "Goodbye Folks," *New Yorker*, May 31, 1931, 41; "Fair Site to Cost the City $11,333,812," *New York Times*, October 2, 1936, 27; August Loeb, "Fine Park from Fair," *New York Times*, October 27, 1940, 141; and Sidney M. Shalett, "Memory of 'Tomorrow,'" *New York Times*, April 27, 1941, SM15.

115. "'Beauty for Ashes': A Statement by the Executive Committee of the New York

World's Fair 1964–1965," June 3, 1967, Folder 230, Box 8, Malcolm Wilson Papers, Rockefeller Archive Center, Sleepy Hollow, NY. See also Marco Duranti, "Utopia, Nostalgia and World War at the 1939–1940 New York World's Fair," *Journal of Contemporary History* 41, no. 4 (October 2006): 663–83.

116. "136 Acres of Lakes and Their Use after the Fair" *Flushing Meadow Improvement* 1, no. 2 (November 1936): 18. For bird species, see "Wild Life Sanctuaries," *Flushing Meadow Improvement* 3, no. 7 (January 1939): 14. On the Atlantic flyway, see Anne-Marie Cantwell and Diana diZerega Wall, *Unearthing Gotham: The Archaeology of New York City* (New Haven, CT: Yale University Press, 2001), 88, and Vileisis, *Discovering the Unknown Landscape*, 161–63.

117. For "unhealable" quote, see *New York World's Fair Preview April 29–30–May 1st* (New York: Junior League of the City of New York, 1939), Special Collections, World's Fairs Collection. See also "Largest Willow Grove," *Flushing Meadow Improvement* 1, no. 1 (October 1936): 8.

118. Robert Moses, "The Heritage of the New York World's Fairs 1939–1940, 1964–1965," February 1972, Folder 248, Box 9, Malcolm Wilson Papers, Rockefeller Archive Center, Sleepy Hollow, NY.

119. On oxygen levels, see Suszkowski, "Sewage Pollution in New York Harbor," 19–21. For "conservation," see Robert Moses, "The Building of Jones Beach," in *Robert Moses: Single-Minded Genius*, ed. Joann P. Krieg (Interlaken, NY: Heart of the Lakes Publishing, 1989), 135–40.

120. "Triborough Bridge a Symbol to Moses," *New York Times*, July 11, 1941, 17; *The Triborough Bridge Authority Fifth Anniversary*; and "Flushing Bay Boat Basin," *Flushing Meadow Improvement* 2, no. 2 (April 1937): 13.

## Epilogue

1. General Motors Corporation, "Futurama: A Comprehensive Description of the GM Highways and Horizons Exhibit at the New York World's Fair," 3, Subseries L, Transportation Zone, Budd Manufacturing–General Motors, Series 3, New York World's Fair, Edward J. Orth Memorial Archives of the New York World's Fair, Archives Center, National Museum of American History, Washington, DC (hereafter Edward J. Orth).

2. Buwei Yang Chao, *Memoir of Buwei Yang Chao, 1947*, in *Autobiography of a Chinese Woman*, trans. Yuenren Chao (New York: John Day, 1947), 327, *North American Women's Letters and Diaries*, Colonial–1950 (Alexandria, VA: Alexander Street Press, 2001–), S9366.

3. Walter Lippmann, *New York Herald Tribune*, June 6, 1939, quoted in Robert W. Rydell, *World of Fairs: The Century-of-Progress Expositions* (Chicago: Chicago University Press, 1993), 134–35. On Bel Geddes as a road pitchman, see Dolores Hayden, "'I Have Seen the Future': Selling the Unsustainable City," *Journal of Urban History* 38, no. 1 (January 2012): 4–5.

4. "An Object-Lesson in Parks," *New York Times*, August 5, 1925, 16.

5. Lewis Mumford to Mr. Branch, August 15, 1939, Subseries L, Transportation Zone, Budd Manufacturing–General Motors, Series 3, New York World's Fair Records, Edward J. Orth. See also Lewis Mumford, "Whither the City?," *American City* 54 (November 1939), 60.

6.  Lewis Mumford, "The Plan of New York," reprinted in *Planning the Fourth Migration: The Neglected Vision of the Regional Planning Association of America*, ed. Carl Sussman (Cambridge, MA: MIT Press, 1976), 208.

7.  Lewis Mumford, "Regions—To Live In," *Survey Graphic* 7 (May 1925), reprinted in Sussman, *Planning the Fourth Migration*, 89.

8.  For "actual 'Planned,'" see Department of Feature Publicity, New York World's Fair, 1939, "The Town of Tomorrow and the Home Building Center, New York World's Fair 1939," 12, Subseries F, Community Interest Zone, Home Building Center, Series 3, New York World's Fair Records, Edward J. Orth. For "intended to represent" quote, see *Official Guide Book New York World's Fair, The World of Tomorrow* (New York: Exposition Publications, 1939), 100, Folder 1, Box 15, Collection 60, World's Expositions 1851–1965, Center, National Museum of American History, Washington, DC (hereafter World's Expositions 1851–1965).

9.  Mumford, "Whither the City?," *American City* 54 (November 1939), 61.

10.  Francis V. O'Connor, "The Unusable Future, the Role of Fantasy in the Promotion of a Consumer Society," in Helen A. Harrison, ed., *Dawn of A New Day: The New York World's Fair 1939–40* (Flushing, NY: Queens Museum, 1980), 62, and Lewis Mumford, "The Sky Line in Flushing: Genuine Bootleg," *New Yorker*, July 29, 1939, 38. An in-depth analysis of the limitations of Futurama and Democracity to imagine creatively a new type of city is available in Jeffrey L. Meikle, *Twentieth Century Limited: Industrial Design in America 1925–1939* (Philadelphia: Temple University Press, 1979), 190–209.

11.  Ira Rosenwaike, *Population History of New York City* (Syracuse, NY: Syracuse University Press, 1972), 137, 140, 174.

12.  Christy Walsh, "Pomonok and the World's Fair," PGA 1939 Championship Souvenir Program, Subseries K, Production and Distribution Zone, Equitable Life Assurance—Petroleum Industry Exhibition Inc. (oversized), Series 3, New York World's Fair Records, Edward J. Orth.

13.  Mildred Adams, "Long Island Has Its Own Land Boom," *New York Times*, June 20, 1926, SM6.

14.  Regional Plan Association, "Requirements for Residential Land in the Region Analyzed by Counties," *Information Bulletin*, June 18, 1934.

15.  Thomas H. Reed, *The Government of Nassau County: A Report Made to the Board of Supervisors* (Mineola, New York: Nassau County, 1934), 2–3.

16.  F. Scott Fitzgerald, *The Great Gatsby* (1925; repr., New York: Scribners, 1992), 9.

17.  Helen Worden, "Where Are the Rich?," *New York World Telegram*, October 20, 1942, 17.

18.  "Long Island Pays Tribute to Moses," *New York Times*, April 4, 1937, 12.

19.  Rosenwaike, *Population History*, 131, 134.

20.  Lee E. Koppelman and Pearl M. Kamer, "Anatomy of the Long Island Economy: Retrospective and Prospective," *Long Island Historical Journal* 6, no. 2 (Spring 1994): 146. For the Munsey estate sale, see the *New York Times*, April 29, 1928, sec. 12, p. 17, and May 1, 1927, sec. 11, p. 2, cited in Jon C. Teaford, *Post-Suburbia: Government and Politics in the Edge Cities* (Baltimore: Johns Hopkins University Press, 1997), 14. See also Virginia L. Bartos, "Robert Weeks De Forest: Forest Hills Gardens and Munsey Park," in *Gardens of Eden: Long Island's Early Twentieth-Century Planned Communities*, ed. Robert B. MacKay (New York: W. W. Norton, 2015), 193–94.

21.  Rosenwaike, *Population History*, 16, 58.

22. Clay McShane, *Down the Asphalt Path: The Automobile and the American City* (New York: Columbia University Press, 1994), 218.

23. Frederick Coburn, "The Five-Hundred Mile City," *World Today* 11 (December 1906): 1253, 1257, 1259.

24. *Long Island: The Sunrise Homeland Brooklyn, Queens, Nassau, and Suffolk Counties* (New York: Long Island-at-the-Fair Committee, 1940), 10, Special Collections World's Fairs Collection.

25. See Christopher W. Wells, *Car Country: An Environmental History* (Seattle: University of Washington Press, 2012). For an environmental history of postwar suburbia, see Adam Rome, *The Bulldozer in the Countryside: Suburban Sprawl and the Rise of American Environmentalism* (New York: Cambridge University Press, 2001), and Christopher C. Sellers, *Crabgrass Crucible: Suburban Nature and the Rise of Environmentalism in Twentieth-Century America* (Chapel Hill: University of North Carolina Press, 2013).

26. Harold M. Lewis, Harold MacLean, Theodore Tremain McCrosky, Henry James, and Thomas Adams, *Physical Conditions and Public Services* (New York: Regional Plan of New York and Its Environs 1929; repr., New York: Arno, 1974), 1.

27. On the acreage of Connecticut marshes, see Richard H. Goodwin, "The Future: A Call for Action," in *Connecticut's Coastal Marshes: A Vanishing Resource*, Bulletin No. 12 (New London, CT: Connecticut Arboretum, Connecticut College, 1961), 30. On wetland protections, see Ann Vileisis, *Discovering the Unknown Landscape: A History of America's Wetlands* (Washington, DC: Island, 1997), 7,185, 222, 275–82. Robert Moses, "Press Release, Dec. 29, 1941," 2, Folder X-118, Soundview Park, Box X-118, Parks Library, New York, NY (hereafter Parks Library).

28. Daniel C. Walsh, "Reconnaissance Mapping of Landfills in New York City," *Proceedings: FOCUS Conference on Eastern Regional Ground Water Issues, October 29–31, 1991* [. . .] *Portland, Maine* (1991), 392, 394, and 398, National Ground Water Association, http://citrix.ngwa.org/gwol/pdf/910155209.pdf (accessed 3 March 2016), and "The History of Waste Landfilling in New York City," *Ground Water* 29 (July–August 1991): 591–92. On the effects of landfill, see also Ted Steinberg, *Gotham Unbound: The Ecological History of Greater New York* (New York: Simon and Schuster, 2014), 242–46, 253. 291–92, 313.

29. "Salt Marshes in New York City Parks: Soundview Park," City of New York Parks and Recreation October 2001, Folder X-118, Soundview Park, Box X-118, New York City Parks, Parks Library.

30. US Army Corps of Engineers New York District, "Soundview Park Bronx, New York Ecosystem Restoration Study Fact Sheet," Folder X-118 Soundview Park [no box], Parks Library.

31. See, for example, US Geological Survey, *Topographic Sheet (Tarrytown), New York, New Jersey, Harlem Quadrangle*, New York City Folio, 83 (Washington, DC: US Department of the Interior, United States Geological Survey, 1902). See also Donald Squires and Kevin Bone, "The Beautiful Lake: The Promise of the Natural Systems," in *The New York Waterfront: Evolution and Building Culture of the Port and Harbor*, ed. Kevin Bone (New York: Monacelli, 1997), 27–31, and Marit Larson and Paul Mankiewicz, "Restoring Soundview Park: HEP Priority Restoration Site LI10," 1, Folder X-118, Soundview Park, Parks Library.

32. Jon C. Teaford, "Caro versus Moses, Round Two: Robert Caro's *The Power Broker*," *Technology and Culture* 49, no. 2 (April 2008): 446–47. Scholarship on Moses's postwar legacy is vast; for leading revisionism, see Joel Schwartz, *The New York Approach: Robert Moses, Urban Liberals, and Redevelopment of the Inner City* (Columbus: Ohio

University Press, 1993), and Samuel Zipp, *Manhattan Projects: The Rise and Fall of Urban Renewal in Cold War New York* (New York: Oxford University Press, 2012). On postwar localism and antigovernment planning movements, see Suleiman Osman, *The Invention of Brownstone Brooklyn: Gentrification and the Search for Authenticity in Postwar New York* (New York: Oxford University Press, 2011).

33.  Regional Plan of New York and Its Environs, Frederic Adrian Delano, Thomas Adams, *The Graphic Regional Plan: Atlas and Description* [. . .], vol. 1 of *Regional Plan of New York and Its Environs* (New York: Regional Plan of New York and Its Environs, 1929), 125. See also Mary Corbin Sies and Christopher Silver, "Planning History and the New American Metropolis," in *Planning the Twentieth-Century American City*, ed. Mary Corbin Sies and Christopher Silver (Baltimore: Johns Hopkins University Press, 1996), 455.

34.  Reed, *Government of Nassau County*, 4, 58–60. On privatism and exclusion, see Gerald Frug, "The Legal Technology of Exclusion in Metropolitan America," in *The New Suburban History*, ed. Kevin H. Kruse and Thomas J. Sugrue (Chicago: University of Chicago Press, 2006), 205, and Sam Bass Warner Jr., *The Private City: Philadelphia in Three Periods of Its Growth*, rev. ed. (Philadelphia: University of Pennsylvania Press, 1987), 213.

35.  John Nolen, "New Cities for the New Age, The Planning Foundation of America," Folder 8, Box 1, John Nolen Pamphlet Collection, John Nolen Papers, 1890–1938, 1954–1960, Division of Rare and Manuscript Collections, Cornell University Library, Ithaca, New York.

# Selected Bibliography

This bibliography lists the chief writings and archival collections referenced in this book. It is not an exhaustive list of the sources consulted. The survey of government reports and archival collections is meant to introduce those who wish to pursue regional history to the range of repositories and government publications that preserve the history of greater New York. The selected bibliography of published scholarship is intended to serve as an introduction to the principal scholarship in the fields of urban and suburban history, the history of planning, and environmental history related to New York City and its coastal setting.

*Archival Collections*

Archives Center, National Museum of American History, Washington, DC
    Edward J. Orth Memorial Archives of the New York World's Fair
    Special Collections, World's Fairs Collection
    World's Expositions 1851–1965
Division of Home and Community Life, Smithsonian Museum of American History, Washington, DC
Division of Rare and Manuscript Collections, Cornell University Library, Ithaca, New York
    John Nolen Papers, 1890–1938, 1954–1960
    Charles Downing Lay Papers, 1898–1956
Fairfield Museum and History Center, Fairfield, Connecticut
    William H. Burr Jr. papers
Knapp House Archives, Rye Historical Society, Rye, New York
La Guardia and Wagner Archives, Long Island City, New York
    Queens Local History Collection
    Steinway and Sons Collection
    Cradle of Aviation Museum Collection
New York City Parks Library, New York, New York
New York Public Library, New York, New York
    Manuscripts and Archives Division
        Robert Moses Papers
        New York World's Fair 1939 and 1940 Incorporated Records 1935–1945
    Miriam and Ira D. Wallach Division of Art, Prints and Photographs
    The Lionel Pincus and Princess Firyal Map Division

New York State Archives, Albany, New York
   Maps of Grants of Lands under Water [ca. 1777–1970]
   Calendars and Minutes of Meetings of the State Council of Parks, 1925–1933
   Governor Smith, Central Subject and Correspondence Files, 1919–1920, 1923–1928
   Pamphlets and Articles Relating to Local History and Historic Sites [ca. 1899–1976]
Rockefeller Archive Center, Sleepy Hollow, New York
   Rockefeller General Files, 1858–1981 (1879–1961)
   Rockefeller Family Archives
   Malcolm Wilson Papers
Special Collections Department, Hofstra University, Hempstead, New York
   Nassau County Museum Reference Library Collection
Special Collections, Lehman College Library, City University of New York (CUNY), The
   Bronx, New York
   The Bronx Institute Oral History Project
   Bronx Regional and Community History Project Photo Record
Special Collections, University Library, American University, Washington, DC
   Frederick Law Olmsted Documentary Editing Project
Westchester County Archives, Elmsford, New York
   Everit Macy Papers
   Parks, Recreation, and Conservation Department Papers
   Parks Department Clippings
Westchester Historical Society Collection, Elmsford, New York
   Hufenland Pamphlet Collection

*Government Reports*

*Board of Commissioners of Central Park Annual Report*
*Board of Health of the Department of Health of the City of New York Annual Report*
*Connecticut State Park and Forest Commission Biennial Report*
*Department of Parks of the Borough of the Bronx Annual Report*
*Department of Public Parks of the City of New York Annual Report*
*Fifteenth Census of the United States*
*Fourteenth Census of the United States*
*Long Island State Park Commission Annual Report*
*New York State Council of Parks Annual Report*
*Westchester County Park Commission Annual Report*

*Published Primary Sources*

Appleton's Hand-Book of American Travel: Northern and Eastern Tours. New York: D. Apple-
   ton, 1870.
Atlas of the City of Bridgeport, Conn, From Official Records and Actual Surveys. Philadelphia:
   G. M. Hopkins, 1888.
Barnum, Phineas Taylor. How I Made Millions: or, The Secret of Success. New York: Belford,
   Clark, 1884.
———. The Life of P. T. Barnum. New York: Redfield, 1855.
———. Struggles and Triumphs, or Forty Years' Recollections of P. T. Barnum. New York:
   American News, 1871.

Basset, Edward M. *The Master Plan, with a Discussion of the Theory of Community Land Planning Legislation.* New York: Russell Sage Foundation, 1938.

Beers, F. W. *Atlas of New York and Vicinity: From Actual Surveys.* New York: F. W. Beers, A. D. Ellis and G. G. Soule, 1868.

Bel Geddes, Norman. *Magic Motorways.* New York: Random House, 1940.

Benton, Joel. "P.T. Barnum, Showman and Humorist." *Century Illustrated Magazine* 64, no. 4 (August 1902): 580.

Bishop, John Peale. "World's Fair Notes." *Kenyon Review* 1, no. 3 (Summer 1939): 239–50.

Bly, Nellie. *Ten Days in a Mad-House.* New York: Ian L. Munro, 1887.

Bromley, George W., and Walter S. Bromley. *Atlas of the City of New York Borough of the Bronx Annexed District: From Actual Surveys and Official Plans.* Vol. 3 of *Atlas of the Borough of the Bronx.* Philadelphia: G. W. Bromley, 1927.

Bronx Board of Trade. *The Bronx: New York City's Fastest Growing Borough.* Bronx, NY: Warontas Press, 1924.

———, ed. *The Great North Side; Or, Borough of The Bronx.* New York: Knickerbocker Press, 1897.

Chapin, Robert Coit. *The Standard of Living among Workingmen's Families in New York City.* New York: Charities Publication Committee, 1909.

*City of New York Office of the President of the Borough of the Bronx, Bureau of Engineer, General Map of the Borough of the Bronx, New York, April 1st 1935.* New York: Hagstrom Map Company, 1935.

Downer, Jay, and James Owen. *Public Parks in Westchester County, 1925.* New York: Lewis Historical, 1925.

Dwight, Timothy. *Travels in New-England and New-York.* Vol. 3. London: William Baynes and Son, 1823.

Evison, Herbert, ed. *A State Park Anthology.* Washington, DC: National Conference on State Parks, 1930.

*The Father of Greater New York: Official Report of the Presentation to Andrew Haswell Green of a Gold Medal* [. . .]. New York: Historical and Memorial Committee of the Mayor's Committee on the Celebration of Municipal Consolidation, 1899.

Foord, John. *The Life and Public Services of Andrew Haswell Green.* Garden City, NY: Doubleday, Page, 1913.

French, Alvah P. *The History of Westchester County.* New York: Lewis Historical, 1925.

*The Graphic Regional Plan: Atlas and Description* [. . .]. Vol. 1 of *Regional Plan of New York and Its Environs.* New York: Regional Plan of New York and Its Environs, 1929.

Green, Andrew H. *Communication to the Commissioners of Central Park, Relative to the Improvement of the Sixth and Seventh Avenues* [. . .] *and Other Subjects.* New York: William Cullen Bryant, 1865.

———. *Public Improvements in the City of New York, Communication from Andrew H. Green to Wm. A. Booth, Esq., and Others. Sept. 28th, 1874.* New York, 1974.

*Hagstrom's Street, Road and Property Ownership Map of Nassau County, Long Island New York.* New York: Hagstrom, 1939.

Hanmer, Lee F., Thomas Adams, Frederick W. Loede Jr., Charles J. Storey, and Frank B. Williams. *Public Recreation: A Study of Parks, Playgrounds, and Other Outdoor Recreational Facilities.* Vol. 5 of *Regional Survey of New York and Its Environs.* New York: Regional Plan of New York and Its Environs, 1928.

Lay, Charles Downing. *A Park System for Long Island: A Report to the Nassau County Committee.* Privately printed, 1925.

Lewis, Harold MacLean, Theodore Tremain McCrosky, Henry James, and Thomas Adams. *Physical Conditions and Public Services*. New York: Regional Plan of New York and Its Environs 1929. Reprint, New York: Arno, 1974.

Lundberg, George A., Mirra Komarovsky, and Mary Alice McInery. *Leisure: A Suburban Study*. New York: Columbia University Press, 1934.

McCabe, James D., Jr. *New York by Sunlight and Gaslight: A Work Descriptive of the Great American Metropolis*. Philadelphia: Douglass Brothers, 1881.

Meyer, Blakeman Quintard. *Views of Rye*. New York: Blakeman Quintard Meyer, 1917.

More, Louise Bolard. *Wage-Earners' Budgets: A Study of Standards and Cost of Living in New York City*. New York: Henry Holt, 1907.

Morris, Fordham. *Address of Mr. Fordham Morris*. New York: North Side Board of Trade, 1895.

Moses, Robert. *The Expanding New York Waterfront*. New York: Triborough Bridge and Tunnel Authority, 1964.

————. *A Half-Century of Achievement: Triborough Bridge and Tunnel Authority Links the Five Boroughs, Parks, Beaches, and Suburbs*. New York: Triborough Bridge and Tunnel Authority, 1948.

Mountain Grove Cemetery. *Lots about Lots; Or, The Great Fair, and What Preceded It: Sold for the Benefit of the Mountain Grove Cemetery*. Farmer Office Presses, 1879.

Mullaly, John. *The New Parks beyond the Harlem*. New York: Record and Guide, 1887.

New York City Department of Parks. *The Reclamation of Park Areas by Sanitation Fill and Synthetic Topsoil*. New York: Comet, 1950.

New York Metropolitan Sewerage Commission. *Sewerage and Sewage Disposal in the Metropolitan District of New York and New Jersey, Report of the New York Metropolitan Sewerage Commission, April 30, 1910*. New York: Martin B. Brown, 1910.

New York State Association, Committee on State Park Plan. *A State Park Plan for New York*. New York: M. B. Brown, 1922.

Nolen, John, and Henry V. Hubbard. *Parkways and Land Values*. Cambridge: Cambridge University Press, 1937.

North Side Board of Trade. *Official Program of the Opening of the Harlem Ship Canal, June 17th, 1895*. New York: Freytag, 1895.

Olmsted, Frederick Law. *The Papers of Frederick Law Olmsted*. 7 vols. Edited by David Schuyler, Jane Turner Censer, Carolyn F. Hoffman, and Kenneth Hawkins. Baltimore: Johns Hopkins University Press, 1992.

Orcutt, Samuel. *A History of the Old Town of Stratford and the City of Bridgeport Connecticut*. Vol. 2. New Haven, CT: Tuttle, Morehouse and Taylor, 1886.

*Present Sanitary Condition of New York Harbor and the Degree of Cleanness Which Is Necessary and Sufficient for the Water Report of the Metropolitan Sewerage Commission of New York*. New York: Wynkoop, Hallenbeck, Crawford, 1912.

Prison Association of New York. *A Study of the Conditions Which Have Accumulated under Many Administrations and Now Exist in the Prisons of Welfare Island* [. . .]. New York: Prison Association of New York, 1924.

Reed, Thomas H. *The Government of Nassau County: A Report Made to the Board of Supervisors*. Mineola, NY: Nassau County, 1934.

*Report of the Metropolitan Sewerage Commission of New York April 30, 1914*. New York: Wynkoop, Hallenbeck, Crawford, 1914.

Scofield, H. G. *Atlas of the City of Bridgeport Connecticut from Actual Surveys*. New York: J. B. Beers, 1876.

Shepp, James W., and Daniel B. Shepp. *Shepp's New York City Illustrated: Scene and Story in the Metropolis of the Western World*. Chicago: Globe Bible, 1894.

Steiner, Jesse Frederick. *Americans at Play: Recent Trends in Recreation and Leisure Time Activities*. New York: McGraw-Hill, 1933.

Stevens, Simon. *Harlem River Ship Canal, Letter from Simon Stevens to the Commissioners of The Sinking Fund of The City Of New York* [ . . . ]. New York: C. G. Burgoyne, 1892.

Todd, Charles B. *A Brief History of the City of New York*. New York: American Book, 1899.

Townshend, Charles Harvey. *The Commercial Interests of Long Island Sound in General, and New Haven in Particular*. New Haven, CT: O. A. Dorman, 1883.

Van Pelt, Daniel. *Leslie's History of the Greater New York*. Vol. 1, *New York to the Consolidation*. New York: Arkell, 1898.

Waldo, George Curtis, Jr. *History of Bridgeport and Vicinity*. 2 vols. New York: S. J. Clarke, 1917.

———. *The Standard's History of Bridgeport*. Bridgeport, CT: Standard Association, 1897.

Walling, Henry F. *Taintor's Route and City Guides, New York to the White Mountains via the Connecticut River*. New York: Taintor Brothers, 1867.

Waring, George E., Jr. *Report on the Social Statistics of Cities, Part I: New England and Middle States, A: New England*. Washington, DC: Department of Interior Census Office, 1885.

Wells, James L., Louis F. Haffen, and Josiah Briggs, eds. *The Bronx and Its People: A History 1909–1927*. New York: Lewis Historical, 1927.

Willis, Walter I. *Queens Borough; Being a Descriptive and Illustrated Book of the Borough of Queens, City of Greater New York* [. . .]. Brooklyn, NY: Brooklyn Eagle, 1913.

### Online Archival Collections

Reps, John W., ed. *Urban Planning, 1794–1918: An International Anthology of Articles, Conference Papers, and Reports*, last modified November 27, 2002. http://urbanplanning.library.cornell.edu/DOCS/homepage.htm.

William Steinway Diary Project, 1861–1891. Smithsonian Institution National Museum of American History. http://americanhistory.si.edu/steinwaydiary.

### Secondary Sources

Archer, Jack H., Donald L. Connors, and Kenneth Laurence. *The Public Trust Doctrine and the Management of America's Coasts*. Amherst: University of Massachusetts Press, 1994.

Aron, Cindy. *Working at Play: A History of Vacations in the United States*. New York: Oxford University Press, 1999.

Ballon, Hilary, and Kenneth Jackson, eds. *Robert Moses and the Modern City: The Transformation of New York*. New York: W. W. Norton, 2008.

Baxandall, Rosalyn, and Elizabeth Ewen. *Picture Windows: How the Suburbs Happened*. New York: Basic Books, 2000.

Beckert, Sven. *The Monied Metropolis: New York City and the Consolidation of the American Bourgeoisie, 1850–1896*. Cambridge: Cambridge University Press, 2001.

Bender, Thomas. "The 'Rural' Cemetery Movement: Urban Travail and the Appeal of Nature." *New England Quarterly* 47, no. 2 (June 1947): 196–211.

———. *Toward an Urban Vision: Ideas and Institutions in Nineteenth-Century America*. Lexington: University Press of Kentucky, 1975.

Blackmar, Elizabeth. *Manhattan for Rent, 1785–1850*. Ithaca, NY: Cornell University Press, 1989.

Bone, Kevin, ed. *The New York Waterfront: Evolution and Building Culture of the Port and Harbor*. New York: Monacelli, 1997.

Borchert, John R. "America's Changing Metropolitan Regions." *Annals of the Association of American Geographers* 62, no. 2 (June 1972): 353–73.

Boyer, M. Christine. *Dreaming the Rational City: The Myth of American City Planning*. Cambridge, MA: MIT Press, 1987.

Bronx Museum of the Arts, ed. *Building a Borough: Architecture and Planning in the Bronx 1890–1940*. Bronx, NY: Bronx Museum of the Arts, 1986.

Burke, Kathryn W. *Playland*. Foreword by Andrew J. Spano. Charleston, SC: Arcadia, 2008.

Burrows, Edwin G., and Mike Wallace. *Gotham: A History of New York City to 1898*. New York: Oxford University Press, 1999.

Campo, Daniel. *The Accidental Playground: Brooklyn Waterfront Narratives of the Undesigned and Unplanned*. New York: Fordham University Press, 2013.

Cantwell, Anne-Marie, and Diana diZerega Wall. *Unearthing Gotham: The Archaeology of New York City*. New Haven, CT: Yale University Press, 2001.

Caro, Robert. *The Power Broker: Robert Moses and the Fall of New York*. New York: Knopf, 1974.

Chiles, Robert. "Working-Class Conservationism in New York: Governor Alfred E. Smith and 'The Property of the People of the State.'" *Environmental History* 18, no. 1 (January 2013): 157–83.

Cranz, Galen. *The Politics of Park Design: A History of Urban Parks in America*. Cambridge, MA: MIT Press, 1982.

Crawford, Margaret. *Building the Workingman's Paradise: The Design of American Company Towns*. London: Verso, 1995.

Currell, Susan. *The March of Spare Time: The Problem and Promise of Leisure in the Great Depression*. Philadelphia: University of Pennsylvania Press, 2005.

Dalbey, Mathew. *Regional Visionaries and Metropolitan Boosters: Decentralization, Regional Planning, and Parkways during the Interwar Years*. Boston: Kluwer Academic, 2002.

Daniels, Tom. *When City and Country Collide: Managing Growth on the Metropolitan Fringe*. Washington, DC: Island, 1999.

Danielson, Michael N., and Jameson W. Doig. *New York: The Politics of Urban Regional Development*. Berkeley: University of California Press, 1983.

Duncan, James S., and Nancy G. Duncan. *Landscapes of Privilege: the Politics of the Aesthetic in an American Suburb*. New York: Routledge, 2004.

Dunne, Collin, John Dunne, Beth Griffin, Ted M. Levine, Robin Russell, and Arlene Weiss. *Rye in the Twenties*. New York: Arno, 1978.

Duranti, Marco. "Utopia, Nostalgia and World War at the 1939–1940 New York World's Fair." *Journal of Contemporary History* 41, no. 4 (October 2006): 663–83.

Eves, Jamie. "Beyond the Dry Highway and the No Trespass Sign: Albert Milford Turner, William H. Burr, Elsie Hill, and the Creation of Connecticut's First Oceanfront State Parks, 1914–1942." *Connecticut League of History Organizations Bulletin* 57, no. 2 (May 2004): 9–12.

Farnham, Thomas J. *Fairfield: The Biography of a Community 1639–2000*. West Kennebunk, ME: Phoenix, 2000.

Fein, Albert ed. *Landscape into Cityscape: Frederick Law Olmsted's Plans for a Greater New York*. New York: Van Nostrand Reinhold, 1981.

Fein, Michael R. *Paving the Way: New York Road Building and the American State, 1880–1956*. Lawrence: University Press of Kansas, 2008.

Fisher, Colin. *Urban Green: Nature, Recreation, and the Working Class in Industrial Chicago*. Chapel Hill: University of North Carolina Press, 2015.

Gilfoyle, Timothy J. *City of Eros: New York City, Prostitution, and the Commercialization of Sex, 1790–1920*. New York: W. W. Norton, 1992.

Gonzalez, Evelyn. *The Bronx*. New York: Columbia University Press, 2004.

Gutman, Marta. "Race, Place, and Play: Robert Moses and the WPA Swimming Pools in New York City." *Journal of the Society of Architectural Historians* 67, no. 4 (December 2008): 532–61.

Hammack, David C. *Power and Society: Greater New York at the Turn of the Century*. New York: Russell Sage Foundation, 1982.

Harris, Richard. "Industry and Residence: The Decentralization of New York City, 1900–1940." *Journal of Historical Geography* 19, no. 2 (1993): 169–90.

———. *Unplanned Suburbs: Toronto's American Tragedy, 1900–1950*. Baltimore: Johns Hopkins University Press, 1996.

Harrison, Helen A. ed. *Dawn of A New Day: The New York World's Fair 1939–40*. Flushing, NY: Queens Museum, 1980.

Hays, Samuel P. "The Social Analysis of American Political History, 1880–1920." *Political Science Quarterly* 80 (September 1965): 373–94.

Hermalyn, Gary. "The Harlem River Ship Canal." *Bronx County Historical Society* 20, no. 1 (1983): 1–23.

Jackson, Kenneth T. *Crabgrass Frontier: The Suburbanization of the United States*. New York: Oxford University Press, 1985.

———, ed. *The Encyclopedia of New York City*. New Haven, CT: Yale University Press, 1995.

Johnson, David A. *Planning the Great Metropolis: The 1929 Regional Plan of New York and Its Environs*. London: E and FN Spon, 1996.

Kahrl, Andrew W. *The Land Was Ours: African American Beaches from Jim Crow to the Sunbelt South*. Cambridge, MA: Harvard University Press, 2012.

Kaplan, Lawrence, and Carol P. Kaplan. *Between Ocean and City: The Transformation of Rockaway, New York*. New York: Columbia University Press, 2003.

Kasson, John. *Amusing the Million: Coney Island at the Turn of the Century*. New York: Hill and Wang, 1978.

Krieg, Joann P., ed. *Long Island Studies, Evoking a Sense of Place*. Interlaken, NY: Heart of the Lakes Publishing, 1988.

———, ed. *Robert Moses: Single-Minded Genius*. Interlaken, NY: Heart of the Lakes Publishing, 1989.

Krieg, Joann P., and Natalie A. Naylor, eds. *Nassau County: From Rural Hinterland to Suburban Metropolis*. Interlaken, NY: Empire State Books, 2000.

Kunstler, James Howard. *The Geography of Nowhere: The Rise of America's Man-Made Landscape*. New York: Simon and Schuster, 1993.

LaFrank, Kathleen. "Real and Ideal Landscapes along the Taconic State Parkway." *Perspectives in Vernacular Architecture* 9 (2003): 247–62.

Lewinnek, Elaine. *The Working Man's Reward: Chicago's Early Suburbs and the Roots of American Sprawl*. New York: Oxford University Press, 2014.

Lewis, Robert, ed. *Manufacturing Suburbs: Building Work and Home on the Metropolitan Edge*. Philadelphia: Temple University Press, 2004.

Lieberman, Richard K. *Steinway and Sons*. New Haven, CT: Yale University Press, 1995.

Low, Setha, and Neil Smith, eds. *The Politics of Public Space*. New York: Routledge, 2006.

Mason, Randall. *The Once and Future New York: Historic Preservation and the Modern City*. Minneapolis: University of Minnesota Press, 2009.

McNamara, John. *History in Asphalt: The Origin of Bronx Street and Place Names*. Harrison, NY: Harbor Hill Books, 1978.

Meikle, Jeffrey L. *Twentieth Century Limited: Industrial Design in America 1925–1939*. Philadelphia: Temple University Press, 1979.

Meyer, Hans. *City and Port: Urban Planning as a Cultural Venture in London, Barcelona, New York, and Rotterdam: Changing Relations between Public Urban Space and Large-Scale Infrastructure*. Utrecht, The Netherlands: International Books, 1999.

Mohl, Raymond A., ed. *The Making of Urban America*. 2nd ed. Wilmington, DE: Scholarly Resources, 1997.

Muller, Peter. "The Evolution of American Suburbs: A Geographical Interpretation." *Urbanism Past and Present*, no. 4 (Summer 1977): 1–10.

Nasaw, David. *Going Out: The Rise and Fall of Public Amusements*. New York: Basic Books, 1993.

Needham, Andrew, and Allen Dieterich-Ward. "Beyond the Metropolis: Metropolitan Growth and Regional Transformation in Postwar America." *Journal of Urban History* 35, no. 7 (November 2009): 943–69.

Newton, Norman T. *Design on the Land: The Development of Landscape Architecture*. Cambridge, MA: Belknap Press, 1971.

O'Shaughnessy, Marjorie. "Childhood Summers at Orchard Beach." *Bronx Historical Society Journal* 12, no. 1 (Spring 1975): 19–23.

Panetta, Roger, ed. *Westchester: The American Suburb*. New York: Fordham University Press, 2006.

Peiss, Kathy. *Cheap Amusements: Working Women and Leisure in Turn-of-the-Century New York*. Philadelphia: Temple University Press, 1986.

Peterson, Jon. *The Birth of City Planning in the United States, 1840–1917*. Baltimore: Johns Hopkins University Press, 2003.

Radde, Bruce. *The Merritt Parkway*. New Haven, CT: Yale University Press, 1996.

Ray, Deborah Wing, and Gloria P. Stewart. *Norwalk: Being an Historical Account of That Connecticut Town*. Canaan, NH: Phoenix, 1979.

Revell, Keith D. *Building Gotham: Civic Culture and Public Policy in New York City, 1898–1938*. Baltimore: Johns Hopkins University Press, 2005.

Robertson, Michael. "Cultural Hegemony Goes to the Fair: The Case of E. L. Doctorow's World's Fair." *American Studies* 33, no. 1 (Spring 1992): 31–45.

Rogers, Martha. "The Bronx Park System: A Faded Design." *Landscape* 27, no. 2 (1983): 13–21.

Roper, Laura Wood. *F.L.O.: A Biography of Frederick Law Olmsted*. Baltimore: Johns Hopkins University Press, 1983.

Rosenzweig, Roy, and Elizabeth Blackmar. *The Park and the People: A History of Central Park*. Ithaca, NY: Cornell University Press, 1992.

Schaffer, Daniel, ed. *Two Centuries of American Planning*. Baltimore: Johns Hopkins University Press, 1988.

Schnitz, Ann, and Robert Loeb. "'More Public Parks!': The First New York Environmental Movement." *Bronx County Historical Society Journal* 21, no. 2 (Fall 1984): 51–63.

Schultz, Stanley K. *Constructing Urban Culture: American Cities and City Planning, 1800–1920.* Philadelphia: Temple University Press, 1989.

Schuyler, David. *The New Urban Landscape: The Redefinition of City Form in Nineteenth-Century America.* Baltimore: Johns Hopkins University Press, 1986.

Scobey, David. *Empire City: The Making and Meaning of the New York City Landscape.* Philadelphia: Temple University Press, 2003.

Seyfried, Vincent F. *Corona: From Farmland to City Suburb, 1650–1935.* [New York?]: Edgian, [1986?].

Seyfried, Vincent F., and William Asadorian. *Old Queens, N.Y. in Early Photographs.* New York: Dover, 1991.

Shultz, Stanley K., and Clay McShane. "To Engineer the Metropolis: Sewers, Sanitation, and City Planning in Late-Nineteenth-Century America." *Journal of American History* 65, no. 2 (September 1978): 389–411.

Sies, Mary Corbin, and Christopher Silver, eds. *Planning the Twentieth-Century American City.* Baltimore: Johns Hopkins University Press, 1996.

Sobin, Dennis P. *Dynamics of Community Change: The Case of Long Island's Declining "Gold Coast."* Port Washington, NY: Ira J. Friedman, 1968.

Spirn, Anne Whiston. *The Granite Garden: Urban Nature and Human Design.* New York: Basic Books, 1984.

Steinberg, Ted. *Gotham Unbound: The Ecological History of Greater New York.* New York: Simon and Schuster, 2014.

Stilgoe, John. *The Metropolitan Corridor: Railroads and the American Scene 1880 to 1935.* New Haven, CT: Yale University Press, 1983.

Stradling, David. *The Nature of New York: An Environmental History of the Empire State.* Ithaca, NY: Cornell University Press, 2010.

Sussman, Carl, ed. *Planning the Fourth Migration: The Neglected Vision of the Regional Planning Association of America.* Cambridge, MA: MIT Press, 1976.

Suszkowski, Dennis J. "Sewage Pollution in New York Harbor: A Historical Perspective." MS thesis, State University of New York at Stony Brook, 1973.

Sutton. S. B., ed. *Civilizing American Cities: Writings on City Landscapes.* New York: Da Capo, 1997.

Teaford, Jon C. *Post-Suburbia: Government and Politics in the Edge Cities.* Baltimore: Johns Hopkins University Press, 1997.

———. *The Unheralded Triumph: City Government in America 1870–1900.* Baltimore: Johns Hopkins University Press, 1984.

Twomey, Bill. "Childhood Memories of Edgewater Park." *Bronx County Historical Society Journal* 33, no. 1 (March 1996): 83–85.

Ultan, Lloyd. *The Beautiful Bronx, 1920–1950.* New Rochelle, NY: Arlington House, 1979.

Ultan, Lloyd, and Gary Hermalyn, eds. *The Bronx in the Innocent Years, 1890–1925.* New York: Harper and Row, 1985.

Van Bloem, Kate. *History of the Village of Lake Success.* Lake Success, NY: Incorporated Village of Lake Success, 1968.

Vileisis, Ann. *Discovering the Unknown Landscape: A History of America's Wetlands.* Washington, DC: Island, 1997.

Walsh, Daniel C. "The History of Waste Landfilling in New York City." *Ground Water Readers' Forum* 29 (July–August 1991): 591–93.

Warner, Sam Bass, Jr. *Streetcar Suburbs: The Process of Growth in Boston, 1870–1900.* Cambridge, MA: Harvard University Press: 1978.

Weigold, Marilyn E. *People and the Parks: A History of Parks and Recreation in Westchester County.* White Plains, NY: Westchester County Dept. of Parks, Recreation and Conservation, 1984.

———. *Pioneering in Parks and Parkways: Westchester County, New York, 1895–1945.* Essays in Public Works History no. 9. Chicago: Public Works Historical Society, 1980.

———, ed. *Westchester County: The Past Hundred Years, 1883–1983.* Valhalla, NY: Westchester County Historical Society, 1983.

Williams, Mason B. *City of Ambition: FDR, La Guardia, and the Making of Modern New York.* New York: W. W. Norton, 2013.

Wiltse, Jeff. *Contested Waters: A Social History of Swimming Pools in America.* Chapel Hill: University of North Carolina Press, 2007.

Witkowski, Mary K. and Bruce Williams. *Bridgeport on the Sound.* Charleston, SC: Arcadia, 2001.

Wolcott, Victoria W. *Race, Riots, and Roller Coasters: The Struggle over Segregated Recreation in America.* Philadelphia: University of Pennsylvania Press, 2012.

Wood, Robert C. *1400 Governments: The Political Economy of the New York Metropolitan Region.* Cambridge, MA: Harvard University Press, 1961.

# Index

Page numbers in *italics* indicate illustrations.